ARM 64-Bit Assembly Language

ARM 64-Bit Assembly Language

Larry D. Pyeatt

with William Ughetta

Newnes is an imprint of Elsevier
The Boulevard, Langford Lane, Kidlington, Oxford OX5 1GB, United Kingdom
50 Hampshire Street, 5th Floor, Cambridge, MA 02139, United States

Notices

Knowledge and best practice in this field are constantly changing. As new research and experience broaden our understanding,
changes in research methods, professional practices, or medical treatment may become necessary.

Practitioners and researchers must always rely on their own experience and knowledge in evaluating and using any
information, methods, compounds, or experiments described herein. In using such information or methods they should be
mindful of their own safety and the safety of others, including parties for whom they have a professional responsibility.

To the fullest extent of the law, neither the Publisher nor the authors, contributors, or editors, assume any liability for any injury
and/or damage to persons or property as a matter of products liability, negligence or otherwise, or from any use or operation of
any methods, products, instructions, or ideas contained in the material herein.

Library of Congress Cataloging-in-Publication Data
A catalog record for this book is available from the Library of Congress

British Library Cataloguing-in-Publication Data
A catalogue record for this book is available from the British Library

ISBN: 978-0-12-819221-4

For information on all Newnes publications
visit our website at https://www.elsevier.com/books-and-journals

Publisher: Mara Conner
Acquisition Editor: Tim Pitts
Editorial Project Manager: Joshua Mearns
Production Project Manager: Anitha Sivaraj
Designer: Greg Harris

Typeset by VTeX

Working together
to grow libraries in
developing countries

www.elsevier.com • www.bookaid.org

Contents

List of tables

List of figures

Preface

This book is intended to be used in a first course in assembly language programming for Computer Science (CS) and Computer Engineering (CE) students. It is assumed that students using this book have already taken courses in programming and data structures, and are competent programmers in at least one high-level language. Many of the code examples in the book are written in C, with an assembly implementation following. The assembly examples can stand on their own, but students who are familiar with C, C++ or Java should find the C· examples helpful.

Computer Science and Computer Engineering are very large fields. It is impossible to cover everything that a student may eventually need to know. There are a limited number of course hours available, so educators must strive to deliver degree programs that make a compromise between the number of concepts and skills that the students learn and the depth at which they learn those concepts and skills. Obviously, with these competing goals it is difficult to reach consensus on exactly what courses should be included in a CS or CE curriculum.

Traditionally, assembly language courses have consisted of a fairly mechanistic learning of a set of instructions, registers, and syntax. Partially because of this tendency, over the years, Assembly language courses have been marginalized in, or removed altogether from, many CS and CE curricula. The author feels that this is unfortunate, because a solid understanding of assembly language leads to better understanding of higher-level languages, compilers, interpreters, architecture, operating systems, and other important CS and CE concepts.

One goal of this book is to make a course in assembly language more valuable by introducing methods (and a bit of theory) that are not covered in any other Computer Science or Computer Engineering course, while using assembly language to implement the methods. In this way, it is intended that the course in assembly language goes far beyond the traditional assembly language course, and once again plays an important role in the overall CS and CE curricula.

Choice of processor family

Because of their ubiquity, x86 based systems have been the platforms of choice for most assembly language courses over the last two decades. This is unfortunate because, in this author's opinion, in every respect other than ubiquity, the x86 architecture is the worst possible choice for learning and teaching assembly language. The newer chips in the family have hundreds of instructions, and irregular rules govern how those instructions can be used. In an attempt to make it possible for students to succeed, typical courses use antiquated assemblers and interface with the antiquated IBM PC BIOS, using only a small subset of the modern x86 instruction set. The programming environment has little or no relevance to modern computing.

Partially because of this tendency to use x86 platforms, and the resulting unnecessary burden placed on students and instructors, as well as the reliance on antiquated and irrelevant development environments, assembly language is often viewed by students as very difficult and lacking in value. The author believes that this situation can be remedied, and hopes that this textbook helps students to realize the value of knowing assembly language. The relatively simple ARM processor family was chosen in hopes that the students also learn that although assembly language programming may be more difficult than high-level languages, it can be mastered.

The recent development of very low-cost ARM based Linux computers has caused a surge of interest in the ARM architecture as an alternative to the x86 architecture, which has become increasingly complex over the years. This book should provide a solution for a growing need. The ARM architecture is a very good choice for teaching and learning assembly language, for several reasons.

Many students have difficulty with the concept that a register can hold variable x at one point in the program, and hold variable y at some other point. They also often have difficulty with the concept that, before it can be involved in any computation, data has to be moved from memory into the CPU. Using a load-store architecture helps the students to more readily grasp these concepts.

Another common difficulty that students have is in relating the concepts of an address and a pointer variable. The design of the ARM architecture makes it easier for students to realize that pointers are just variables that hold an address. Students who learn assembly using this book will understand pointers at a fundamental level.

Many students also struggle with the concept of recursion, regardless of what language is used. In assembly, the mechanisms involved are exposed and directly manipulated by the programmer. Examples of recursion are scattered throughout this textbook.

Some students have difficulty understanding the flow of a program, and tend to put many unnecessary branches into their code. Many assembly language courses spend so much time and space on learning the instruction set that they never have time to teach good programming practices. This textbook puts strong emphasis on using structured programming concepts. The relative simplicity of the ARM architecture makes this possible.

One of the major reasons to learn and use assembly language is that it allows the programmer to create very efficient mathematical routines. The concepts introduced in this book will enable students to perform efficient non-integral math on any processor. These techniques are rarely taught because of the time that it takes to cover the x86 instruction set. With the ARM processor, less time is spent on the instruction set, and more time can be spent teaching how to optimize the code.

The combination of the ARM processor and the Linux operating system provide the least costly hardware platform and development environment available. A cluster of ten Raspberry Pis, or similar hosts, with power supplies and networking, can be assembled for 500 US dollars or less. This cluster can support up to 50 students logging in through secure shell (ssh). If their client platform supports the X window system, then they can run GUI enabled applications on the host, and have the GUI display directed to their client. Alternatively, most low-cost ARM systems can directly drive a display and take input from a keyboard and mouse. With the addition of an NFS server (which itself could be a low-cost ARM system and a hard drive), an entire Linux ARM based laboratory of 20 workstations could be built for 250 US dollars per seat or less. Admittedly, it would not be a high-performance laboratory, but could be used to teach C, assembly, and other languages. The author would argue that beginning programmers *should* learn to program on low-performance machines, because it reinforces a life-long tendency towards efficiency.

General approach

The ARMv8-A architecture has two architectural modes: AArch32 and AArch64. In moving from 32-bit machines to 64-bit machines, ARM chose to completely re-design the architecture. As a result, the two processing modes have completely different assembly languages. The AArch32 mode allows backwards-compatible execution of ARMv7 (and older) code. This book focuses exclusively on the 64-bit AArch64 execution state and assembly language.

The approach of this book is to present concepts in different ways throughout the book, slowly building from simple examples towards complex programming on bare-metal embedded systems. Students who don't understand a concept when it is explained in a certain way may easily grasp the concept when it is presented later from a different viewpoint.

The main objective of this book is to provide an improved course in assembly language by replacing the x86 platform with one that is less costly, more ubiquitous, well-designed, powerful, and easier to learn. Since students are able to master the basics of assembly language quickly, it is possible to teach a wider range of topics, such as fixed and floating point mathematics, ethical considerations, performance tuning, and interrupt processing. The author hopes that courses using this book will better prepare students for the junior and senior level courses in operating systems, computer architecture, and compilers.

## CHAPTER 1

Introduction

An executable computer program is, ultimately, just a series of numbers that have very little or no meaning to a human being. We have developed a variety of human-friendly languages in which to express computer programs, but in order for the program to execute, it must eventually be reduced to a stream of numbers. Assembly language is one step above writing the stream of numbers. The stream of numbers is called the *instruction stream*. Each number in the instruction stream instructs the computer to perform one (usually small) operation. Although each instruction does very little, the ability of the programmer to specify any *sequence* of instructions and the ability of the computer to perform billions of these small operations every second makes modern computers very powerful and flexible tools. In Assembly language, one line of code usually gets translated into one machine instruction. In higher level languages, a single line of code may generate *many* machine instructions.

A simplified model of a computer system, as shown in Fig. 1.1, consists of memory, input/output devices, and a Central Processing Unit (CPU), connected together by a system bus. The bus can be thought of as a roadway that allows data to travel between the components of the computer system. The CPU is the part of the system where most of the computation occurs, and the CPU controls the other devices in the system.

Memory can be thought of as a series of mailboxes. Each mailbox can hold a single postcard with a number written on it, and each mailbox has a unique numeric identifier. The identifier, x is called the memory address, and the number stored in the mailbox is called the contents of address x. Some of the mailboxes contain data, and other contain instructions which tell the CPU what actions to perform.

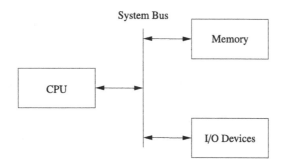

Figure 1.1: Simplified representation of a computer system.

ARM 64-Bit Assembly Language
https://doi.org/10.1016/B978-0-12-819221-4.00008-0
1

The CPU also contains a much smaller set of mailboxes, which we call registers. Data can be copied from cards stored in memory to cards stored in the CPU, or vice-versa. Once data has been moved into one of the CPU registers, it can be used in computation. For example, in order to add two numbers in memory, they must first be moved into registers on the CPU. The CPU can then add the numbers together and store the result in one of the CPU registers. The result of the addition can then be moved back into one of the mailboxes in memory.

Modern computers execute instructions sequentially. In other words, the next instruction to be executed is at the memory address immediately following the current instruction. One of the registers in the CPU, the program counter (PC) keeps track of the location from which the next instruction is to be fetched. The CPU follows a very simple sequence of actions. It fetches an instruction from memory, increments the program counter, executes the instruction, and then repeats the process with the next instruction. However some instructions may change the program counter, so that the next instruction is fetched from a non-sequential address.

1.1 Reasons to learn assembly

There are many high-level *programming languages*, such as Java, Python, C, and C++ that have been designed to allow programmers to work at a high level of abstraction, so that they do not need to understand exactly what instructions are needed by a particular CPU. For *compiled* languages, such as C and C++, a compiler handles the task of translating the program, written in a high-level language into *assembly language* for the particular CPU on the system. An assembler then converts the program from assembly language into the binary codes that the CPU reads as instructions.

High-level languages can greatly enhance programmer productivity. However, there are some situations where writing assembly code directly is desirable or necessary. For example, assembly language may be the best choice when writing

- the first steps in booting the computer,
- code to handle interrupts,
- low-level locking code for multi-threaded programs,
- code for machines where no compiler exists,
- in situations where the compiler cannot generate code that is optimal (or efficient enough),
- on computers with very limited memory, and
- code that requires low-level access to architectural and/or processor features.

Aside from sheer necessity, there are several other reasons why it is still important for computer scientists to learn assembly language.

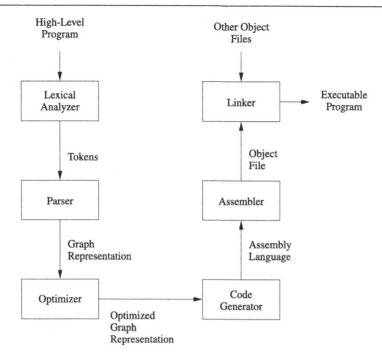

Figure 1.2: Stages of a typical compilation sequence.

One example where knowledge of assembly is indispensable is when designing and implementing compilers for high-level languages. As shown in Fig. 1.2, a typical compiler for a high level language must generate assembly language as its output. Most compilers are designed to have multiple stages. In the input stage, the source language is read and converted into a graph representation. The graph may be optimized before being passed to the output, or code generation, stage where it is converted to assembly language. The assembly is then fed into the system's assembler to generate an object file. The object file is linked with other object files (which are often combined into *libraries*) to create an executable program.

The code generation stage of a compiler must traverse the graph and emit assembly code. The quality of the assembly code that is generated can have a profound influence on the performance of the executable program. Therefore, the programmer responsible for the code generation portion of the compiler must be well-versed in assembly programming for the target CPU.

Some people believe that a good optimizing compiler will generate better assembly code than a human programmer. This belief is not justified. Highly optimizing compilers have lots of clever algorithms, but like all programs, they are not perfect. Outside of the cases that they were designed for, they do not optimize well. Many newer CPUs have instructions which operate on multiple items of data at once. However, compilers rarely make use of these powerful Single Instruction Multiple Data (SIMD) instructions. Instead, it is common for programmers to write functions in assembly language to take advantage of SIMD instructions. The assembly functions are assembled into object file(s), then linked with the object file(s) generated from the high-level language compiler.

Many modern processors also have some support for processing vectors (arrays). Compilers are usually not very good at making effective use of the vector instructions. In order to achieve excellent vector performance for audio or video codecs and other time-critical code, it is often necessary to resort to small pieces of assembly code in the performance-critical inner loops. A good example of this type of code is when performing vector and matrix multiplies. Such operations are commonly needed in processing images and in graphical applications. The vector instructions are explained in Chapter 10.

Another reason for assembly is when writing certain parts of an operating system. Although modern operating systems are mostly written in high-level languages, there are some portions of the code that can only be done in assembly. Typical uses of assembly language are when writing device drivers, saving the state of a running program so that another program can use the CPU, restoring the saved state of a running program so that it can resume executing, and managing memory and memory protection hardware. There are many other tasks that are central to a modern operating system which can only be accomplished in assembly language. Careful design of the operating system can minimize the amount of assembly required, but cannot eliminate it completely.

Another good reason to learn assembly is for debugging. Simply understanding what is going on at a fundamental level can be very valuable when trying to debug programs. If there is a problem in a call to a third party library, sometimes the only way a developer can isolate and diagnose the problem is to run the program under a debugger and step through it one machine instruction at a time. This does not require a deep knowledge of assembly language coding but at least a passing familiarity with assembly is helpful in that particular case. Analysis of assembly code is an important skill for C programmers and C++ programmers, who may occasionally have to diagnose a fault by looking at the contents of CPU registers and single-stepping through machine instructions.

Assembly language is an important part of the path to understanding how the machine works. Even though only a small percentage of computer scientists will be lucky enough to work on the code generator of a compiler, they all can benefit from the deeper level of understanding that they gain by learning assembly language. Many programmers do not really understand pointers until they have written assembly language.

Without first learning assembly language, it is impossible to learn advanced concepts such as microcode, pipelining, instruction scheduling, out-of-order execution, threading, branch prediction, and speculative execution. There are many other concepts dealing with operating systems and computer architecture, which require some understanding of assembly language. The best programmers understand why some language constructs perform better than others, how to reduce cache misses, and how to prevent buffer overruns that destroy security.

Every program is meant to run on a real machine. Even though there are many languages, compilers, virtual machines, and operating systems to enable the programmer to use the machine more conveniently, the strengths and weaknesses of that machine still determine what is easy and what is hard. Learning assembly is a fundamental part of understanding enough about the machine to make informed choices about how to write efficient programs, even when writing in a high-level language.

As an analogy, most people do not need to know a lot about how an internal combustion engine works in order to operate an automobile. A race car driver needs a much better understanding of exactly what happens when he or she steps on the accelerator pedal, in order to be able to judge precisely when (and how hard) to do so. Also, who would trust their car to a mechanic who could not tell the difference between a spark plug and a brake caliper? Worse still, should we trust an engineer to *build* a car without that knowledge? Even in this day of computerized cars, someone needs to know the gritty details, and they are paid well for that knowledge. Knowledge of assembly language is one of the things that defines the computer scientist and engineer.

When learning assembly language, the specific instruction set is not critically important, because what is really being learned is the fine detail of how a typical stored-program machine uses different storage locations and logic operations to convert a string of bits into a meaningful calculation. However, when it comes to learning assembly languages, some processors make it more difficult than it needs to be. Because some processors have an instruction set is extremely irregular, non-orthogonal, large, and poorly designed, they are not a good choice for learning assembly. The author feels that teaching students their first assembly language on one of those processors should be considered a crime, or at least a form of mental abuse. Luckily, there are processors that are readily available, low-cost, and relatively easy to learn assembly with. This book uses one of them as the model for assembly language.

1.2 The ARM processor

In the late 1970's, the microcomputer industry was a fierce battleground, with several companies competing to sell computers to small business and home users. One of those companies, based in the United Kingdom, was Acorn Computers Ltd. Acorn's flagship product, the BBC

Micro, was based on the same processor that Apple Computer had chosen for their Apple II line of computers; the 8-bit 6502 made by MOS Technology. As the 1980's approached, microcomputer manufacturers were looking for more powerful 16-bit and 32-bit processors. The engineers at Acorn considered the processor chips that were available at the time, and concluded that there was nothing available that would meet their needs for the next generation of Acorn computers.

The only reasonably-priced processors that were available were the Motorola 68000, (a 32-bit processor used in the Apple Macintosh and most high-end Unix workstations), and the Intel 80286 (a 16-bit processor used in less powerful personal computers such as the IBM PC). During the previous decade, a great deal of research had been conducted on developing high-performance computer architectures. One of the outcomes of that research was the development of a new paradigm for processor design, known as Reduced Instruction Set Computing (RISC). One advantage of RISC processors was that they could deliver higher performance with a much smaller number of transistors than the older Complex Instruction Set Computing (CISC) processors such as the 68000 and 80286. The engineers at Acorn decided to design and produce their own processor. They used the BBC Micro to design and simulate their new processor, and in 1987, they introduced the Acorn Archimedes. The Archimedes was arguably the most powerful home computer in the world at that time, with graphics and audio capabilities that IBM PC and Apple Macintosh users could only dream about. Thus began the long and successful dynasty of the Acorn RISC Machine (ARM) processor.

Although their technology was superb, Acorn never made a big impact on the global computer market. Acorn eventually went out of business, but the processor that they created has lived on. It was re-named to the Advanced RISC Machine, and is now known simply as ARM. Stewardship of the ARM processor belongs to ARM Holdings, LLC which manages the design of new ARM architectures and licenses the manufacturing rights to other companies. ARM Holdings does not manufacture any processor chips, yet more ARM processors are produced annually than all other processor designs combined. Most ARM processors are used as components for embedded systems and portable devices. If you have a smart phone or similar device, then there is a very good chance that it has an ARM processor in it. Because of its enormous market presence, clean architecture, and small, orthogonal instruction set, the ARM is a very good choice for learning assembly language.

Although it dominates the portable device market, the ARM processor family has almost no presence in the desktop or server market. However, that may change. In 2012, ARM Holdings announced the ARMv8-A architecture, which introduced a new 64-bit instruction set, referred to as AArch64, or ARM64. This represents the first major redesign of the ARM architecture in 30 years. Most processor families have been extended to include 64-bit instructions, while simultaneously supporting the older instruction set(s). ARM chose another route. The processor operates either in ARM64 mode or ARM32 mode, and the two modes are completely

**Table 1.1: Values
represented by two bits.**

Bit 1	Bit 0	Value
0	0	0
0	1	1
1	0	2
1	1	3

separate, with different (thought similar) assembly languages. This separation of modes results in a much simpler assembly language which is easier to learn. AArch64 is intended to compete for the desktop and server market with other high-end processors such as the Sun SPARC and Intel Xeon, and as of 2018, is the most common processor in high-end smartphones and other portable devices. Because of its ubiquity and clean design, the AArch64 assembly language was chosen as the target for this book.

1.3 Computer data

The basic unit of data in a digital computer is the binary digit, or *bit*. A bit can have a value of zero or one. In order to store numbers larger than 1, bits are combined into larger units. For instance, using two bits, it is possible to represent any number between zero and three. This is shown in Table 1.1. When stored in the computer, all data is simply a string of binary digits. There is more than one way that such a fixed-length string of binary digits can be interpreted.

Computers have been designed using many different bit group sizes, including 4, 8, 10, 12, and 14 bits. Today most computers recognize a basic grouping of 8 bits, which we call a *byte*. Some computers can work in units of 4 bits, which is commonly referred to as a *nibble* (sometimes spelled "nybble"). A nibble is a convenient size because it can exactly represent one hexadecimal digit. Additionally, most modern computers can also work with groupings of 16, 32, and 64 bits. The CPU is designed with a *default word size*. For most modern CPUs, the default word size is 32 bits. Many processors support 64 bit words, which is increasingly becoming the default size.

1.3.1 Representing natural numbers

A numeral system is a writing system for expressing numbers. The most common system is the Hindu-Arabic number system, which is now used throughout the world. Almost from the first day of formal education, children begin learning how to add, subtract, and perform other operations using the Hindu-Arabic system. After years of practice, performing basic mathe-

Table 1.2: The first 21 integers (starting with 0) in various bases.

Base									
16	10	9	8	7	6	5	4	3	2
0	0	0	0	0	0	0	0	0	0
1	1	1	1	1	1	1	1	1	1
2	2	2	2	2	2	2	2	2	10
3	3	3	3	3	3	3	3	10	11
4	4	4	4	4	4	4	10	11	100
5	5	5	5	5	5	10	11	12	101
6	6	6	6	6	10	11	12	20	110
7	7	7	7	10	11	12	13	21	111
8	8	8	10	11	12	13	20	22	1000
9	9	10	11	12	13	14	21	100	1001
A	10	11	12	13	14	20	22	101	1010
B	11	12	13	14	15	21	23	102	1011
C	12	13	14	15	20	22	30	110	1100
D	13	14	15	16	21	23	31	111	1101
E	14	15	16	20	22	24	32	112	1110
F	15	16	17	21	23	30	33	120	1111
10	16	17	20	22	24	31	100	121	10000
11	17	18	21	23	25	32	101	122	10001
12	18	20	22	24	30	33	102	200	10010
13	19	21	23	25	31	34	103	201	10011
14	20	22	24	26	32	40	110	202	10100

matical operations using strings of digits between 0 and 9 seems natural. However, there are other ways to count and perform arithmetic, such as Roman numerals, unary systems, and Chinese numerals. With a little practice, it is possible to become as proficient at performing mathematics with other number systems as with the Hindu-Arabic system.

The Hindu-Arabic system is a *base ten* or *radix ten* system, because it uses the ten digits 0, 1, 2, 3, 4, 5, 6, 7, 8, and 9. For our purposes, the words radix and base are equivalent, and refer to the number of individual digits available in the numbering system. The Hindu-Arabic system is also a *positional system*, or a place-value notation, because the value of each digit in a number depends on its position in the number. The radix ten Hindu-Arabic system is only one of an infinite family of closely related positional systems. The members of this family differ only in the radix used (and therefore, the number of characters used). For bases greater than base ten, characters are borrowed from the alphabet and used to represent digits. For example, the first column in Table 1.2 shows the character "A" being used as a single digit representation for the number ten.

In base ten, we think of numbers as strings of the ten digits, "0" through "9". Each digit counts 10 times the amount of the digit to its right. If we restrict ourselves to integers, then the digit furthest to the right is always the ones digit. It is also referred to as the *least significant digit*. The digit immediately to the left of the ones digit is the tens digit. To the left of that is the hundreds digit, and so on. The leftmost digit is referred to as the *most significant digit*. The following equation shows how a number can be decomposed into its constituent digits:

$$57839_{10} = 5 \times 10^4 + 7 \times 10^3 + 8 \times 10^2 + 3 \times 10^1 + 9 \times 10^0.$$

Note that the subscript of "10" on 57839_{10} indicates that the number is given in base ten.

Imagine that we only had seven digits: 0, 1, 2, 3, 4, 5, and 6. We need ten digits for base ten. If we only have seven digits, then we are limited to base seven. In base seven, each digit in the string represents a power of seven rather than a power of ten. We can represent any integer in base seven, but it may take more digits than in base ten. Other than using a different base for the power of each digit, the math works exactly the same as for base ten. For example, suppose we have the following number in base seven: 330425_7. We can convert this number to base ten as follows:

$$
\begin{aligned}
330425_7 &= 3 \times 7^5 + 3 \times 7^4 + 0 \times 7^3 + 4 \times 7^2 + 2 \times 7^1 + 5 \times 7^0 \\
&= 50421_{10} + 7203_{10} + 0_{10} + 196_{10} + 14_{10} + 5_{10} \\
&= 57839_{10}
\end{aligned}
$$

Base two, or binary is the "native" number system for modern digital systems. The reason for this is mainly because it is relatively easy to build circuits with two stable states: on and off (or 1 and 0). Building circuits with more than two stable states is much more difficult and expensive, and any computation that can be performed in a higher base can also be performed in binary. The least significant (rightmost) digit in binary is referred to as the *least significant bit*, or LSB, while the leftmost binary digit is referred to as the *most significant bit*, or MSB.

1.3.2 Base conversion

The most common bases used by programmers are base two (binary), base ten (octal), base ten (decimal), and base sixteen (hexadecimal). Octal and hexadecimal are common because, as we shall see later, they can be translated quickly and easily to and from base two, and are often easier for humans to work with than base two. Note that for base 16, we need 16 characters. We use the digits 0 through 9 plus the letters A through F. Table 1.2 shows the equivalents for all numbers between 0 and 20 in base two through base ten, and base sixteen.

Before learning assembly language it is essential to know how to convert from any base to any other base. Since we are already comfortable working in base ten, we will use that as an intermediary when converting between two arbitrary bases. For instance, if we want to convert a number in base three to base five, we will do it by first converting the base three number to base ten, then from base ten to base five. By using this two-stage process, we will only need to learn to convert between base ten and any arbitrary base b.

1.3.2.1 Base b to decimal

Converting from an arbitrary base b to base ten simply involves multiplying each base b digit d by b^n, where n is the significance of digit d, and summing all of the results. For example, converting the base five number 3421_5 to base ten is performed as follows:

$$
\begin{aligned}
3421_5 &= 3 \times 5^3 + 4 \times 5^2 + 2 \times 5^1 + 1 \times 5^0 \\
&= 375_{10} + 100_{10} + 10_{10} + 1_{10} \\
&= 486_{10}
\end{aligned}
$$

This conversion procedure works for converting any integer from any arbitrary base b to its equivalent representation in base ten. Example 1 gives another specific example of how to convert from base b to base ten.

Example 1. Converting from an arbitrary base to base ten.

Converting 4362_7 to base ten is accomplished
by expanding and summing the terms:

$$
\begin{aligned}
4362_7 &= 4 \times 7^3 + 3 \times 7^2 + 6 \times 7^1 + 2 \times 7^0 \\
&= 4 \times 343 + 3 \times 49 + 6 \times 7 + 2 \times 1 \\
&= 1372 + 147 + 42 + 2 \\
&= 1563_{10}
\end{aligned}
$$

1.3.2.2 Decimal to base b

Converting from base ten to an arbitrary base b involves repeated division by the base, b. After each division, the remainder is used as the next *more significant* digit in the base b number, and the quotient is used as the dividend for the next iteration. The process is repeated until the quotient is zero. For example, converting 56_{10} to base 4 is accomplished as follows:

$$\begin{array}{r} 14 \\ 4\overline{)56} \\ 40 \\ \hline 16 \\ 16 \\ \hline 0 \end{array} \searrow \begin{array}{r} 3 \\ 4\overline{)14} \\ 12 \\ \hline 2 \end{array} \searrow \begin{array}{r} 0 \\ 4\overline{)3} \end{array}$$

Reading the remainders from right to left yields: 320_4. This result can be double-checked by converting it back to base ten as follows:

$$
\begin{aligned}
320_4 &= 3 \times 4^2 + 2 \times 4^1 + 0 \times 4^0 \\
&= 48 + 8 + 0 \\
&= 56_{10}.
\end{aligned}
$$

Since we arrived at the same number we started with, we have verified that $56_{10} = 320_4$. This conversion procedure works for converting any integer from base ten to any arbitrary base b. Example 2 gives another example of converting from base ten to another base b.

Example 2. Converting from base ten to an arbitrary base.

Converting 8341_{10} to base seven is accomplished as follows:

$$\begin{array}{r} 1191 \\ 7\overline{)8341} \\ 7000 \\ \hline 1341 \\ 700 \\ \hline 641 \\ 630 \\ \hline 11 \\ 7 \\ \hline 4 \end{array} \searrow \begin{array}{r} 170 \\ 7\overline{)1191} \\ 700 \\ \hline 491 \\ 490 \\ \hline 1 \end{array} \searrow \begin{array}{r} 24 \\ 7\overline{)170} \\ 140 \\ \hline 30 \\ 28 \\ \hline 2 \end{array} \searrow \begin{array}{r} 3 \\ 7\overline{)24} \\ 21 \\ \hline 3 \end{array} \searrow \begin{array}{r} 0 \\ 7\overline{)3} \end{array}$$

$$8341_{10} = 33214_7$$

1.3.2.3 Bases that are powers-of-two

In addition to the methods above, there is a simple method for quickly converting between base two, base eight, and base sixteen. These shortcuts rely on the fact that two, eight, and sixteen are all powers of two. Because of this, it takes exactly four binary digits (bits) to represent exactly one hexadecimal digit. Likewise, it takes exactly three bits to represent an octal

Base 2	Base 16
0000	0
0001	1
0010	2
0011	3
0100	4
0101	5
0110	6
0111	7
1000	8
1001	9
1010	A
1011	B
1100	C
1101	D
1110	E
1111	F

Base 2	Base 8
000	0
001	1
010	2
011	3
100	4
101	5
110	6
111	7

Figure 1.3: Tables used for converting between binary, octal, and hexadecimal.

digit. Conversely, each hexadecimal digit can be converted to exactly four binary digits, and each octal digit can be converted to exactly three binary digits. This relationship makes it possible to do very fast conversions using the tables shown in Fig. 1.3.

When converting from hexadecimal to binary, all that is necessary is to replace each hex digit with the corresponding binary digits from the table. For example, to convert $5AC4_{16}$ to binary, we just replace "5" with "0101," replace "A" with "1010," replace "C" with "1100," and replace "4" with "0100." So, just by referring to the table, we can immediately see that $5AC4_{16} = 0101101011000100_2$. This method works exactly the same for converting from octal to binary, except that it uses the table on the right side of Fig. 1.3.

Converting from binary to hexadecimal is also very easy using the table. Given a binary number, n, take the four least significant digits of n and find them in the table on the left side of Fig. 1.3. The hexadecimal digit on the matching line of the table is the least significant hex digit. Repeat the process with the next set of four bits and continue until there are no bits remaining in the binary number. For example, to convert 0011100101010111_2 to hexadecimal, just divide the number into groups of four bits, starting on the right, to get: $0011|1001|0101|0111_2$. Now replace each group of four bits by looking up the corresponding hex digit in the table on the left side of Fig. 1.3, to convert the binary number to 3957_{16}. In the case where the binary number does not have enough bits, simply pad with zeros in the high-order bits. For example, dividing the number 100110001001_2 into groups of four yields $1|0011|0001|0011_2$ and padding with zeros in the high-order bits results in $0001|0011|0001|0011_2$. Looking up the four groups in the table reveals that $0001|0011|0001|0011_2 = 1313_{16}$.

1.3.2.4 Conversion between arbitrary bases

It is possible to perform the division and multiplication steps in any base and directly convert from any base to any other base. However, most people are much better at working in base ten. For that reason, the easiest and least error-prone way to convert from any base a to any other base b is to use a two step process. The first step is to convert from base a to base ten. The second step is to convert from base ten to base b.

Example 3. Converting 42301_5 to base 11 is accomplished with two steps.

The number is first converted to base ten as follows:

$$
\begin{aligned}
42301_5 &= 4 \times 5^4 + 2 \times 5^3 + 3 \times 5^2 + 0 \times 5^1 + 1 \times 5^0 \\
&= 4 \times 625 + 2 \times 125 + 3 \times 25 + 0 \times 5 + 1 \times 1 \\
&= 2500 + 250 + 75 + 0 + 1 \\
&= 2826_{10}
\end{aligned}
$$

Then the result is converted to base 11:

$$
\begin{array}{cccc}
256 \searrow & 23 \searrow & 2 \searrow & 0 \\
11)\overline{2826} & 11)\overline{256} & 11)\overline{23} & 11)\overline{2} \\
\underline{2200} & \underline{220} & \underline{22} & \\
626 & 36 & 1 & \\
\underline{550} & \underline{33} & & \\
76 & 3 & & \\
\underline{66} & & & \\
10 & & &
\end{array}
$$

$$42301_5 = 2826_{10} = 213A_{11}$$

1.3.3 Representing integers

The computer stores groups of bits, but the bits by themselves have no meaning. The programmer gives them meaning by deciding what the bits represent, and how they are interpreted. Interpreting a group of bits as unsigned integer data is relatively simple. Each bit is weighted by a power-of-two, and the value of the group of bits is the sum of the non-zero bits multiplied by their respective weights. However, programmers often need to represent negative as well as non-negative numbers, and there are many possibilities for storing and

Binary	Unsigned	Sign Magnitude	Excess-127	Two's Complement
00000000	0	0	−127	0
00000001	1	1	−126	1
⋮	⋮	⋮	⋮	⋮
01111110	126	126	−1	126
01111111	127	127	0	127
10000000	128	−0	1	−128
10000001	129	−1	2	−127
⋮	⋮	⋮	⋮	⋮
11111110	254	−126	127	−2
11111111	255	−127	128	−1

Figure 1.4: Four different representations for binary integers.

interpreting integers whose value can be both positive and negative. Programmers and hardware designers have developed several standard schemes for encoding such numbers The three main methods for storing and interpreting signed integer data are two's complement, sign-magnitude, and excess-N, Fig. 1.4 shows how the same binary pattern of bits can be interpreted as a number in four different ways.

1.3.3.1 Sign-magnitude representation

The sign-magnitude representation simply reserves the most significant bit to represent the sign of the number, and the remaining bits are used to store the magnitude of the number. This method has the advantage that it is easy for humans to interpret, with a little practice. However, addition and subtraction are slightly complicated. The addition/subtraction logic must compare the sign bits, complement one of the inputs if they are different, implement an end-around carry, and complement the result if there was no carry from the most significant bit. Complements are explained in Section 1.3.3.3. Because of the complexity, most integer CPUs do not directly support addition and subtraction of integers in sign-magnitude form. However, this method is commonly used for mantissa in floating-point numbers, as will be explained in Chapter 8. Another drawback to sign-magnitude is that it has two representations for zero, which can cause problems if the programmer is not careful.

1.3.3.2 Excess-$(2^{n-1} - 1)$ representation

Another method for representing both positive and negative numbers is by using an *excess-N* representation. With this representation, the number that is stored is N greater than the actual value. This representation is relatively easy for humans to interpret. Addition and subtraction are easily performed using the complement method, which is explained in Section 1.3.3.3. This representation is just the same as unsigned math, with the addition of a *bias* which is usually $(2^{n-1} - 1)$. So, zero is represented as zero plus the bias. In $n = 12$ bits, the bias is $2^{12-1} - 1 = 2047_{10}$, or 011111111111_2. This method is commonly used to store the exponent in floating-point numbers, as will be explained in Chapter 8.

1.3.3.3 Complement representation

A very efficient method for dealing with signed numbers involves representing negative numbers as the *radix complements* of their positive counterparts. The complement is the amount that must be added to something to make it "whole." For instance, in geometry, two angles are complementary if they add to $90°$. In radix mathematics, the complement of a digit x in base b is simply $b - x - 1$. For example, in base ten, the complement of 4 is $10 - 4 - 1 = 5$.

In complement representation, the most significant digit of a number is reserved to indicate whether or not the number is negative. If the first digit is less than $\frac{b}{2}$ (where b is the radix), then the number is positive. If the first digit is greater than or equal to $\frac{b}{2}$, then the number is negative. The first digit is not part of the magnitude of the number, but only indicates the sign of the number. For example, numbers in ten's complement notation are positive if the first digit is less than 5, and negative if the first digit is greater than 4. This works especially well in binary, since the number is considered positive if the first bit is zero and negative if the first bit is one. The magnitude of a negative number can be obtained by taking the radix complement. Because of the nice properties of the complement representation, it is the most common method for representing signed numbers in digital computers.

Finding the complement

The *radix complement* of an n digit number y in radix (base) b is defined as

$$C(y_b) = b^n - y_b. \tag{1.1}$$

For example, the ten's complement of the four digit number 8734_{10} is $10^4 - 8734 = 1266$. In this example, we directly applied the definition of the radix complement from Eq. (1.1). That is easy in base ten, but not so easy in an arbitrary base, because it involves performing a subtraction. However, there is a very simple method for calculating the complement which does not require subtraction. This method involves finding the *diminished radix complement*, which is $(b^n - 1) - y$ by substituting each digit with its complement from a *complement table*. The radix complement is found by adding one to the diminished radix complement. Fig. 1.5 shows the complement tables for base ten and base two. Examples 4 and 5 show how

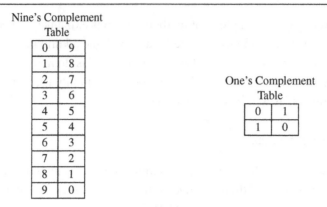

Figure 1.5: Complement tables for base ten and base two.

the complement is obtained in bases ten and two respectively. Example 6 and Example 7 show additional conversions between binary and decimal.

Example 4. The nine's and ten's complements.

> The nine's complement of the base ten number 593 is found by finding the digit '5' in the complement table, and replacing it with its complement, which is the digit '4.' The digit '9' is replaced with '0,' and '3' is replaced with '6.' Therefore the nine's complement of 593_{10} is 406. Likewise, the nine's complement of 1000_{10} is 8999_{10} and the nine's complement of 0999_{10} is 9000_{10}.

> The ten's complement is the nine's complement plus one, so the ten's complement of 726_{10} is $273_{10} + 1 = 274_{10}$.

Example 5. The one's and two's complements.

> The one's complement of a binary number is found in the same way as the nine's complement of a decimal number, but using the one's complement table instead of the nine's complement table. The one's complement of 01001101_2 is 10110010_2 and the one's complement of 0000000010110110 is 1111111101001001_2. Note that the one's complement of a base two number is equivalent to the bit-wise logical "not" (Boolean complement) operator. This operator is very easy to implement in digital hardware.

> The two's complement is the one's complement plus one. The two's complement of 1010100_2 is $0101011_2 + 1 = 0101100_2$.

Example 6. Conversion from binary to decimal.

  1. If the most significant bit is '1', then
       a. Find the two's complement
       b. Convert the result to base 10
       c. Add a negative sign
  2. else
       a. Convert the result to base 10

Number	one's Complement	two's Complement	Base 10	Negative
11010010	00101101	00101110	46	−46
1111111100010110	0000000011101001	0000000011101010	234	−234
01110100	Not negative		116	
1000001101010110	0111110010101001	0111110010101010	31914	−31914
0101001111011011	Not negative		21467	

Example 7. Conversion from decimal to binary.

  1. Remove the negative sign
  2. Convert the number to binary
  3. Take the two's complement

Base 10	Positive Binary	one's Complement	two's Complement
−46	00101110	11010001	11010010
−234	0000000011101010	1111111100010101	1111111100010110
−116	01110100	10001011	10001100
−31914	0111110010101010	1000001101010110	1000001101010111
−21467	0101001111011011	1010110000100100	1010110000100101

Subtraction using complements

One very useful feature of complement notation is that it can be used to perform subtraction by using addition. Given two numbers in base b, x_b and y_b, the difference can be computed as:

$$z_b = x_b - y_b \tag{1.2}$$
$$= x_b + (b^n - y_b) - b^n \tag{1.3}$$
$$= x_b + C(y_b) - b^n, \tag{1.4}$$

where $C(y_b)$ is the radix complement of y_b. Assume that x_b and y_b, both positive and $y_b \le x_b$ and both numbers have the same number of digits n (y_b may have leading zeros). In this case, the result of $x_b + C(y_b)$ will always be greater than or equal to b^n, but less than $2 \times b^n$. This means that the result of $x_b + C(y_b)$ will always begin with a '1' in the $n + 1$ digit position. Dropping the initial '1' is equivalent to subtracting b^n, making the result $x - y + b^n - b^n$ or just $x - y$, which is the desired result. This can be reduced to a simple procedure. When y and x are both positive and $y \leq x$, there are four steps to be performed:

1. pad the subtrahend (y) with leading zeros, as necessary, so that both numbers have the same number of digits (n)
2. find the b's complement of the subtrahend,
3. add the complement to the minuend, and
4. discard the leading '1'.

The complement notation provides a very easy way to represent both positive and negative integers using a fixed number of digits, and to perform subtraction by using addition. Since modern computers typically use a fixed number of bits, complement notation provides a very convenient and efficient way to store signed integers and perform mathematical operations on them. Hardware is simplified because there is no need to build a specialized subtractor circuit. Instead a very simple complement circuit is built and the adder is used to perform subtraction as well as addition.

1.3.4 Representing characters

In the previous section, we discussed how the computer stores information as groups of bits, and how we can interpret those bits as numbers in base two. Given that the computer can only store information using groups of bits, how can we store textual information? The answer is that we create a table, which assigns a numerical value to each character in our language.

Early in the development of computers, several computer manufacturers developed such tables, or character coding schemes. These schemes were incompatible and computers from different manufacturers could not easily exchange textual data without the use of translation software to convert the character codes from one coding scheme to another.

Eventually, a standard coding scheme, known as the American Standard Code for Information Interchange (ASCII) was developed. Work on the ASCII standard began on October 6, 1960, with the first meeting of the American Standards Association's (ASA) X3.2 subcommittee. The first edition of the standard was published in 1963. The standard was updated in 1967 and again in 1986. Within a few years of its development, ASCII was accepted by all major computer manufacturers, although some continue to support their own coding schemes as well.

ASCII was designed for American English, and does not support some of the characters that are used by non-English languages. For this reason, ASCII has been extended to create more comprehensive coding schemes. Most modern multi-lingual coding schemes are compatible with ASCII, though they support a wider range of characters.

At the time that it was developed, transmission of digital data over long distances was very slow, and usually involved converting each bit into an audio signal which was transmitted over a telephone line using an acoustic modem. In order to maximize performance, the standards committee chose to define ASCII as a 7-bit code. Because of this decision, all textual data could be sent using seven bits rather than eight, resulting in approximately 10% better overall performance when transmitting data over a telephone modem. A possibly unforeseen benefit was that this also provided a way for the code to be extended in the future. Since there are 128 possible values for a 7-bit number, the ASCII standard provides 128 characters. However, 33 of the ASCII characters are *non-printing control characters*. These characters, shown in Table 1.3, are used to send information about how the text is to be displayed and/or printed. The remaining 95 *printable characters* are shown in Table 1.4.

1.3.4.1 Non-printing characters

The non-printing characters are used to provide hints or commands to the device that is receiving, displaying, or printing the data. The FF character, when sent to a printer, will cause the printer to eject the current page and begin a new one. The LF character causes the printer or terminal to end the current line and begin a new one. The CR character causes the terminal or printer to move to the beginning of the current line. Many text editing programs allow the user to enter these non-printing characters by using the control key on the keyboard. For instance, to enter the BEL character, the user would hold the control key down and press the G key. This character, when sent to a character display terminal, will cause it to emit a beep. Many of the other control characters can be used to control specific features of the printer, display, or other device that the data is being sent to.

1.3.4.2 Converting character strings to ASCII codes

Suppose we wish to convert a string of characters, such as "Hello World" to an ASCII representation. We can use an 8-bit byte to store each character. Also, it is common practice to include an additional byte at the end of the string. This additional byte holds the ASCII NUL character, which indicates the end of the string. Such an arrangement is referred to as a null-terminated string.

To convert the string "Hello World" into a null terminated string, we can build a table with each character on the left and its equivalent binary, octal, hexadecimal, or decimal value (as defined in the ASCII table) on the right. Table 1.5 shows the characters in "Hello World" and their equivalent binary representations, found by looking in Table 1.4. Since most modern

Table 1.3: The ASCII control characters.

Binary	Oct	Dec	Hex	Abbr	Glyph	Name
000 0000	000	0	00	NUL	^@	Null character
000 0001	001	1	01	SOH	^A	Start of Header
000 0010	002	2	02	STX	^B	Start of Text
000 0011	003	3	03	ETX	^C	End of Text
000 0100	004	4	04	EOT	^D	End of Transmission
000 0101	005	5	05	ENQ	^E	Enquiry
000 0110	006	6	06	ACK	^F	Acknowledgment
000 0111	007	7	07	BEL	^G	Bell
000 1000	010	8	08	BS	^H	Backspace
000 1001	011	9	09	HT	^I	Horizontal Tab
000 1010	012	10	0A	LF	^J	Line feed
000 1011	013	11	0B	VT	^K	Vertical Tab
000 1100	014	12	0C	FF	^L	Form feed
000 1101	015	13	0D	CR	^M	Carriage return[g]
000 1110	016	14	0E	SO	^N	Shift Out
000 1111	017	15	0F	SI	^O	Shift In
001 0000	020	16	10	DLE	^P	Data Link Escape
001 0001	021	17	11	DC1	^Q	Device Control 1 (oft. XON)
001 0010	022	18	12	DC2	^R	Device Control 2
001 0011	023	19	13	DC3	^S	Device Control 3 (oft. XOFF)
001 0100	024	20	14	DC4	^T	Device Control 4
001 0101	025	21	15	NAK	^U	Negative Acknowledgment
001 0110	026	22	16	SYN	^V	Synchronous idle
001 0111	027	23	17	ETB	^W	End of Transmission Block
001 1000	030	24	18	CAN	^X	Cancel
001 1001	031	25	19	EM	^Y	End of Medium
001 1010	032	26	1A	SUB	^Z	Substitute
001 1011	033	27	1B	ESC	^[	Escape
001 1100	034	28	1C	FS	^\	File Separator
001 1101	035	29	1D	GS	^]	Group Separator
001 1110	036	30	1E	RS	^^	Record Separator
001 1111	037	31	1F	US	^_	Unit Separator
111 1111	177	127	7F	DEL	^?	Delete

computers use 8-bit bytes (or multiples thereof) as the basic storage unit, an extra zero bit is shown in the most significant bit position.

Reading the Binary column from top to bottom results in the following sequence of bytes:
01001000 01100101 01101100 01101100 01101111 00100000 01010111 01101111

Table 1.4:   The ASCII printable characters.

Binary	Oct	Dec	Hex	Glyph	Binary	Oct	Dec	Hex	Glyph
010 0000	040	32	20	␢	101 0000	120	80	50	P
010 0001	041	33	21	!	101 0001	121	81	51	Q
010 0010	042	34	22	"	101 0010	122	82	52	R
010 0011	043	35	23	#	101 0011	123	83	53	S
010 0100	044	36	24	$	101 0100	124	84	54	T
010 0101	045	37	25	%	101 0101	125	85	55	U
010 0110	046	38	26	&	101 0110	126	86	56	V
010 0111	047	39	27	'	101 0111	127	87	57	W
010 1000	050	40	28	(	101 1000	130	88	58	X
010 1001	051	41	29	)	101 1001	131	89	59	Y
010 1010	052	42	2A	*	101 1010	132	90	5A	Z
010 1011	053	43	2B	+	101 1011	133	91	5B	[
010 1100	054	44	2C	,	101 1100	134	92	5C	\
010 1101	055	45	2D	-	101 1101	135	93	5D	]
010 1110	056	46	2E	.	101 1110	136	94	5E	^
010 1111	057	47	2F	/	101 1111	137	95	5F	_
011 0000	060	48	30	0	110 0000	140	96	60	`
011 0001	061	49	31	1	110 0001	141	97	61	a
011 0010	062	50	32	2	110 0010	142	98	62	b
011 0011	063	51	33	3	110 0011	143	99	63	c
011 0100	064	52	34	4	110 0100	144	100	64	d
011 0101	065	53	35	5	110 0101	145	101	65	e
011 0110	066	54	36	6	110 0110	146	102	66	f
011 0111	067	55	37	7	110 0111	147	103	67	g
011 1000	070	56	38	8	110 1000	150	104	68	h
011 1001	071	57	39	9	110 1001	151	105	69	i
011 1010	072	58	3A	:	110 1010	152	106	6A	j
011 1011	073	59	3B	;	110 1011	153	107	6B	k
011 1100	074	60	3C	<	110 1100	154	108	6C	l
011 1101	075	61	3D	=	110 1101	155	109	6D	m
011 1110	076	62	3E	>	110 1110	156	110	6E	n
011 1111	077	63	3F	?	110 1111	157	111	6F	o
100 0000	100	64	40	@	111 0000	160	112	70	p
100 0001	101	65	41	A	111 0001	161	113	71	q
100 0010	102	66	42	B	111 0010	162	114	72	r
100 0011	103	67	43	C	111 0011	163	115	73	s
100 0100	104	68	44	D	111 0100	164	116	74	t
100 0101	105	69	45	E	111 0101	165	117	75	u
100 0110	106	70	46	F	111 0110	166	118	76	v
100 0111	107	71	47	G	111 0111	167	119	77	w
100 1000	110	72	48	H	111 1000	170	120	78	x
100 1001	111	73	49	I	111 1001	171	121	79	y
100 1010	112	74	4A	J	111 1010	172	122	7A	z
100 1011	113	75	4B	K	111 1011	173	123	7B	{
100 1100	114	76	4C	L	111 1100	174	124	7C		
100 1101	115	77	4D	M	111 1101	175	125	7D	}
100 1110	116	78	4E	N	111 1110	176	126	7E	~
100 1111	117	79	4F	O					

Table 1.5: Binary equivalents
for each character in
"Hello World".

Character	Binary
H	01001000
e	01100101
l	01101100
l	01101100
o	01101111
	00100000
W	01010111
o	01101111
r	01110010
l	01101100
d	01100100
NUL	00000000

Table 1.6: Binary, hexadecimal, and decimal equivalents
for each character in "Hello World".

Character	Binary	Hexadecimal	Decimal
H	01001000	48	72
e	01100101	65	101
l	01101100	6C	108
l	01101100	6C	108
o	01101111	6F	111
	00100000	20	32
W	01010111	57	87
o	01101111	6F	111
r	01110010	62	98
l	01101100	6C	108
d	01100100	64	100
NUL	00000000	00	0

01110010 01101100 01100100 0000000. To convert the same string to a hexadecimal representation, we can use the shortcut method that was introduced previously to convert each 4-bit nibble into its hexadecimal equivalent, or read the hexadecimal value from the ASCII table. Table 1.6 shows the result of extending Table 1.5 to include hexadecimal and decimal equivalents for each character. The string can now be converted to hexadecimal or decimal simply by reading the correct column in the table. So "Hello World" expressed as a null-terminated

Table 1.7: Interpreting a hexadecimal string as ASCII.

Hexadecimal	ASCII
46	F
61	a
62	b
75	u
6C	l
6F	o
75	u
73	s
21	!
00	NUL

string in hexadecimal is "48 65 6C 6C 6F 20 57 6F 62 6C 64 00" and in decimal it is "72 101 108 108 111 32 87 111 98 108 100 0".

1.3.4.3 Interpreting data as ASCII strings

It is sometimes necessary to convert a string of bytes in hexadecimal into ASCII characters. This is accomplished simply by building a table with the hexadecimal value of each byte in the left column, then looking in the ASCII table for each value and entering the equivalent character representation in the right column. Table 1.7 shows the table used to interpret the hexadecimal string "466162756C6F75732100" as an ASCII string.

1.3.4.4 ISO extensions to ASCII

ASCII was developed to encode all of the most commonly used characters in North American English text. The encoding uses only 128 of the 256 codes that are available in an 8-bit byte. ASCII does not include symbols frequently used in other countries, such as the British pound symbol (£) or accented characters (ü). However, the International Standards Organization (ISO) has created several extensions to ASCII to enable the representation of characters from a wider variety of languages.

The ISO extended ASCII standards are known collectively as ISO 8859. Several variations of the ISO 8859 standard exist for supporting different language families. Each ISO 8859 character set is an eight-bit extension to ASCII which includes the 128 ASCII characters along with 128 additional characters, such as the British Pound symbol and the American cent symbol. Table 1.8 provides a brief description of the various ISO standards.

Table 1.8:   Variations of the ISO 8859 standard.

Name	Alias	Languages
ISO8859-1	Latin-1	Western European languages
ISO8859-2	Latin-2	Non-Cyrillic Central and Eastern European languages
ISO8859-3	Latin-3	Southern European languages and Esperanto
ISO8859-4	Latin-4	Northern European and Baltic languages
ISO8859-5	Latin/Cyrillic	Slavic languages that use a Cyrillic alphabet
ISO8859-6	Latin/Arabic	Common Arabic language characters
ISO8859-7	Latin/Greek	Modern Greek language
ISO8859-8	Latin/Hebrew	Modern Hebrew languages
ISO8859-9	Latin-5	Turkish
ISO8859-10	Latin-6	Nordic languages
ISO8859-11	Latin/Thai	Thai language
ISO8859-12	Latin/Devanagari	Never completed. Abandoned in 1997
ISO8859-13	Latin-7	Some Baltic languages not covered by Latin-4 or Latin-6
ISO8859-14	Latin-8	Celtic languages
ISO8859-15	Latin-9	Update to Latin-1 that replaces some characters. Most notably, it includes the euro symbol (€), which did not exist when Latin-1 was created
ISO8859-16	Latin-10	Covers several languages not covered by Latin-9 and includes the euro symbol (€)

Although the ISO extensions helped to standardize text encodings for several languages that were not covered by ASCII, there were still some issues. The first issue is that the input devices must be configured to correctly encode the text, and output devices must be configured for displaying or printing documents with one of the multiple encodings. This often requires some mechanism for changing the encoding on-the-fly. Another issue has to do with the lexicographical ordering of characters. Although two languages may share a character, that character may appear in a different place in the alphabets of the two languages. This leads to issues when programmers need to sort strings into lexicographical order. The ISO extensions help to unify character encodings across multiple languages, but do not solve all of the issues involved in defining a universal character set.

1.3.4.5 Unicode and UTF-8

In the late 1980's, there was growing interest in developing a universal character encoding for all languages. People from several computer companies worked together, and by 1990, had developed a draft standard for Unicode. In 1991, the Unicode Consortium was formed and charged with guiding and controlling the development of Unicode. The Unicode Consortium has worked closely with the ISO to define, extend, and maintain the international standard

Table 1.9: UTF-8 encoding of the ISO/IEC 10646 code points.

UCS bits	First code point	Last code point	Bytes	Byte 1	Byte 2	Byte 3	Byte 4
7	U+0000	U+007F	1	0xxxxxxx			
11	U+0080	U+07FF	2	110xxxxx	10xxxxxx		
16	U+0800	U+FFFF	3	1110xxxx	10xxxxxx	10xxxxxx	
21	U+10000	U+10FFFF	4	11110xxx	10xxxxxx	10xxxxxx	10xxxxxx

for a Universal Character Set (UCS). This standard is known as the ISO/IEC 10646 standard. The ISO/IEC 10646 standard defines the mapping of code points (numbers) to glyphs (characters), but does not specify character collation or other language-dependent properties. UCS code points are commonly written in the form U+XXXX, where XXXX in the numerical code point in hexadecimal. For example, the code point for the ASCII DEL character would be written as U+007F. Unicode extends the ISO/IEC standard and specifies language-specific features.

Originally, Unicode was designed as a 16-bit encoding. It was not fully backward-compatible with ASCII, and could encode only 65,536 code points. Eventually, the Unicode character set grew to encompass 1,112,064 code points, which requires 21 bits per character for a straightforward binary encoding. By early 1992, it was clear that some clever and efficient method for encoding character data was needed.

UTF-8 (UCS Transformation Format-8-bit) was proposed and accepted as a standard in 1993. UTF-8 is a variable-width encoding that can represent every character in the Unicode character set using between one and four bytes. It was designed to be backward compatible with ASCII and to avoid the major issues of previous encodings. Code points in the Unicode character set with lower numerical values tend to occur more frequently than code points with higher numerical values. UTF-8 encodes frequently occurring code points with fewer bytes than those which occur less frequently. For example the first 128 characters of the UTF-8 encoding are exactly the same as the ASCII characters, requiring only seven bits to encode each ASCII character. Thus any valid ASCII text is also valid UTF-8 text. UTF-8 is now the most common character encoding for the World Wide Web, and is the recommended encoding for email messages.

In November 2003, UTF-8 was restricted by RFC 3629 to end at code point $10FFFF_{16}$. This allows UTF-8 to encode 1,114,111 code points, which is slightly more than the 1,112,064 code points defined in the ISO/IEC 10646 standard. Table 1.9 shows how ISO/IEC 10646 code points are mapped to a variable-length encoding in UTF-8. Note that the encoding allows each byte in a stream of bytes to be placed in one of three distinct categories as follows:

1. If the most significant bit of a byte is zero, then it is a single-byte character, and is completely ASCII-compatible.
2. If the two most significant bits in a byte are set to one, then the byte is the beginning of a multi-byte character.
3. If the most significant bit is set to one, and the next bit is set to zero, then the byte is part of a multi-byte character, but is not the first byte in that sequence.

The UTF-8 encoding of the UCS characters has several important features:

Backwards compatible with ASCII: This allows the vast number of existing ASCII documents to be interpreted as UTF-8 documents without any conversion.

Self synchronization: Because of the way code points are assigned, it is possible to find the beginning of each character by looking only at the top two bits of each byte. This can have important performance implications when performing searches in text.

Encoding of code sequence length: The number of bytes in the sequence is indicated by the pattern of bits in the first byte of the sequence. Thus, the beginning of the next character can be found quickly. This feature can also have important performance implications when performing searches in text.

Efficient code structure: UTF-8 efficiently encodes the UCS code points. The high-order bits of the code point go in the lead byte. Lower-order bits are placed in continuation bytes. The number of bytes in the encoding is the minimum required to hold all the significant bits of the code point.

Easily extended to include new languages: This feature will be greatly appreciated when we contact intelligent species from other star systems.

With UTF-8 encoding The first 128 characters of the UCS are each encoded in a single byte. The next 1,920 characters require two bytes to encode. The two-byte encoding covers almost all Latin alphabets, and also Arabic, Armenian, Cyrillic, Coptic, Greek, Hebrew, Syriac, and Tāna alphabets. It also includes combining diacritical marks, which are used in combination with another character, such as á, ñ, and ö. Most of the Chinese, Japanese, and Korean (CJK) characters are encoded using three bytes. Four bytes are needed for the less common CJK characters, various historic scripts, mathematical symbols, and emoji (pictographic symbols).

Consider the UTF-8 encoding for the British Pound symbol (£), which is UCS code point U+00A3. Since the code point is greater than $7F_{16}$, but less than 800_{16}, it will require two bytes to encode. The encoding will be 110xxxxx 10xxxxxx, where the x characters are replaced with the 11 least-significant bits of the code point, which are 00010100011. Thus, the character £ is encoded in UTF-8 as 11000010 10100011 in binary, or C2 A3 in hexadecimal.

The UCS code point for the Euro symbol (€) is U+20AC. Since the code point is between 800_{16} and $FFFF_{16}$, it will require three bytes to encode in UTF-8. The three-byte encoding is 1110xxxx 10xxxxxx 10xxxxxx where the x characters are replaced with the 16 least-significant bits of the code point. In this case the code point, in binary is 0010000010101100. Therefore, the UTF-8 encoding for € is 11100010 10000010 10101100 in binary, or E2 82 AC in hexadecimal.

In summary, there are three components to modern language support. The ISO/IEC 10646 defines a mapping from code points (numbers) to glyphs (characters). UTF-8 defines an efficient variable-length encoding for code points (text data) in the ISO/IEC 10646 standard. Unicode adds language specific properties to the ISO/IEC 10646 character set. Together, these three elements currently provide support for textual data in almost every human written language, and they continue to be extended and refined.

1.4 Memory layout of an executing program

Computer memory consists of number of storage locations, or cells, each of which has a unique numeric *address*. Addresses are usually written in hexadecimal. Each storage location can contain a fixed number of binary digits. The most common size is one byte. Most computers group bytes together into words. A computer CPU that is capable of accessing a single byte of memory is said to have *byte addressable* memory. Some CPUs are only capable of accessing memory only in word-sized groups. They are said to have *word addressable* memory.

Fig. 1.6A shows a section of memory containing some data. Each byte has a unique address that is used when data is transferred to or from that memory cell. Most processors can also move data in word-sized chunks. On a 32-bit system, four bytes are grouped together to form a word. There are two ways that this grouping can be done. Systems that store the most significant byte of a word in the smallest address, and the least significant byte in the largest address, are said to be *big-endian*. The big-endian interpretation of a region of memory is shown in Fig. 1.6B. As shown in Fig. 1.6C, *little-endian* systems store the least significant byte in the lowest address and the most significant byte in the highest address. Some processors, such as the ARM, can be configured as either little-endian or big-endian. The Linux operating system, by default, configures the ARM processor to run in little-endian mode.

The memory layout for a typical program is shown in Fig. 1.7. The program is divided into six major memory regions, or *sections*. The programmer specifies the contents of the `.text`, `.data`, `.rodata`, and `.bss` sections. The Stack and Heap sections are defined when the program is loaded for execution. The Stack and Heap may grow and shrink as the program executes, while the other sections are set to fixed sizes by the compiler, linker, and loader. The `.text` section contains the executable instructions. The `.data`, `.rodata`, and `.bss` sections

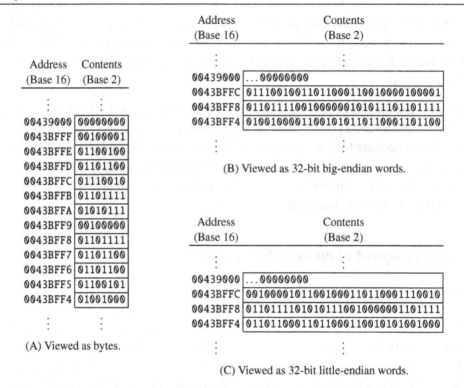

Figure 1.6: A section of memory can be viewed in different ways.

contain constants and static variables. The size of the `.text`, `.data`, `.rodata`, and `.bss` sections depend on how large the program is, and how much static data storage has been declared by the programmer. The heap contains variables that are allocated dynamically, and the stack is used to store parameters for function calls, return addresses, and local (automatic) variables.

In a high level language, storage space for a variable can be allocated in one of three ways: statically, dynamically, and automatically. Statically allocated variables are allocated in the `.data`, `.rodata`, or `.bss` section. The storage space is reserved, and usually initialized, when the program is loaded and begins execution. The address of a statically allocated variable is fixed at the time the program begins running, and cannot be changed. Automatically allocated variables, often referred to as local variables, are stored on the stack. The stack pointer is adjusted down to make space for the newly allocated variable. The address of an automatic variable is always computed as an offset from the stack pointer. Dynamic variables are allocated from the heap, using `malloc`, `new`, or a language-dependent equivalent. The address of a dynamic variable is always stored in another variable, known as a pointer, which may be an automatic or static variable, or even another dynamic variable. The six major sections of

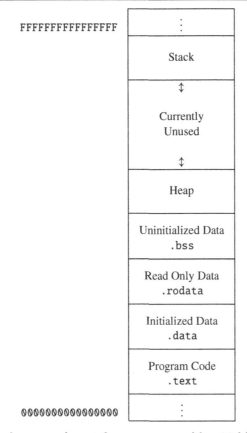

FFFFFFFFFFFFFFFF

0000000000000000

Stack

Currently
Unused

Heap

Uninitialized Data
.bss

Read Only Data
.rodata

Initialized Data
.data

Program Code
.text

Figure 1.7: Typical memory layout for a program with a 32-bit address space.

program memory correspond to executable code, statically allocated variables, dynamically allocated variables, and automatically allocated variables.

1.5 Chapter summary

There are several reasons for Computer Scientists and Computer Engineers to learn at least one assembly language. There are programming tasks that can only be performed using assembly language, and some tasks can be written to run much more efficiently and/or quickly if written in assembly language. Programmers with assembly language experience tend to write better code even when using a high-level language, and are usually better at finding and fixing bugs.

Although it is possible to construct a computer capable of performing arithmetic in any base, it is much cheaper to build one that works in base two. It is relatively easy to build an electrical circuit with two states, using two discrete voltage levels, but much more difficult to build a stable circuit with ten discrete voltage levels. Therefore, modern computers work in base two.

Computer data can be viewed as simple bit strings. The programmer is responsible for supplying interpretations to give meaning to those bit strings. A set of bits can be interpreted as a number, a character, or anything that the programmer chooses. There are standard methods for encoding and interpreting characters and numbers. Fig. 1.4 shows some common methods for encoding integers. The most common encodings for characters are UTF-8 and ASCII.

Computer memory can be viewed as a sequence of bytes. Each byte has a unique address. A running program has several regions of memory. One region holds the executable code. The other regions hold different types of variables.

Exercises

1.1.   What is the two's complement of 11011101?

1.2.   Perform the base conversions to fill in the blank spaces in the following table:

Base 10	Base 2	Base 16	Base 21
23			
	010011		
		ABB	
			2HE

1.3.   What is the 8-bit ASCII binary representation for the following characters?
   a.   "A"
   b.   "a"
   c.   "!"

1.4.   What is \ minus ! given that \ and ! are ASCII characters? Give your answer in binary.

1.5.   Representing characters:
   a.   Convert the string "Super!" to its ASCII representation. Show your result as a sequence of hexadecimal values.
   b.   Convert the hexadecimal sequence into a sequence of values in base four.

1.6.   Suppose that the string "This is a nice day" is stored beginning at address $4B3269AC_{16}$. What are the contents of the byte at address $4B3269B1_{16}$ in hexadecimal?

1.7.   Perform the following:
   a.   Convert 101101_2 to base ten.
   b.   Convert 1023_{10} to base nine.
   c.   Convert 1023_{10} to base two.
   d.   Convert 301_{10} to base 16.

e. Convert 301_{10} to base 2.

f. Represent 301_{10} as a null-terminated ASCII string. (Write your answer in hexadecimal.)

g. Convert 3420_5 to base ten.

h. Convert 2314_5 to base nine.

i. Convert 116_7 to base three.

j. Convert 1294_{11} to base 5.

1.8. Given the following binary string:

01001001 01110011 01101110 00100111 01110100 00100000 01000001 01110011
01110011 01100101 01101101 01100010 01101100 01111001 00100000 01000110
01110101 01101110 00111111 00000000

a. Convert it to a hexadecimal string.

b. Convert the first four bytes to a string of base ten numbers.

c. Convert the first (little-endian) halfword to base ten.

d. Convert the first (big-endian) halfword to base ten.

e. If this string of bytes were sent to an ASCII printer or terminal, what would be printed?

1.9. The number 1,234,567 is stored as a 32-bit word starting at address $F04390000_{16}$. Show the address and contents of each byte of the 32-bit word on a

a. little-endian system.

b. big-endian system.

1.10. Really good assembly programmers can convert small numbers between binary, hexadecimal, and decimal in their heads. Without referring to any tables or using a calculator or pencil, fill in the blanks in the following table:

Binary	Decimal	Hexadecimal
	5	
1010		
		C
	23	
0101101		
		4B

1.11. UTF-8 is often referred to as Unicode. Why is this not correct?

1.12. What are the differences between a CPU register and a memory location?

1.13. The ISO/IEC 10646 standard defines 1,112,064 code points (glyphs). Each code point could be encoded using 24 bits, or three bytes. The UTF-8 encoding uses up to four bytes to encode a code point. Give three reasons why UTF-8 is preferred over a simple 3-byte per code point encoding.

## CHAPTER 2

GNU assembly syntax

All modern computers consist of three main components: the central processing unit (CPU), memory, and devices. It can be argued that the major factor that distinguishes one computer from another is the CPU architecture. The architecture determines the set of instructions that can be performed by the CPU. The human-readable language which is closest to the CPU architecture is assembly language. When a new processor architecture is developed, its creators also define an assembly language for the new architecture. In most cases, a precise assembly language syntax is defined and an assembler is created by the processor developers. Because of this, there is no single syntax for assembly language, although most assembly languages are similar in many ways, and have certain elements in common. The GNU assembler (GAS) is a highly portable re-configurable assembler. GAS uses a simple and general syntax that works for a wide variety of architectures.

2.1 Structure of an assembly program

An assembly program consists of four basic elements: assembler directives, labels, assembly instructions, and comments. Assembler directives allow the programmer to reserve memory for the storage of variables, control which program section is being used, define macros, include other files, and perform other operations that control the conversion of assembly instructions into machine code. The assembly instructions are given as *mnemonics*, or short character strings that are easier for human brains to remember than sequences of binary, octal, or hexadecimal digits. Each assembly instruction may have an optional label, and most assembly instructions require the programmer to specify one or more operands.

Most assembly language programs are written in lines of 80 characters organized into four columns. The first column is for optional labels. The second column is for assembly instructions or assembler directives. The third column is for specifying operands, and the fourth column is for comments. Traditionally, the first two columns are 8 characters wide, the third column is 16 characters wide, and the last column is 48 characters wide. However, most modern assemblers (including GAS) do not require fixed column widths. Listing 2.2 shows a basic "Hello World" program written in AArch64 Assembly to run under Linux. For comparison,

Listing 2.1 "Hello World" program in C.

```
 1   #include <stdio.h>
 2   /*
 3   * Prints "Hello World\n" and returns 0.
 4   */
 5   int main(void) {
 6      printf("Hello World\n");
 7      return 0;
 8   }
```

Listing 2.1 shows an equivalent program written in C. The assembly language version of the program is significantly more lines than the C version, and will only work on an AArch64 processor or emulator. The C version is at a higher level of abstraction, and can be compiled to run on any system that has a C compiler. Thus, C is referred to as a high-level language, and assembly as a low-level language.

2.1.1 Labels

Most modern assemblers are called two-pass assemblers because they read the input file twice. On the first pass, the assembler keeps track of the location of each piece of data and each instruction, and assigns an address or numerical value to each label and symbol in the input file. The main goal of the first pass is to build a symbol table, which maps each label or symbol to a numerical value.

On the second pass, the assembler converts the assembly instructions and data declarations into binary, using the symbol table to supply numerical values whenever they are needed. In Listing 2.2, there are two labels: `main` and `message`. During assembly, those labels are assigned the value of the *address counter* at the point where they appear. Labels can be used anywhere in the program to refer to the address of data, functions, or blocks of code. In GNU assembly syntax, label definitions always end with a colon (`:`) character.

2.1.2 Comments

In the GNU assembler, there are two basic comment styles: multi-line and single-line. Multi-line comments start with /* and everything is ignored until a matching sequence of */ is found. These comments are exactly the same as multi-line comments in C and C++. Single line comments begin with //, and all remaining characters on the line are ignored. This is the same as single line comments in C++. Listing 2.2 shows both types of comment.

Listing 2.2 "Hello World" program in AArch64 assembly.

```
        .section .rodata        // Read-only data section
mesg:   .asciz   "Hello World\n" // Define null-terminated string

        .text                   // Text section
        .global  main
        /*
        * Prints "Hello World\n" and returns 0.
        */
main:   stp      x29, x30, [sp, #-16]!

        // printf("Hello World\n")
        adr      x0, mesg
        bl       printf

        // return 0
        mov      w0, #0
        ldp      x29, x30, [sp], #16
        ret
        .size    main,(. - main)
```

2.1.3 Directives

Directives are used mainly to define symbols, allocate storage, and control the behavior of the assembler. Directives allow the programmer to control how the assembler does its job. The GNU assembler has many directives, but assembly programmers typically need to know only a few of them. All assembler directives begin with a period '.' which is followed by a sequence of letters, usually in lower case. Listing 2.2 uses the .section, .rodata, .asciz, .text, and .global directives. The most commonly used directives are discussed later in this chapter. There are many other directives available in the GNU Assembler which are not covered here. Complete documentation is available on-line as part of the GNU Binutils package.

2.1.4 Assembly instructions

Assembly instructions are the program statements that will be executed on the CPU. Most instructions cause the CPU to perform one low-level operation. In most assembly languages, operations can be divided into a few major types. Some instructions move data from one location to another. Others perform addition, subtraction, and other computational operations. Another class of instructions are used to perform comparisons, and control which part of the

Listing 2.3 "Hello World" assembly listing.

```
Line Addr Code/Data    Label    Instruction/Directive
------------------------------------------------------------------
1                               .data
2    0000 48656C6C     mesg:    .asciz  "Hello World\n"
2         6F20576F
2         726C640A
2         00
3
4                               .text
5                               .global  main
6                      /*
7                      * Prints "Hello World\n" and returns 0.
8                      */
9    0000 FD7BBFA9     main:    stp     x29, x30, [sp, #-16]!
10
11                              // printf("Hello World\n")
12   0004 00000010              adr     x0, mesg
13   0008 00000094              bl      printf
14
15                              // return 0
16   000c 00008052              mov     w0, #0
17   0010 FD7BC1A8              ldp     x29, x30, [sp], #16
18   0014 C0035FD6              ret

DEFINED SYMBOLS
HelloWorld.S:2     .data:0000000000000000 mesg
HelloWorld.S:9     .text:0000000000000000 main
HelloWorld.S:9     .text:0000000000000000 $x

UNDEFINED SYMBOLS
printf
```

program is to be executed next. Chapter 3 and Chapter 4 explain most of the assembly instructions that are available in AArch64.

2.2 What the assembler does

Listing 2.3 shows how the GNU assembler will assemble the "Hello World" program from Listing 2.2. The assembler converts the string on input line 2 into the binary representation of the string. The results are shown in hexadecimal in the Code/Data column of the listing. The first byte of the string is stored at address 0000, in the .data section of the program, as shown in the Addr column on line 2. This line also has a label, which shows up in the DEFINED SYMBOLS table at the end of the listing.

On line 4, the assembler switches to the .text section of the program and begins converting instructions into binary. The first instruction, on line 9, is converted into its 4-byte machine code, $FD7BBFA9_{16}$, and stored at location 0000 in the .text section of the program, as shown in the Code/Data and Addr columns on line 9. This line also has a label, which shows up in the DEFINED SYMBOLS table at the end of the listing.

Next, the assembler converts the `adr` instruction on line 12 into the four-byte machine instruction 00000010_{16} and stores it at address `0004`. It repeats this process with each remaining instruction until the end of the program. The assembler writes the resulting data into a specially formatted file, called an object file. Note that the assembler was unable to locate the `printf` function. The linker will take care of that. The object file created by the assembler, `hello.o`, contains the data in the `Code` column of Listing 2.3, along with information to help the linker to link (or "patch") the instruction on line 13 so that `printf` is called correctly.

After creating the object file, the next step in creating an executable program would be to invoke the linker and request that it link `hello.o` with the C standard library. The linker will generate the final executable file, containing the code assembled from `hello.s`, along with the `printf` function and other start-up code from the C standard library. The GNU C compiler is capable of automatically invoking the assembler for files that end in `.s` or `.S`, and can also be used to invoke the linker. For example, if Listing 2.2 is stored in a file named `hello.S` in the current directory, then the command

```
gcc -o hello hello.S
```

will run the GNU C compiler, telling it to create an executable program file named `hello`, and to use `hello.S` as the source file for the program. The C compiler will notice the `.S` extension, and invoke the assembler to create an object file which is stored in a temporary file, possibly named `hello.o`. Then the C compiler will invoke the linker to link `hello.o` with the C standard library, which provides the `printf` function and some start-up code which calls the `main` function. The linker will create an executable file named `hello`. When the linker has finished, the C compiler will remove the temporary object file.

2.3 GNU assembly directives

Each processor architecture has its own assembly language, created by the designers of the architecture. Although there are many similarities between assembly languages, the designers may choose different names for various directives. The GNU assembler supports a relatively large set of directives, and some of them have more than one name. This is because it is designed to handle assembling code for many different processors, without drastically changing the assembly language designed by the processor manufacturers. We will now cover some of the most commonly used directives for the GNU assembler.

2.3.1 Selecting the current section

The instructions and data that make up a program are stored in different *sections* of the program file. There are several standard sections that the programmer can choose to put code and data in. Sections can also be further divided into numbered subsections. Each section has its own address counter, which is used to keep track of the location of bytes within that section. When a label is encountered, it is assigned the value of the current address counter for the currently active section.

Selecting a section and subsection is done by using the appropriate assembly directive. Once a section has been selected, all of the instructions and/or data will go into that section until another section is selected. The most important directives for selecting a section are:

`.data subsection`
> Instructs the assembler to append the following instructions or data to the data subsection numbered `subsection`. If the subsection number is omitted, it defaults to zero. This section is normally used for global variables and constants, which have labels.

`.text subsection`
> Tells the assembler to append the following statements to the end of the text subsection numbered `subsection`. If the subsection number is omitted, subsection number zero is used. This section is normally used for executable instructions.

`.bss subsection`
> The bss (short for Block Started by Symbol) section is used for defining data storage areas that should be initialized to zero at the beginning of program execution. The `.bss` directive tells the assembler to append the following statements to the end of the bss subsection numbered `subsection`. If the subsection number is omitted, subsection number zero is used. This section is normally used for global variables which need to be initialized to zero. Regardless of what is placed into the section at compile-time, all bytes will be set to zero when the program begins executing. This section does not actually consume any space in the object or executable file. It is really just a request for the loader to reserve some space when the program is loaded into memory.

`.section name`
> In addition to the three common sections, the programmer can create other sections using this directive. However in order for custom sections to be linked into a program, the linker must be made aware of them. With the GNU tools, this can be accomplished by providing a *linker script*. That topic is outside the scope of this book, but interested readers are referred to the GNU linker documentation.

2.3.2 Allocating space for variables and constants

There are several directives that allow the programmer to allocate and initialize static storage space for variables and constants. The assembler supports bytes, integer types, floating point types, and strings. These directives are used to allocate a fixed amount of space in memory and optionally initialize the memory. Some of these directives allow the memory to be initialized using an expression. An expression can be a simple integer, or a C-style expression. The directives for allocating storage are as follows:

`.byte expressions`

> `.byte` expects zero or more expressions, separated by commas. Each expression produces one byte of data. The first byte is placed at the current address counter, and then the counter is incremented so that the second expression is placed at the next byte, and so on. If no expressions are given, then the address counter is not advanced and no bytes are reserved.

`.2byte expressions`
`.hword expressions`
`.short expressions`

> For the AArch64 processor, these directives all do exactly the same thing. They expect zero or more expressions, separated by commas, and emit a 16-bit number for each expression. If no expressions are given, then the address counter is not advanced and no bytes are reserved.

`.4byte expressions`
`.word expressions`
`.long expressions`

> For the AArch64 processor, these three directives do exactly the same thing. They expect zero or more expressions, separated by commas, and will emit four bytes for each expression given. If no expressions are given, then the address counter is not advanced and no bytes are reserved.

`.8byte expressions`
`.quad expressions`

> For the AArch64 processor, these two directives do exactly the same thing. They expect zero or more expressions, separated by commas, and will eight bytes for each expression given. If no expressions are given, then the address counter is not advanced and no bytes are reserved.

`.ascii "string"`

> The `.ascii` directive expects zero or more string literals, each enclosed in quotation marks and separated by commas. It assembles each string (with no trailing ASCII NULL character) into consecutive addresses.

`.asciz "string"`
`.string "string"`

> The `.asciz` directive is similar to the `.ascii` directive, but each string is followed by an ASCII NULL character (zero). The "z" in `.asciz` stands for zero. `.string` is just another name for `.asciz`.

```
        .data
i:          .word    0
j:          .word    1
fmt:        .asciz   "Hello\n"
ch:         .byte    'A','B',0
ary:        .word    0,1,2,3,4
```

```
static int i = 0;
static int j = 1;
static char fmt[] = "Hello\n";
static char ch[] = {'A','B',0};
static int ary[] = {0,1,2,3,4};
```

(A) Declarations in Assembly

(B) Declarations in C

Figure 2.1: Equivalent static variable declarations in Assembly and C.

`.float flonums`
`.single flonums`

> This directive assembles zero or more floating point numbers, separated by commas.
> In AArch64, they are 4-byte IEEE standard single precision numbers. `.float` and
> `.single` are synonyms.

`.double flonums`

> The `.double` directive expects zero or more floating point numbers, separated by com-
> mas. In AArch64, they are stored as 8-byte IEEE standard double precision numbers.

Fig. 2.1A shows how these directives are used to declare variables and constants. Fig. 2.1B
shows the equivalent statements for creating global variables in C or C++. Note that in both
cases, the variables created will be visible anywhere within the file that they are declared, but
not visible in other files which are linked into the program.

In C, the declaration of an array can be performed by leaving out the number of elements and
specifying an initializer, as shown in the last three lines of Fig. 2.1B. In assembly, the equivalent is accomplished by providing a label, a type, and a list of values, as shown in the last
three lines of Fig. 2.1A. The syntax is different, but the result is precisely the same.

Listing 2.4 shows how the assembler assigns addresses to these labels. The second column
of the listing shows the address (in hexadecimal) that is assigned to each label. The variable
i is assigned the first address. Since it is a word variable, the address counter is incremented
by four bytes and the next address is assigned to the variable j. The address counter is incremented again, and fmt is assigned the address 0008. The fmt variable consumes seven bytes,
so the ch variable gets address 000f. Finally, the array of words named ary begins at address
0012. Note that $12_{16} = 18_{10}$ is not evenly divisible by four, which means that the word variables in ary are not aligned on word boundaries.

Listing 2.4 A listing with mis-aligned data.

```
ARM GAS   variable1.S                    page 1

line addr value          code
  1                              .data
  2 0000 00000000       i:      .word    0
  3 0004 01000000       j:      .word    1
  4 0008 48656C6C       fmt:    .asciz   "Hello\n"
  4      6F0A00
  5 000f 414200         ch:     .byte    'A','B',0
  6 0012 00000000       ary:    .word    0,1,2,3,4
  6      01000000
  6      02000000
  6      03000000
  6      04000000
```

2.3.3 Filling and aligning

On the AArch64 CPU, data can be moved to and from memory one byte at a time, two bytes at a time (half-word), four bytes at a time (word), or eight bytes at a time (double-word). Moving a word between the CPU and memory takes significantly more time if the address of the word is not aligned on a four-byte boundary (one where the least significant two bits of the address are zero). Similarly, moving a half-word between the CPU and memory takes significantly more time if the address of the half-word is not aligned on a two-byte boundary (one where the least significant bit of the address is zero), and moving a double-word takes more time if it is not aligned on an eight-byte boundary (one where the least significant three bits of the address are zero). Therefore, when declaring storage, it is important that double-words, words, and half-words are stored on appropriate boundaries. The following directives allow the programmer to insert as much space as necessary to align the next item on any boundary desired.

.align abs-expr, abs-expr, abs-expr
> Pad the location counter (in the current subsection) to a particular storage boundary. For the AArch64 processor, the first expression specifies the number of low-order zero bits the location counter must have after advancement. The second expression gives the fill value to be stored in the padding bytes. It (and the comma) may be omitted. If it is omitted, then the fill value is assumed to be zero. The third expression is also optional. If it is present, it is the maximum number of bytes that should be skipped by this alignment directive.

`.balign[lw] abs-expr, abs-expr, abs-expr`

These directives adjust the location counter to a particular storage boundary. The first expression is the byte-multiple for the alignment request. For example `.balign 16` will insert fill bytes until the location counter is an even multiple of 16. If the location counter is already a multiple of 16, then no fill bytes will be created. The second expression gives the fill value to be stored in the fill bytes. It (and the comma) may be omitted. If it is omitted, then the fill value is assumed to be zero. The third expression is also optional. If it is present, it is the maximum number of bytes that should be skipped by this alignment directive.

The `.balignw` and `.balignl` directives are variants of the `.balign` directive. The `.balignw` directive treats the fill pattern as a two byte word value, and `.balignl` treats the fill pattern as a four byte longword value. For example, "`.balignw 4,0x368d`" will align to a multiple of four bytes. If it skips two bytes, they will be filled in with the value `0x368d` (the exact placement of the bytes depends upon the endianness of the processor).

`.skip size, fill`

`.space size, fill`

Sometimes it is desirable to allocate a large area of memory and initialize it all to the same value. This can be accomplished by using these directives. These directives emit `size` bytes, each of value `fill`. Both `size` and `fill` are absolute expressions. If the comma and `fill` are omitted, `fill` is assumed to be zero. For the ARM processor, the `.space` and `.skip` directives are equivalent. This directive is very useful for declaring large arrays in the `.bss` section.

Listing 2.5 shows how the code in Listing 2.4 can be improved by adding an alignment directive at line 6. The directive causes the assembler to emit two zero bytes between the end of the ch variable and the beginning of the ary variable. These extra "padding" bytes cause the following word data to be word aligned, thereby improving performance when the word data is accessed. It is good practice to always put an appropriate alignment directive when the size of the data on the next line is larger than the size of the data on the current line. For example, if the previous line declares string, byte, half-word, or word data, and the current line declares double-word data, then an `.align 3` directive should be placed between them. Likewise, if the current line declares word data and the previous line declares string, byte, or half-word data, then an `.align 2` directive should be placed between them. For a line declaring half-word data following a line declaring string or byte data, an `.align 1` directive should be placed between them.

2.3.4 Setting and manipulating symbols

The assembler provides support for setting and manipulating symbols which can then be used in other places within the program. The labels that can be assigned to assembly statements

Listing 2.5 A listing with properly aligned data.

```
ARM GAS   variable2.S                        page 1

line addr value          code
  1                             .data
  2 0000 00000000    i:         .word   0
  3 0004 01000000    j:         .word   1
  4 0008 48656C6C    fmt:       .asciz  "Hello\n"
  4      6F0A00
  5 000f 414200      ch:        .byte   'A','B',0
  6 0012 0000                   .align  2
  7 0014 00000000    ary:       .word   0,1,2,3,4
  7      01000000
  7      02000000
  7      03000000
  7      04000000
```

and directives are one type of symbol. The programmer can also declare other symbols and use them throughout the program. Such symbols may not have an actual storage location in memory, but they are included in the assembler's symbol table, and can be used anywhere that their value is required. The most common use for defined symbols is to allow numerical constants to be declared in one place and easily changed. The .equ directive allows the programmer to use a label instead of a number throughout the program. This contributes to readability, and has the benefit that the constant value can then be easily changed every place that it is used, just by changing the definition of the symbol. The most important directives related to symbols are:

.equ symbol, expression
.set symbol, expression

> This directive sets the value of symbol to expression. It is similar to the C language #define directive.

.equiv symbol, expression

> The .equiv directive is like .equ and .set, except that the assembler will signal an error if the symbol is already defined.

.global symbol
.globl symbol

> This directive makes the symbol visible to the linker. If symbol is defined within a file, and this directive is used to make it global, then it will be available to any file that is

Listing 2.6 Defining a symbol for the number of elements in an array.

```
ARM GAS   variable3.S                    page 1

line addr value          code
  1                           .data
  2 0000 00000000      i:    .word    0
  3 0004 01000000      j:    .word    1
  4 0008 48656C6C      fmt:  .asciz   "Hello\n"
  4      6F0A00
  5 000f 414200        ch:   .byte    'A','B',0
  6 0012 0000                .align   2
  7 0014 00000000      ary:  .word    0,1,2,3,4
  7      01000000
  7      02000000
  7      03000000
  7      04000000
  8                          .equ     arysize,(. - ary)/4
  9


DEFINED SYMBOLS
        variable3.S:2    .data:0000000000000000 i
        variable3.S:3    .data:0000000000000004 j
        variable3.S:4    .data:0000000000000008 fmt
        variable3.S:5    .data:000000000000000f ch
        variable3.S:6    .data:0000000000000012 $d
        variable3.S:7    .data:0000000000000014 ary
        variable3.S:8    *ABS*:0000000000000005 arysize

NO UNDEFINED SYMBOLS
```

linked with the one containing the symbol. Without this directive, symbols are visible only within the file where they are defined.

.comm symbol, length

This directive declares symbol to be a common symbol, meaning that if it is defined in more than one file, then all instances should be merged into a single symbol. If the symbol is only defined in one file, then the linker will allocate length bytes of uninitialized memory. If there are multiple definitions for symbol, and they have different sizes, the linker will merge them into a single instance using the largest size defined.

Listing 2.6 shows how the .equ directive can be used to create a symbol holding the number of elements in an array. The symbol arysize is defined as the value of the current address counter (denoted by the .) minus the value of the ary symbol, divided by four (each word in the array is four bytes). The listing shows all of the symbols defined in this program segment.

Note that the four variables are shown to be in the data segment, and the `arysize` symbol is marked as an "absolute" symbol, which simply means that it is a number and not an address. It will have no storage location in the running program, but it's value can be used whereever needed while the file is being assembled. The programmer can now use the symbol `arysize` to control looping when accessing the array data. If the size of the array is changed by adding or removing constant values, the value of `arysize` will change automatically, and the programmer will not have to search through the code to change the original value, 5, to some other value in every place it is used.

2.3.5 Functions and objects

There are a few assembler directives that are used for defining the size and type of labels. This information is stored in the object file along with the code and data, and is used by the linker and/or debugger.

`.size name,expression`

>   The `.size` directive is used to set the size associated with a symbol. This information helps the linker to exclude unneeded code and/or data when creating an executable file, and helps the debugger to keep track of where it is in the program.

`.type name,type_description`

>   The `.type` directive sets the type of a symbol name to be either a function or an object. Valid values for `type_desription` in GNU AArch64 assembly include:
>
>   `%function`   The symbol is a function name.
>
>   `%gnu_indirect_function`   The symbol is an indirect function (may be called through a function pointer table).
>
>   `%object`   The symbol is a data object.
>
>   `%tls_object`   The symbol is a thread-local data object.
>
>   `%common`   The symbol is a common (shared) object.
>
>   `%notype`   The symbol has no type.
>
>   **Note:** Some assemblers, including some versions of the GNU assembler, a may require the @ character instead of the % character.

The following example shows how a typical function is declared.

```
        .type   myfunc, %function
        .global myfunc
myfunc: // an example function named myfunc

        // The function code goes here

        .size myfunc,(. - myfunc)
```

The period "." in the expression (. - myfunc) is a reference to the current location counter value. The expression (. - myfunc) means "Subtract the location of myfunc from the current location." This directive calculates how many bytes there are between the label myfunc and the location of the .size directive, and provides that information for the linker and/or debugger.

2.3.6 Conditional assembly

Sometimes it is desirable to skip assembly of portions of a file. The assembler provides some directives to allow conditional assembly. One use for these directives is to optionally assemble code as an aid for debugging.

.if expression
> .if marks the beginning of a section of code which is only considered part of the source program being assembled if the argument (which must be an absolute expression) is non-zero. The end of the conditional section of code must be marked by the .endif directive. Optionally, code may be included for the alternative condition by using the .else directive.

.ifdef symbol
> Assembles the following section of code if the specified symbol has been defined.

.ifndef symbol
> Assembles the following section of code if the specified symbol has not been defined.

.else
> Assembles the following section of code only if the condition for the preceding .if or .ifdef was false.

.endif
> Marks the end of a block of code that is only assembled conditionally.

2.3.7 Including other source files

.include "file"
> This directive provides a way to include supporting files at specified points in the source program. The code from the included file is assembled as if it followed the point of the .include directive. When the end of the included file is reached, assembly of the original file continues. The search paths used can be controlled with the '-I' command line parameter when running the assembler. Quotation marks are required around file. This assembler directive is similar to including header files in C and C++ using the #include compiler directive.

2.3.8 Macros

The directives .macro and .endm allow the programmer to define *macros* that the assembler
expands to generate assembly code. The GNU assembler supports simple macros. Some other
assemblers have much more powerful macro capabilities.

.macro macname
.macro macname macargs ...

> Begin the definition of a macro called macname. If the macro definition requires argu-
> ments, their names are specified after the macro name, separated by commas or spaces.
> The programmer can supply a default value for any macro argument by following the
> name with '=deflt'.

The following begins the definition of a macro called reserve_str, with two arguments. The
first argument has a default value, but the second does not:

```
.macro reserve_str p1=0 p2
```

When a macro is called, the argument values can be specified either by position, or by keyword. For example, reserve_str 9,17 is equivalent to reserve_str p2=17,p1=9. After
the definition is complete, the macro can be called either as
reserve_str x,y
(with \p1 evaluating to x and \p2 evaluating to y), or as
reserve_str ,y
(with \p1 evaluating as the default, in this case 0, and \p2 evaluating to y). Other examples of
valid .macro statements are:

```
// Begin the definition of a macro called comm,
// which takes no arguments:
.macro comm
```

```
// Begin the definition of a macro called plus1,
// which takes two arguments:
.macro plus1 p, p1
// Write \p or \p1 to use the arguments.
```

.endm
> End the current macro definition.

.exitm
> Exit early from the current macro definition. This is usually used only within a .if or
> .ifdef directive.

\\@

This is a pseudo-variable used by the assembler to maintain a count of how many macros it has executed. That number can be accessed with '\\@', but only within a macro definition.

2.3.8.1 Macro example

The following definition specifies a macro SHIFT that will emit the instruction to shift a given register left by a specified number of bits. If the number of bits specified is negative, then it will emit the instruction to perform a right shift instead of a left shift.

```
.macro SHIFT a,b
.if \b < 0
asr \a, \a, #-\b
.else
lsl \a, \a, #\b
.endif
.endm
```

After that definition, the following code:

```
SHIFT   x1, 3
SHIFT   x4, -6
```

will generate these instructions:

```
lsl     x1, x1, #3
asr     x4, x4, #6
```

The meaning of these instructions will be covered in Chapter 3 and Chapter 4.

2.3.8.2 Recursive macro example

The following definition specifies a macro enum that puts a sequence of numbers into memory by using a recursive macro call to itself:

```
.macro  enum first=0, last=5
.long   \first
.if     \last-\first
enum    "(\first+1)",\last
.endif
.endm
```

With that definition, 'enum 0,5' is equivalent to this assembly input:

```
    .long   0
    .long   1
    .long   2
    .long   3
    .long   4
    .long   5
```

2.4 Chapter summary

There are four elements to assembly syntax: labels, directives, instructions, and comments. Directives are used mainly to define symbols, allocate storage, and control the behavior of the assembler. The most common assembler directives were introduced in this chapter, but there are many other directives available in the GNU Assembler. Complete documentation is available on-line as part of the GNU Binutils package.

Directives are used to declare statically allocated storage, which is equivalent to declaring global static variables in C. In assembly, labels and other symbols are visible only within the file that they are declared, unless they are explicitly made visible to other files with the .global directive. In C, variables which are declared outside of any function are visible to all files in the program, unless the static keyword is used to make them visible only within the file where they are declared. Thus, both C and assembly support file and global scope for static variables, but with the opposite defaults.

Directives can also be used to declare macros. Macros are expanded by the assembler and may generate multiple statements. Careful use of macros can automate some simple tasks, allowing several lines of assembly code to be replaced with a single macro invocation.

Exercises

2.1. What is the difference between
   a. the .data section and the .bss section?
   b. the .ascii and the .asciz directives?
   c. the .word and the .long directives?
2.2. What is the purpose of the .align assembler directive? What does ".align 2" do in GNU AArch64 assembly?
2.3. Assembly language has four main elements. What are they?
2.4. Using the directives presented in this chapter, show three different ways to create a null-terminated string containing the phrase "segmentation fault".

2.5. What is the total memory, in bytes, allocated for the following variables?

```
var1:   .word    23
var2:   .long    0xC
expr:   .ascii   ">>"
```

2.6. Identify the directive(s), label(s), comment(s), and instruction(s) in the following code:

```
        .text
        .global main
main:
        mov     w0,#1    // the program return code is 1
        ret              // return and exit the program
```

2.7. Write assembly code to declare variables equivalent to the following C code:

```
/* these variables are declared outside of any function */
static int foo[3];    /* visible anywhere in the current file */
static char bar[4];   /* visible anywhere in the current file */
char barfoo;          /* visible anywhere in the program */
int foobar;           /* visible anywhere in the program */
```

2.8. Show how to store the following text as a single string in assembly language, while making it readable and keeping each line shorter than 80 characters:

```
The three goals of the mission are:
1) Keep each line of code under 80 characters,
2) Write readable comments,
3) Learn a valuable skill for readability.
```

2.9. Insert the minimum number of `.align` directives necessary in the following code so that all double-word variables are aligned on double-word boundaries, all word variables are aligned on word boundaries, and all halfword variables are aligned on halfword boundaries, while minimizing the amount of wasted space.

```
        .data
a:      .byte    7
b:      .word    32
c:      .byte    7
d:      .quad    0x1234567890abcdef
e:      .byte    7
f:      .hword   0xffff
```

2.10. Re-order the directives in the previous problem so that no `.align` directives are necessary to ensure proper alignment. How many bytes of storage were wasted by the original ordering of directives, compared to the new one?

2.11. What are the most important directives for selecting a section?

2.12. Why are `.ascii` and `.asciz` directives usually followed by an `.align` directive, but `.word` directives are not?

2.13. Using the "Hello World" program shown in Listing 2.2 as a template, write a program that will print your last name.

2.14. Listing 2.3 shows that the assembler will assign the location 00000000_{16} to the `main` symbol and also to the `mesg` symbol. Why does this not cause problems?

## CHAPTER 3

Load/store and branch instructions

The part of the computer architecture related to programming is referred to as the *instruction set architecture* (ISA). The ISA is a contract between the hardware and the software. It defines the *set of instructions* and the *set of registers* that the hardware must support. How the hardware actually implements the ISA is called the *microarchitecture*, and it is not defined by the ISA. Assembly programmers can write code that will work across a spectrum of different processor implementations. As long as the software and the hardware are both built around the same ISA, they will be compatible. The ISA is an effective interface between hardware–the *data paths* and *processing elements*–and software, because it creates an *abstraction* that allows each side to be imagined to function independently.

The first step in learning a new assembly language is to become familiar with the ISA. For most modern computer systems, data must be loaded in a register before it can be used for any data processing instruction, but there are a limited number of registers. Memory provides a place to store data that is not currently needed. Program instructions are also stored in memory and fetched into the CPU as they are needed. This chapter introduces A64, the 64-bit ISA for AArch64 processors.

3.1 CPU components and data paths

The CPU is composed of data storage and computational components connected together by a set of buses, or wires. The most important components of the CPU are the *registers*, where data is stored, and the *arithmetic and logic unit* (ALU), where arithmetic and logical operations are performed on the data. Many CPUs also have dedicated hardware units for multiplication and/or division. Fig. 3.1 shows a conceptual view of the major components of an AArch64 CPU and the buses that connect the components together. These buses provide pathways for the data to move between the computational and storage components. The organization of the components and buses in a CPU govern what types of operations can be performed.

The set of instructions and addressing modes available on the AArch64 processor is closely related to the architecture shown in Fig. 3.1. The architecture provides for certain operations to be performed efficiently, and this has a direct relationship to the types of instructions that are supported.

ARM 64-Bit Assembly Language
https://doi.org/10.1016/B978-0-12-819221-4.00010-9
Copyright © 2020 Elsevier Inc. All rights reserved.

ARM CPU

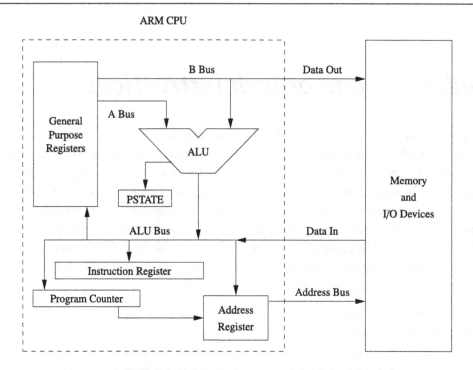

Figure 3.1: A simplified view of the AArch64 architecture.

The AArch64 architecture provides a set of registers that can be used by programmers to store data. There are a limited number of registers, so large programs must store data in memory most of the time, and bring that data into a CPU register to be processed. Results of the processing must be moved back to memory to free up registers to process more data.

3.2 AArch64 user registers

As shown in Fig. 3.2, the AArch64 ISA provides 31 general-purpose registers, which are called X0 through X30. These registers can each store 64 bits of data. To use all 64 bits, they are referred to as x0 through x30 (capitalization is optional). To use only the lower (least significant) 32 bits, they are referred to as w0-w30. Since each register has a 64-bit name and a 32-bit name, we use R0 through R30 to specify a register without specifying the number of bits. For example, when we refer to R12, we are really referring to either x12 or w12.

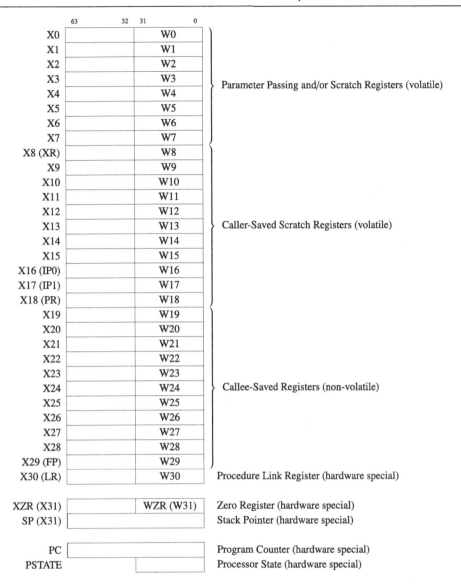

Figure 3.2: AArch64 general purpose registers (R0–R30) and special registers.

3.2.1 General purpose registers

The general-purpose registers are each used according to specific *conventions*. These rules are defined in the *application binary interface (ABI)*. The AArch64 ABI is called *AAPCS64*. The difference between callee saved and caller saved registers will also be explained in Section 5.4.4.

31	30	29	28	27	26	25	24	23	22	21	20	19	18	17	16	15	14	13	12	11	10	9	8	7	6	5	4	3	2	1	0
N	Z	C	V					UAO	PAN	SS	IL											D	A	I	F		0		cEL		SPs

Figure 3.3: Fields in the PSTATE register.

Registers R0-R7 are used for passing arguments when calling a procedure or function Registers R9-R15 are scratch registers and can be used at any time because no assumptions are made about what they contain. They are called scratch registers because they are useful for holding temporary results of calculations. Registers R16-R29 can also be used as scratch registers, but their contents must be saved before they are used, and restored to their original contents before the procedure exits.

Some of the registers have alternate names. For example, X16 is also known as IP0. Most of these alternate names are only of interest to people writing compilers and operating systems. However, two of these registers are of interest to all AArch64 programmers.

3.2.2 Frame pointer

The frame pointer, x29, is used by high-level language compilers to track the current *stack frame*. This register can be helpful when the program is running under a debugger, and can sometimes help the compiler to generate more efficient code for returning from a subroutine. The GNU C compiler can be instructed to use x29 as a general-purpose register by using the -fomit-frame-pointer command line option. The use of x29 as the frame pointer is a *programming convention*. Some instructions (e.g. branches) implicitly modify the program counter, the link register, and even the stack pointer, so they are considered to be *hardware special* registers. As far as the hardware is concerned, the frame pointer is exactly the same as the other general-purpose registers, but AArch64 programmers use it for the frame pointer because of the ABI.

3.2.3 PSTATE register

The PSTATE register contains bits that indicate the status of the current process, including information about the results of previous operations. Fig. 3.3 shows all of its bits. The dashed lines indicate unused space that may be reserved for future AArch64 architectural extensions. The PSTATE register is actually a collection of independent fields, most of which are only used by the operating system. User programs make use of the first four bits, N, Z, C, and V. These are referred to as the condition *flags* field. Most instructions can modify these flags, and later instructions can use the flags to control their operation. Their meaning is as follows:

Negative: This bit is set to one if the signed result of an operation is negative, and set to zero if the result is positive or zero.

Zero: This bit is set to one if the result of an operation is zero, and set to zero if the result is non-zero.

Carry: This bit is set to one if an add operation results in a carry out of the most significant bit, or if a subtract operation results in a borrow. For shift operations, this flag is set to the last bit shifted out by the shifter.

oVerflow: For addition and subtraction, this flag is set if a signed overflow occurred.

3.2.4 Link register

The procedure link register, x30, is used to hold the *return address* for subroutines. Certain instructions cause the program counter to be copied to the link register, then the program counter is loaded with a new address. These *branch-and-link* instructions are briefly covered in Section 3.5 and in more detail in Section 5.4. The link register could theoretically be used as a scratch register, but its contents are modified by hardware when a subroutine is called, in order to save the correct return address. Using x30 as a general-purpose register is dangerous and is strongly discouraged.

3.2.5 Stack pointer

The program stack was introduced in Section 1.4. The stack pointer, sp, is used to hold the address where the stack ends. This is commonly referred to as the *top* of the stack, although on most systems the stack grows downwards and the stack pointer really refers to the *lowest* address in the stack. The address where the stack ends may change when registers are pushed onto the stack, or when temporary local variables (*automatic variables*) are allocated or deleted. The use of the stack for storing automatic variables is described in Chapter 5. The stack pointer can only be modified or read by a small set of instructions.

3.2.6 Zero register

The zero register, zr, can be referred to as a 64-bit register, xzr, or a 32-bit register, wzr. It always has the value zero. Most instructions can use the zero register as an operand, even as a destination register. If this is the case, the instruction will not change the destination register. However, it can still have side effects, including updating the PSTATE flags based on the ALU operation and incrementing a register in pre-indexed or post-indexed addressing. The zero register cannot always be used as an operand. It shares the same binary encoding with the stack pointer register, sp, which is the value 31. Some instructions can access the zero register, while others can access the stack pointer.

3.2.7 Program counter

The program counter, pc, always contains the *address* of the next instruction that will be executed. The processor increments this register by four, automatically, after each instruction is fetched from memory. By moving an address into this register, the programmer can cause the processor to fetch the next instruction from the new address. This gives the programmer the ability to jump to any address and begin executing code there. Only a small number of instructions can access the pc directly. For example instructions that create a PC-relative address, such as adr, and instructions which load a register, such as ldr, are able to access the program counter directly.

3.3 Instruction components

The AArch64 processor supports a relatively small set of instructions grouped into four basic instruction types. Many instructions have optional modifiers, which results in a very rich programming language. The following sections give a brief overview of the components which are common to instructions in each category. The individual instructions are explained later in this chapter, and in the following chapter.

3.3.1 Setting and using condition flags

As mentioned previously, PSTATE contains four flag bits (bits 28–31), which can be used to control whether or not certain instructions are executed. Most of the data processing instructions have an optional modifier to control whether or not the flag bits are set when the instruction is executed. For example, the basic instruction for addition is add. When the add instruction is executed, the result is stored in the destination register, but the flag bits in PSTATE are not affected.

However, the programmer can add the *s modifier* to the add instruction to create the adds instruction. When it is executed, this instruction will affect the NZCV flag bits in PSTATE. The flag bits can be used to control subsequent branch instructions. The meaning of the flags depends on the type of instruction that was most recently used to set the flags. Table 3.1 shows the names and meanings of the four bits depending on the type of instruction that set or cleared them. Many logical and arithmetic instructions support the s modifier to control setting the flags.

Table 3.1: Flag bits NZCV in PSTATE.

Name		Logical instruction	Arithmetic instruction
N	(Negative)	No meaning	Bit 31 of the result is set. Indicates a negative number in signed operations.
Z	(Zero)	Result is all zeroes	Result of operation was zero
C	(Carry)	After Shift operation, '1' was left in carry flag	Result was greater than 64 bits
V	(oVerflow)	No meaning	The signed two's complement result requires more than 64 bits. Indicates a possible corruption of the result.

Table 3.2: AArch64 condition modifiers.

Condition code	Meaning	Condition flags	Binary encoding
EQ	Equal	$Z = 1$	0000
NE	Not Equal	$Z = 0$	0001
HI	Unsigned Higher	$(C = 1) \wedge (Z = 0)$	1000
HS	Unsigned Higher or Same	$C = 1$	0010
LS	Unsigned Lower or Same	$(C = 0) \vee (Z = 1)$	1001
LO	Unsigned Lower	$C = 0$	0011
GT	Signed Greater Than	$(Z = 0) \wedge (N = V)$	1100
GE	Signed Greater Than or Equal	$N = V$	1010
LE	Signed Less Than or Equal	$(Z = 1) \vee (N \neq V)$	1101
LT	Signed Less Than	$N \neq V$	1011
CS	Unsigned Overflow (Carry Set)	$C = 1$	0010
CC	No Unsigned Overflow (Carry Clear)	$C = 0$	0011
VS	Signed Overflow	$V = 1$	0110
VC	No Signed Overflow	$V = 0$	0111
MI	Minus, Negative	$N = 1$	0100
PL	Plus, Positive or Zero	$N = 0$	0101
AL	Always Executed	*Any*	1110
NV	Never Executed	*Any*	1111

The branch instruction, b, is one instruction that can have a *condition modifier* appended to the mnemonic. If present, the modifier controls, at run-time, whether or not the instruction is actually executed. Table 3.2 shows the condition modifiers that can be attached to the branch instruction. For example, to create an instruction that only branches if the Z flag is set, the programmer would add the eq condition modifier to the basic b instruction to create the beq instruction. There are a few other instructions that can use the condition modifier to control their execution using the flags.

3.3.2 Immediate values

An immediate value in assembly language is a constant value that is specified by the programmer. Some assembly languages encode the immediate value as part of the instruction. Other assembly languages create a table of immediate values in a *literal pool*, stored in memory, and insert appropriate instructions to access them. ARM assembly language provides both methods.

Immediate values can be specified in decimal, octal, hexadecimal, or binary. Octal values must begin with a zero, and hexadecimal values must begin with "`0x`". Likewise immediate values that start with "`0b`" are interpreted as binary numbers. Any value that does not begin with zero, `0x`, or `0b` will be interpreted as a decimal value.

There are two ways that immediate values can be specified in GNU ARM assembly. The first way is as a literal immediate value. This can be optionally prefixed with a pound sign for clarity: `#<immediate|symbol>`. The second way is the `=<immediate|symbol>` syntax, which can only be used with the `ldr` pseudo-instruction. The `=<immediate|symbol>` syntax can be used to specify any immediate value (up to 64-bits), or to specify the 64-bit value of any symbol in the program. Symbols include program labels (such as `main`) and symbols that are defined using `.equ` and similar assembler directives. This syntax is a pseudo-instruction because of the way the AArch64 machine instructions are encoded. For data processing instructions, there are a limited number of bits that can be devoted to storing immediate data as part of the instruction. In order to store a full 64-bit immediate, the linker must create a literal pool.

The `#<immediate|symbol>` syntax is used to specify immediate data values for data processing instructions. The `#<immediate|symbol>` syntax has some restrictions. The assembler must be able to construct the specified value using only a certain number of bits of data, a shift or rotate, and/or a complement. For example, arithmetic immediate instructions can only use an immediate value that fits within 12 bits. For immediate values that can cannot be constructed by shifting or rotating and complementing an 12-bit value, the programmer must use an `ldr` instruction with the `=<immediate|symbol>` to specify the value. That method is covered in Section 3.4.

Some examples of immediate values are shown in Table 3.3. When an immediate value is necessary, the programmer can always use the `#<immediate|symbol>` syntax. The assembler will emit an error message when an invalid immediate value is attempted, and the programmer can change their code to use the `ldr Rx,=<immediate|symbol>` method instead.

Table 3.3: Summary of valid immediate values.

Immediate type	Bits	Description	Legal	Illegal
Arithmetic	12	Unsigned immediate. Optionally shifted left by 12 bits.	`0xfff`	`0x1fff`
Logical	13	Pattern immediate. Series of one or more 1's followed by one or more 0's is repeated as an element 1, 2, 4, 8, 16, or 32 times to fill the register, and may be rotated.	`0xf0f0f0f0 f0f0f0f0`	`0xfeedfeed feedfeed`
Move	16	Optionally shifted left by 16, 32, or 48. Alternatively, may use a Logical pattern.	`0x8765`	`0x88888`
PC-relative addressing	21	Range is ±1 MB. `adrp` shifts left by 12 and has a range of ±4 GB.	`-0x100000`	`-0x100001`
Shift	6	Range is [0, 63].	`7`	`0x4f`
Load/Store	12	Unsigned immediate. Left shifted by 3, 2, 1, or 0 depending on data size.	`0x7ff8`	`0x7ff4`
Load/Store Indexed	9	Signed immediate.	`-0xff`	`0x1ff`

3.3.3 Addressing modes

The AArch64 architecture has a strict separation between instructions that perform computation and those that move data between the CPU and memory. Computational instructions can only modify registers, not main memory. Because of this separation between load/store operations and computational operations, it is a classic example of a *load-store architecture*. The programmer can transfer bytes (8 bits), half-words (16 bits), words (32 bits), and double-words (64 bits) from memory into a register, or from a register into memory. The programmer can also perform computational operations (such as adding) using two source operands and one register as the destination for the result. All *computational* instructions assume that the registers already contain the data. *Load* instructions are used to move data from memory into the registers, and *store* instructions are used to move data from the registers to memory.

Most of the load/store instructions use an `<address>` which is one of the six options shown in Table 3.4. The brackets used in the modes denote a memory access. There are three fundamental addressing modes in AArch64 instructions: register offset, immediate offset, and

Table 3.4: Load/Store memory addressing modes.

Name	Syntax	Range
Register Address	[Xn|sp]	
Signed Immediate Offset	[Xn|sp, #±<imm9>]	[−256, 255]
Unsigned Immediate Offset	[Xn|sp, #<imm12>]	[0, 0x7ff8]
Pre-indexed Immediate Offset	[Xn|sp, #±<imm9>]!	[−256, 255]
Post-indexed Immediate Offset	[Xn|sp], #±<imm9>	[−256, 255]
Register Offset	[Xn|sp, Xm, (U|S)XTW]	(or LSL #1-3)
Literal	label	±1 MB
Pseudo Load	=<immediate|symbol>	64 bits

literal. Immediate has two important variants: pre-indexed and post-indexed. The pseudo addressing mode allows an immediate data value or the address of a label to be loaded into a register, and may result in the assembler generating more than one instruction. The following section describes each addressing mode in detail.

Register Address: [Xn|sp]

This addressing method is used to access the memory address that is contained in the register Xn or sp. The brackets around Xn denote that it is a memory access using the contents of the register as the address in memory.

For example, the following line of code:

        ldr   x3, [x2]

uses the contents of register x2 as a memory address and loads eight bytes of data, starting at that address, into register x3. Likewise,

        str   x3, [x2]

copies the contents of x3 to the eight bytes of memory starting at the address that is in x2. This is really encoded as an unsigned immediate offset. [Xn] or [sp] is just shorthand notation for [Xn, #0] or [sp, #0], respectively.

Signed Immediate Offset: [Xn|sp, <±imm9>]

The signed immediate offset (which may be negative or positive) is added to the contents of Xn or sp. The result is used as the address of the item to be loaded or stored. For example, the following line of code:

        ldur   x0, [x1, #0x50]

calculates a memory address by adding 0x50 to the contents of register x1. It then loads eight bytes of data, starting at the calculated memory address, into register x0. Similarly, the line:

        stur   x0, [x1, #-0x50]

adds negative 0x50 to the contents of x1 and uses that as the address where it stores the eight bytes of x0 into memory.

Unsigned Immediate Scaled Offset: [Xn|sp, <imm12>]

The unsigned immediate offset (which may only be zero or positive) is *scaled* and then added to the contents of Xn or sp. If the register being loaded or stored is a 64-bit register, then the immediate value is scaled by shifting it left three bits. Likewise, if the load or store is 32-bits, the immediate value is scaled by shifting it left two bits. For half-word loads and stores, the offset is scaled by shifting left by one bit, and for byte loads and stores, no scaling occurs.

Note that the syntax for this addressing mode is the same as the syntax for Signed Immediate Offset mode, but the set of possible immediate values is different. The programmer does not need to worry about which mode is used. The programmer just specifies the offset as an immediate value. The Assembler will automatically select whether to use Signed Immediate Offset or Unsigned Immediate Scaled Offset mode depending on the immediate offset value that is specified.

The result of adding the scaled offset to the base register is used as the address of the item to be loaded or stored. For example, the following line of code:

```
ldr    x0, [x1, #0x7ff8]
```

calculates a memory address by adding 0x7ff8 to the contents of register x1. It then loads eight bytes of data, starting at the calculated memory address, into register x0. Similarly, the line:

```
str    w0, [x1, #0x3ffc]
```

adds 0x3ffc to the contents of x1 and uses that as the address where it stores the 4 bytes of x0 in memory.

Pre-indexed Immediate Offset: [Xn|sp, #±<imm9>]!

The memory address is computed by adding the unshifted, signed 9-bit immediate to the number stored in Xn or sp. Then, Xn is set to contain the memory address. This mode can be used to step through elements in an array, updating a pointer to the next array element before each element is accessed.

Post-indexed Immediate Offset: [Xn|sp], #±<imm9>

Register Xn or sp is used as the address of the value to be loaded or stored. After the value is loaded or stored, the value in Xn is updated by adding the unshifted immediate offset, which may be negative or positive. This mode can also be used to step through elements in an array, updating a pointer to point at the next array element after each one is accessed.

Register Offset: [Xn|sp, Rm, <option>]

Rm is extended or shifted, then added to Xn or sp. The result is used as the address of the item to be loaded or stored. For example,

```
ldr    x3, [x2, x1, lsl #3]
```

shifts the contents of x1 left three bits, adds the result to the contents of x2 and uses the sum as an address in memory from which it loads eight bytes into x3. Recall that shifting a binary number left by three bits is equivalent to multiplying that number by eight. This addressing mode is typically used to access an array, where x2 contains the address of the beginning of the array, and x1 is an integer index. The integer shift amount depends on the size of the objects in the array.

This is convenient when the size of the items in an array are powers of two. For example, the shift would be lsl #3 for double-words, lsl #2 for words, and lsl #1 for half-words. For an array of structures, this method is only appropriate if the size of the structures in the array is a power of two. Many programs use 32-bit integers (words). For example, int in C is often 32-bits. The following instruction illustrates how to access an array of words:

```
ldr  w14, [x2, x3, lsl #2]
```

where w14 is the register to which the array element indexed by x3 is saved.

To store an item from register x0 into an array of half-words, the following instruction could be used:

```
strh  w0, [x4, w5, lsl #1]
```

where x4 holds the 64-bit address of the first byte of the array, and w5 holds the integer index for the desired array item.

Subroutines often keep data on the stack, including their return addresses and local variables, if they use them. The following instruction shows how to store a double-word variable on the stack:

```
str   x0, [sp, x1, lsl #3]
```

In this instruction x1 is an offset to the local variable, starting from the stack pointer as the base address, and x0 is the value used to overwrite the local variable on the stack.

If Rm is specified as a 32-bit register (Wm), then the <option> for sign extension can be applied. The programmer can choose either sign extend word (sxtw) or unsigned extend word (uxtw). Sign extension and unsigned extension are used to preserve the values of binary numbers when more bits are used to represent them. Sign extension replicates the sign bit while unsigned extension uses only zeros to extend the number. If a 32-bit negative register offset is used to calculate a memory address, then it should be sign extended:

```
ldr   w1, [x2, w3, sxtw]
```

In this case, w3 is sign extended to become a 64-bit value, and then that sign-extended value is added to x3 to form the memory address. w1 is loaded with the word in memory at the calculated address.

Literal: label

When using a literal load instruction, an address in memory within one megabyte of the program counter can be calculated. This is possible because the label address is encoded as a signed offset from the load instruction. Since instructions are four bytes long, the label will be at an address that is a multiple of four bytes. On a binary level, the label's offset is encoded in 19 bits. It is then multiplied by four (shifted left by two) and added to the program counter to obtain the label's address.

Pseudo load: =<immediate|symbol>

This is a pseudo-instruction. The assembler will generate a mov instruction if possible. Otherwise it will store the value of immediate or the address of symbol in a "literal pool", or "literal table", and generate a load instruction, using one of the previous addressing modes, to load the value into a register. This addressing mode can only be used with the ldr instruction. An example pseudo-instruction and its disassembly are shown in Listing 3.1 and Listing 3.2.

Listing 3.1 LDR pseudo-instruction.

```
    .text
    ldr    x0, =0x123456789abcdef0
    ret
```

Listing 3.2 Disassembly of LDR pseudo-instruction

```
0: 58000040  ldr x0, 8 <.text+0x8>
4: d65f03c0  ret
8: 9abcdef0  .word 0x9abcdef0
c: 12345678  .word 0x12345678
```

3.4 Load and store instructions

The load and store instructions allow the programmer to move data from memory to registers or from registers to memory. The load/store instructions can be grouped into the following types:

- single register,
- register pair,
- atomic.

3.4.1 Load/store single register

These instructions transfer a double-word, single word, half-word, or byte from a register to memory or from memory to a register:

ldr   Load Register, and
str   Store Register.

3.4.1.1 Syntax

```
<op>{<size>}  Rd, <addr>
```

- <op> is either ldr or str.
- The optional <size> is one of:
  - b   unsigned byte
  - h   unsigned half-word
  - sb   signed byte
  - sh   signed half-word
  - sw   signed word
- <addr> is one of the address specifiers described in Section 3.3.3.
- str cannot use a signed <size>. It also cannot use the literal addressing mode.

3.4.1.2 Operations

Name	Effect	Description
ldr	Rd ← Mem[addr]	Load register from memory at addr
str	Mem[addr] ← Rd	Store register in memory at addr

3.4.1.3 Examples

Load the word (4 byte) value from Mem[x4] into w8, and set the upper four bytes of x8 to zero.

```
        ldr     w8, [x4]
```

Store the least-significant byte from register x12 into Mem[x2].

```
        strb    x12, [x2]
```

Load the double-word (8 byte) value from Mem[x3 + 7] into x5. Then set x3 = x3 + 7:

```
        ldr     x5, [x3, #7]!
```

Store the half-word (2 byte) value in w9 to Mem[x6]. Then set x6 = x6 + 7:

```
        strh    w9, [x6], #7
```

Load the half-word value from Mem[x0 + 8] into x5 and sign extend it:

```
    ldrsh   x5, [x0, 8]
```

Store the least significant byte in `w1` at `Mem[x9]`:

```
    strb    w1, [x9]
```

3.4.2 Load/store single register (unscaled)

These instructions are the same as Load/Store Single Register, except that they only use an unscaled, signed addressing mode with an offset range of $[-256, 256]$. Programmers rarely need to write `ldur` or `stur` explicitly. The programmer can just use `ldr` or `str`, and the assembler will almost always automatically convert them to `ldur` or `stur` when appropriate.

ldur   Load Register (Unscaled), and
stur   Store Register (Unscaled).

3.4.2.1 Syntax

```
    <op>{<size>}   Rd, [Xn, #imm9]
```

- <op> is either `ldur` or `stur`.
- The optional <size> is one of:
  - b   unsigned byte
  - h   unsigned half-word
  - sb   signed byte
  - sh   signed half-word
  - sw   signed word
- The addressing mode is signed immediate with 9 bits.
- `stur` cannot use a signed <size>.

3.4.2.2 Operations

Name	Effect	Description
ldur	Rd ← Mem[addr]	Load register from memory at addr
stur	Mem[addr] ← Rd	Store register in memory at addr

3.4.2.3 Examples

Load the byte value from `Mem[x5 + 255]`. Sign extend it and store the value in `x4`:

```
        ldursb  x4, [x5, #255]
```

Store the double-word value in x1 to Mem[x2 - 256]:

```
        stur    x1, [x2, #-256]
```

3.4.3 Load/store pair

These instructions are used to store or load two registers at a time. This can be useful for moving registers onto the stack or for copying data. These two instructions are particularly useful for transferring data in a load-store architecture because each instruction can move twice as much information as the ldr and str instructions.

ldp   Load Pair, and
stp   Store Pair.

3.4.3.1 Syntax

```
    <op>{<size>}    Rt, Rt2, <addr>
```

- <op> is either ldp or stp.
- <size> is optionally sw for signed words.
- <addr> is 7 bits Pre-indexed, Post-indexed, or Signed immediate.
- Signed immediate Xt range: [-0x200, 0x1f8]. Wt range: [-0x100, 0xfc].

3.4.3.2 Operations

Name	Effect	Description
ldp	Rt ← Mem[addr] Rt2 ← Mem[addr+size(Rt)]	Load register pair from memory at addr where sizeof(Rt) is 4 for Wt registers and 8 for Xt registers
stp	Mem[addr] ← Rt Mem[addr+size(Rt)] ← Rt2	Store register pair in memory at addr

3.4.3.3 Examples

The stp and ldp instructions are commonly used at the beginning of a function, and at the end of a function. For example, the following instruction allocates sixteen bytes on the stack by decrementing the stack pointer. Then it stores the frame pointer (x29) at Mem[sp] (the bottom of the stack), and stores the procedure link register (x30) at Mem[sp + 8], which completely fills the 16 bytes that were allocated on the stack.

```
        stp     x29, x30, [sp, #-16]!
```

The opposite of the previous instruction can be used to restore the values of the frame pointer and procedure link registers:

```
        ldp     x29, x30, [sp], #16
```

Here x29 is set to Mem[sp] and x30 is set to Mem[sp + 8]. Then, sp is incremented by 16, essentially deleting the values of x29 and x30 from the stack. Although the bits are still there, they may be overwritten by subsequent instructions that store data on the stack.

The following assembly program demonstrates the two instructions used together to save and restore a pair of registers.

```
        .text
        .type   main, %function
        .global main
main:
        stp     x29, x30, [sp, #-16]!    // Store x29 and x30 on the stack
        mov     x29, #0x0                // change their contents
        mov     x30, #0x0
        ldp     x29, x30, [sp], #16      // Restore x29 and x30, and sp
        ret
        .size   main, (. - main)
```

The following function demonstrates how to use the stp and ldp instructions to save, and later restore, all of the non-volatile registers.

```
        .text
        .type   exmpl, %function
        .global exmpl
exmpl:
        stp     x19, x20, [sp, #-16]!    // Store x19 and x20 on the stack
        stp     x21, x22, [sp, #-16]!    // Store x21 and x22 on the stack
        stp     x23, x24, [sp, #-16]!    // Store x23 and x24 on the stack
        stp     x25, x26, [sp, #-16]!    // Store x25 and x26 on the stack
        stp     x27, x28, [sp, #-16]!    // Store x27 and x28 on the stack
        stp     x29, x30, [sp, #-16]!    // Store x29 and x30 on the stack

        // The function can now use x19 through x30. If it calls another
        // function, then it is guaranteed that these registers will not
        // be modified by the function that was called.
        // INSERT FUNCTION BODY HERE
```

```
16
17        ldp      x29, x30, [sp], #16    // Restore x29 and x30
18        ldp      x27, x28, [sp], #16    // Restore x27 and x28
19        ldp      x25, x26, [sp], #16    // Restore x25 and x26
20        ldp      x23, x24, [sp], #16    // Restore x23 and x24
21        ldp      x21, x22, [sp], #16    // Restore x21 and x22
22        ldp      x19, x20, [sp], #16    // Restore x19 and x20
23        ret
24        .size    exmpl, (. - exmpl)
```

3.5 Branch instructions

Branch instructions allow the programmer to change the address of the next instruction to be executed. They are used to implement loops, if-then structures, subroutines, and other flow control structures. There are five instructions related to branching:

- Branch,
- Branch to Register,
- Branch and Link (subroutine call),
- Compare and Branch, and
- Form program-counter-relative Address.

3.5.1 Branch

This instruction is used to perform conditional and unconditional branches in program execution:

b   Branch.

It is used for creating loops and if-then-else constructs. This is the only instruction that may have a conditional suffix attached to the mnemonic.

3.5.1.1 Syntax

```
b{<cond>}   <target_label>
```

- The optional <cond> can be any of the codes from Table 3.2 specifying conditional execution.
- The <target_label> can be any label in the current file, or any label that is defined as .global or .globl in any file that is linked in (within a certain range).

- The <target_label> is encoded as a program-counter-relative offset with 19 bits for a conditional branch and 26 bits for an unconditional branch. To calculate the correct address, the immediate value is first shifted left by two bits and then added to the program counter.
- The range of the <target_label> is ±1 MB for a conditional branch and ±128 MB for an unconditional branch.

3.5.1.2 Operations

Name	Effect	Description
b	pc ← target_address	Unconditionally move new address to the program counter (pc).

b<cond>	**if** <cond> **then**
	    pc ← target_address
	end if
		Conditionally move new address to the program counter (pc).

3.5.1.3 Examples

Branch to label overflow if the signed overflow flag V is set (meaning if it has a value of 1) in the PSTATE register:

```
        bvs     overflow
```

An unconditional branch is always taken. One example is an endless loop:

```
        .text
        .type   main, %function
        .global main
main:   b       main    // Endless loop branching to main
        ret             // Does not return
        .size   main, (. - main)
```

Conditional branching is often used to implement if statements and other control-flow logic:

```
        .text
        .global main
main:
        stp     x29, x30, [sp, #-16]!

        // if (w0 = 0) goto endif1
        cmp     w0, wzr
        beq     endif1

        // w0 = 1
```

```
11          mov     w0, #1
12  endif1:
13
14          // return w0
15          ldp     x29, x30, [sp], #16
16          ret
```

3.5.2 Branch register

These instructions allow the program counter to be set to the value contained in a 64-bit general-purpose register (x0-x30):

br   Branch to Register and
ret   Return from Subroutine.

3.5.2.1 Syntax

```
br      Xn
ret     {Xn}
```

- ret uses the procedure link register (LR), which is x30, if no register is specified.
- ret is only used to return from a subroutine.
- br can be used for implementing a computed goto (switch statement).

3.5.2.2 Operations

Name	Effect	Description
br	pc ← Xn	Copy register Xn to the program counter (pc).
ret	pc ← X30 or pc ← Xn	Copy the link register (X30), or any other register (Xn) to the program counter (pc).

3.5.2.3 Examples

The branch register instruction will set the pc register to the contents of a register:

```
br      x10
```

This can generate a Segmentation Fault if the contents of x10 are invalid. If not, the processor will begin executing instructions at the address contained in x10.

The return from procedure instruction, `ret`, uses the procedure link register `x30` to return when no register is specified:

```
        .text
        .type   main, %function
        .global main
main:
        // Save values of registers x29, x30 to the stack
        stp     x29, x30, [sp, #-16]!

        // Program code
        // ...

        // Restore values of registers x29, x30 from the stack
        ldp     x29, x30, [sp], #16
        // Return from subroutine to address in x30
        ret     x30
        .size   main, (. - main)
```

The return instruction `ret    x30` is the same as just `ret`.

3.5.3 Branch and link

The branch and link instructions are used to call subroutines:

bl   Branch and Link and
blr   Branch to Register and Link.

The branch and link instruction is identical to the branch instruction, except that it copies the current program counter to the link register before performing the branch. The branch to register and link instruction does the same thing, except that it branches to the address contained in a specified register. This allows the programmer to copy the link register back into the program counter at some later point with the `ret` instruction. This is how subroutines are called, and how subroutines return and resume executing at the next instruction after the one that called them.

3.5.3.1 Syntax

```
    bl      <target_address>
    blr     Xn
```

- The `target_address` can be any label in the current file, or any label that is defined as `.global` or `.globl` in any file that is linked in, within a range of ±128 MB.

3.5.3.2 Operations

Name	Effect	Description
bl	$X30 \leftarrow pc + 4$ $pc \leftarrow$ target_address	Save address of next instruction in link register (X30), then load pc with new address
blr	$X30 \leftarrow pc + 4$ $pc \leftarrow Xn$	Save address of next instruction in link register (X30), then load pc with Xn

3.5.3.3 Examples

The following C code uses the `printf` function from the C Standard Library to print a message to standard output.

```c
int main(void)
{
  printf("hello, world!\n");
  return 0;
}
```

The equivalent ARM assembly code uses the `bl` instruction call `printf`:

```
        .section .rodata
msg:
        .asciz  "Hello, world!\n"      // Declare our message
        .text
        .type   main, %function
        .global main
main:
        stp     x29, x30, [sp, #-16]!  // push link register (and x30) to stack

        // printf("hello, world\n")
        // Note: the following call to printf will change the link
        // register, but we saved it on the stack.
        ldr     x0, =msg
        bl      printf                 // Branch and Link to printf


        mov     w0, 0                  // return 0

        ldp     x29, x30, [sp], #16    // pop link register (and x30) from stack
        ret
        .size   main, (. - main)
```

The C compiler will automatically create a constant array of characters and initialize it to hold the message in the read-only data, .rodata, section. Note that strings in C must be null-terminated, meaning the last character must be 0. The compiler does this automatically. The compiler directive .asciz allocates a null-terminated string in memory. The program will load the address of the first character in the array into register x0 before calling printf(). The printf() function will expect to see an address in x0, which it will assume is the address of the format string to be printed. By convention, when a function is called, it will expect to find its first argument in R0. There are other rules which AArch64 programmers must follow regarding how registers are used when passing arguments to functions and procedures. Those rules will be explained fully in Chapter 5, Section 5.4.

Example 8 shows how the bl instruction can be used to call a function from the C standard library to read a single character from standard input. By convention, when a function is called, it will leave its return value in R0.

Example 8. Using the bl instruction to read a character. Suppose we want to read a single character from standard input. This can be accomplished in C by calling the getchar() function from the C standard library as follows:

```
c = getchar();
```

The above C code assumes that the variable c has been declared to hold the result of the function. In A64 assembly language, functions always return their results in x0. The assembly programmer may then move the result to any register or memory location they choose. In the following example, it is assumed that w9 was chosen to hold the value of the variable c:

```
        bl      getchar // Call the getchar function
        mov     w9,w0   // Move the result to register 9
```

The full assembly program below accepts one character of input and then prints it, utilizing getchar and bl:

```
        .section .rodata
formatString:
        .string "char: %c\n"
        .text
        .type   main, %function
        .global main
main:
```

```
8        stp     x29, x30, [sp, #-16]!
9
10       // w9 = getchar()
11       bl      getchar
12       mov     w9, w0
13
14       // printf("char: %c\n", x9)
15       ldr     x0, =formatString
16       mov     w1, w9
17       bl      printf
18
19       // return 0
20       mov     w0, 0
21       ldp     x29, x30, [sp], #16
22       ret
23       .size   main, (. - main)
```

3.5.4 Compare and branch

These instructions are used to branch conditionally if a register is zero or nonzero (or if a bit is zero or nonzero):

cbz   Compare and Branch if Zero,
cbnz Compare and Branch if Nonzero,
tbz   Test Bit and Branch if Zero, and
tbnz Test Bit and Branch if Nonzero.

3.5.4.1 Syntax

```
   cb{n}z  Rt, <label>
   tb{n}z  Rt, #imm6, <label>
```

- The cbz and cbnz instructions have a range of ±1 MB (encoded in 19 bits).
- The range of the tbz and tbnz instructions is ±32 KB (encoded in 14 bits).
- imm6 specifies which bit to test (0 to 63).
- The <label> must always be word-aligned.

3.5.4.2 Operations

Name	Effect	Description

cbz	**if** Rt = 0 **then**
	    pc ← label
	end if
		Conditionally branch to label if Rt is zero.

continued on next page

Name	Effect	Description

cbnz	**if** Rt $\neq 0$ **then**
	 pc $\leftarrow$ label
	end if
		Conditionally branch to label if Rt is nonzero.

tbz	**if** Rt[imm6] $= 0$ **then**
	 pc $\leftarrow$ label
	end if
		Conditionally branch to label if the specified bit in Rt is zero.

tbnz	**if** Rt[imm6] $\neq 0$ **then**
	 pc $\leftarrow$ label
	end if
		Conditionally branch to label if the specified bit in Rt is nonzero.

3.5.4.3 Examples

The following program reads one character at a time until it reads in a character with an odd ASCII value. For example, it will stop when it reads 'a' or 'c', which have ASCII values 97 and 99 respectively. When it reads an odd ASCII value, it stops and prints out the character, then returns zero to its caller:

```
        .section .rodata
format: .string "You entered an odd char: %c.\n"
        .text
        .type   main, %function
        .global main
main:
        stp     x29, x30, [sp, #-16]!

        // do { c = getchar(); } while (c % 2 == 0)
loop:
        bl      getchar
        tbz     x0, #0, loop

        // printf("You entered an odd char: %c\n", c)
        mov     x1, x0
        ldr     x0, =format
        bl      printf

        // return 0
        mov     w0, #0
        ldp     x29, x30, [sp], #16
        ret
        .size   main, (. - main)
```

3.5.5 Form PC-relative address

These instructions are used to calculate the address associated with a label:

adr   Form PC-Relative Address
adrp   Form PC-Relative Address to 4 KB Page

They are more efficient than the `ldr Rx,=label` pseudo instruction, because they can calculate a 64-bit address in one or two instructions without performing a memory access. They can be used to load any address that is within range. If the label is out of range, then the assembler or linker will emit an error, and the programmer can change their code to use the `ldr Rx,=label` syntax.

3.5.5.1 Syntax

```
    <op>    Rd, <label>
```

- `<op>` is either `adr` or `adrp`.
- `adr` has a range of ±1 MB. (21 bit immediate).
- `adrp` has a range of ±4 GB to the nearest 4 KB page (4096 bytes). The 21-bit immediate is shifted left by 12 bits and added to the `pc`.
- The lower 12 bits of a label's address can be added to `adrp` to exactly address a label.

3.5.5.2 Operations

Name	Effect	Description
adr	Rd ← Address of `label`	Load address with pc-relative immediate addressing.
adrp	Rd ← Page address of `label`	Load address of the beginning of the 4-Kilobyte memory page which contains the label using pc-relative immediate addressing.

3.5.5.3 Examples

The `adr` instruction is helpful for calculating the address of labels at run-time. This is particularly useful when the address of a label must be passed to a function as an argument, but the address cannot be determined at compile time. For example, the address of some system libraries may not be set by the linker, but are set when the program is loaded and prepared to run. The addresses of labels in these libraries cannot be loaded with the `ldr Rx,=label` syntax, because the assembler and linker cannot predict the location of the label. The `adr` instruction provides a way to get the address of the label.

```
        .section .rodata
format: .string "hello, world\n"

        .text
        .type   main, %function
        .global main
main:
        stp     x29, x30, [sp, #-16]!

        // printf("hello, world\n")
        adr     x0, format      // Load address of format string
        bl      printf

        // return 0
        mov     w0, 0

        ldp     x29, x30, [sp], #16
        ret
        .size   main, (. - main)
```

The range of `adr` is just as limited as an unconditional `b` or a `bl`. To address a label that is a greater distance away, yet within 4 GB in either direction, the `adrp` instruction can be used. Note that the following code allocates 4 GB of data. If that allocation were done in the `.data` section, then the executable file would contain the 4 GB of data. By allocating the data in the `.bss` section, the program size is reduced by 4 GB. As an added benefit, it takes much less time to assemble the program, and less time to load and run it.

```
        .bss
closeBy:
        .word 0x00000000
        .skip 4000000000
farFarAway:
        .word 0xffffffff
        .text
        .type   main, %function
        .global main
main:
        adrp    x0, farFarAway
        add     x0, x0, #:lo12:farFarAway

        ret
        .size   main, (. - main)
```

A special notation is used to add only the lowest 12 bits of the label to the `adrp` address. This fully calculates the label address because the 4 KB page is addressable with 12 bits.

3.6 Chapter summary

The AArch64 Instruction Set Architecture, known as A64, includes 32 general-purpose registers and a four basic instruction types. This chapter explained the instructions used for

- moving data between memory and registers, and
- branching and calling subroutines.

The load and store operations are used to move data between memory and registers. The basic load and store operations, ldr and str, have a very powerful set of addressing modes. The assembler provides pseudo-instructions for loading addresses and loading immediate values.

The AArch64 processor provides both conditional and unconditional branch instructions. The b instruction can be used to create loops and to create if-then-else constructs and other types of control-flow. The bl instruction is used to call subroutines (functions). It sets the register x30 to the pc of the next instruction before branching to the subroutine. At the end of the called subroutine, the programmer can use the ret instruction to copy register x30 back to the pc, thereby resuming the calling function at its next instruction.

Exercises

3.1. Which registers hold the stack pointer, return address, and program counter?
3.2. Which is more efficient for loading a constant value, the ldr pseudo-instruction, or the mov instruction? Explain.
3.3. The str/ldr and stp/ldp instructions include an optional '!' after the address register. What does it do?
3.4. The following C statement declares an array of four integers, and initializes their values to 7, 3, 21, and 10, in that order.

```
int nums[]={7,3,21,10};
```

   a. Write the equivalent in GNU ARM assembly.
   b. Write the ARM assembly instructions to load all four numbers into registers w3, w5, w6, and w9, respectively, using:
      i. two ldp instructions, and
      ii. four ldr instructions.
3.5. What is the difference between a memory location and a CPU register?
3.6. How many registers are provided by the ARM Instruction Set Architecture?
3.7. Assume that x is an array of integers. Convert the following C statements into ARM assembly language.
   a. x[8] = 100;
   b. x[10] = x[0];

     c. `x[9] = x[3];`

3.8. Assume that `x` is an array of integers, and `i` and `j` are integers. Convert the following C statements into arm64 assembly language.

     a. `x[i] = j;`

     b. `x[j] = x[i];`

     c. `x[i] = x[j*2];`

3.9. What is the difference between the `b` instruction and the `bl` instruction? What is each used for?

3.10. What is the meaning of the following instructions?

     a. `ldaxr`

     b. `sttr`

     c. `bgt`

     d. `bne`

     e. `bge`

     f. `bcc`

Data processing and other instructions

The AArch64 processor provides a rich set of data processing instructions, most of which perform arithmetic or bitwise logical operations. There are also a few special instructions that are used infrequently to perform operations that are not classified as load/store, branch, or data processing.

The data processing instructions cannot access data stored in memory. They operate only on CPU registers, and possibly immediate data that is encoded as part of the instruction. In general, data must first be moved from memory into a CPU register before processing can be performed. The most common way to move data from memory into registers and vice versa is by using the load and store instructions, which are covered in Chapter 3. Each data processing instruction performs one basic arithmetic or logical operation. The instructions are grouped in the following categories of operations:

- Arithmetic
- Logical,
- Data Movement,
- Multiplication,
- Division,
- Comparison,
- Conditional, and
- Special.

4.1 Operand2

Most of the data processing instructions require the programmer to specify two *source operands* and one *destination register*. Because three items must be specified for these instructions, they are known as *three address instructions*. The use of the word *address* in this case has nothing to do with memory addresses. The term *three address instruction* comes from earlier processor architectures that allow arithmetic operations to be performed with data that is stored in memory rather than registers.

Referring back to Fig. 3.1, the first source operand specifies a register whose contents will be supplied from the register file on the A Bus and enter the ALU. The second source

Table 4.1: Formats for Operand2.

Operand2	Meaning
#<immediate|symbol>	A 12 bit immediate value
#<pattern>	A 13 bit pattern immediate used only for logical instructions.
Rm	Any of the 31 registers R0-R30
Rm, <shift_op> #<shift_imm>	The contents of a register shifted or rotated by an immediate amount between 0 and 64
Rm, <extend_op> #<shift_imm>	The contents of a register are extended and then are shifted left by 0-4 bits. Note that this mode cannot be used with logical instructions.

operand will enter the ALU through the B Bus and will either be the contents of a register, or an immediate value which is encoded in the instruction currently being executed. This second operand entering the ALU is referred to as Operand2, and can be one of four things:

- a register (R0-R30),
- a register (R0-R30) and a *shift operation* to modify it, or
- a register (R0-R30) and a *sign extend operation* that also shifts the register left by zero to four bits, or
- a 12-bit *immediate value*.

The options for Operand2 allow a great deal of flexibility. Many operations that would require two instructions on most processors can be performed using a single AArch64 instruction. Table 4.1 shows all of the possible forms for Operand2.

4.1.1 Shift and rotate operations

Table 4.2 shows the mnemonics used for specifying shift operations, which we refer to as <shift_op>. All of the shift operations require the programmer to specify a shift amount, which we will refer to as n. The lsl operation shifts each bit left. Zero is shifted into the n least significant bits, and the most significant n bits are lost. The lsr operation shifts each bit right. Zero is shifted into the n most significant bits, and the least significant n bits are lost. The asr operation shifts each bit right. The n most significant bits become copies of the sign bit (bit 63), and the least significant n bits are lost. The ror operation rotates the bits to the

Table 4.2: **Shift and rotate operations in Operand2.**

`<shift_op>`	Meaning
lsl	Logical Shift Left
lsr	Logical Shift Right
asr	Arithmetic Shift Right
ror	Rotate Right (Bitwise Instructions Only)

Table 4.3: **Extension operations in Operand2.**

`<extend_op>`	Meaning
uxtb	Unsigned extend byte
uxth	Unsigned extend half-word
uxtw	Unsigned extend word
uxtx	Unsigned extend double-word
sxtb	Sign-extend byte
sxth	Sign-extend half-word
sxtw	Sign-extend word
sxtx	Sign-extend double-word
lsl	Logical Shift Left

right. The n least significant bits become the most significant n bits, and all other bits are shifted right by n. The `ror` option can only be used with bitwise instructions, such as and, orr, mvn, or tst. It cannot be used with arithmetic instructions such as add or cmp. All of these instructions will be covered later in this chapter.

4.1.2 Extend operations

Table 4.3 shows the mnemonics available for the extension operations, referred to as `<extend_op>`. There is an extension for each size: double-word, word, half-word, and byte, which can be signed or unsigned. Unsigned extension is the same as zero extension fills the higher bits with zeroes. Signed extension copies the sign bit into the higher bits. Note that two extension operations are always redundant depending on the size of the register specified. For example, `sxtx` and `uxtx` are redundant for 64-bit registers. However, if Rm is a 32-bit register (Wm), then `sxtw` and `uxtw` are redundant. In each of these cases, the unsigned extension is always preferred. If the `<shift_imm>` amount is not zero and the `<extend_op>` is redundant, then the `lsl` mnemonic should be used as the `<extend_op>` for clarity. Extension cannot be used with logical instructions, which are able to use the `ror` `<shift_op>`.

4.1.3 Immediate data

There are two types of immediate values: 12 bit unsigned immediate, with a range of [0, 4095], and 13 bit pattern immediate. The first immediate is used for every instruction type that is not a logical instruction. Logical instructions use pattern immediates, which were introduced in Table 3.3. When choosing a pattern immediate, only values that can be specially encoded in the thirteen bits are valid. When choosing a pattern immediate value, a bit pattern of consecutive one's must be used. This pattern is then repeated either 1, 2, 4, 8, 16, or 32 times across the register, and it may also be rotated. For example, the immediate 5 would not work because its binary representation, 0b101, is not a consecutive pattern of ones. A consecutive pattern of 0b10101010101010101010101010101010 (hexadecimal: 0xaaaaaaaa) would work for a 32-bit Wn. An immediate value of 7 is valid because its binary representation, 0b111, is three ones in a row. Furthermore, 7 can be rotated to create 0x700, which is also valid.

4.2 Data processing instructions

The following instructions use the ALU to perform arithmetic and logical operations. Most of them require two source operands, and one destination register. However there are some exceptions. For example, the instruction for calculating the two's complement only requires one source operand, and the comparison operations do not require a destination register. Most of these instructions have an optional s modifier, which allows the programmer to choose whether or not the instruction will affect the flags in the PSTATE register.

4.2.1 Arithmetic operations

There are six basic arithmetic operations:

add   Add,
adc   Add with Carry,
sub   Subtract,
sbc   Subtract with Carry,
neg   Negate, and
ngc   Negate with Carry.

All of the non-carry instructions involve two source operands, and a destination register. The carry instructions may only use unshifted, unextended registers as the operands.

4.2.1.1 Syntax

```
<op>{s}      Rd, Rn, <Operand2>
<op2>{s}     Rd, Rn, Rm
neg{s}       Rd, <Operand2>
ngc{s}       Rd, Rm
```

- <op> is one of add, sub.
- <op2> is one of adc, sbc. The carry bit from PSTATE is added to the two operands.
- neg and ngc are aliases of sub and sbc where Rn is ZR. Therefore, neither syntax uses Rn.
- The optional s specifies whether or not the instruction should affect the bits in PSTATE.

4.2.1.2 Operations

Name	Effect	Description
add	Rd ← Rn + Operand2	Add.
adc	Rd ← Rn + Rm + *carry*	Add with carry.
sub	Rd ← Rn − Operand2	Subtract.
sbc	Rd ← Rn − Rm + *carry* − 1	Subtract with carry.
neg	Rd ← −Operand2	Negate.
ngc	Rd ← −Rn + *carry* − 1	Negate with carry.

4.2.1.3 Examples

```
    add     x0, x1, x2      // x0 = x1 + x2 and don't set PSTATE

    subs    x3, x3, #1      // x3 = x3 - 1 and set nzcv flags in PSTATE

    neg     x4, x5, lsl #4  // x4 = -(x5 << 4)
```

Example 9. The following listings show a complete program for adding the contents of two statically allocated variables and printing the result. The printf() function expects to find the address of a string in x0, as it prints the string, it finds the %d formating command, which indicates that the value of an integer variable should be printed. It expects the variable to be stored in w1. Note that the variable sum does not need to be stored in memory. It is stored in w1, where printf() expects to find it.

The following C program will add together two numbers stored in memory and print the result.

```c
#include <stdio.h>
static int x = 5;
static int y = 8;
int main(void)
{
  int sum;
  sum = x + y;
  printf("The sum is %d\n",sum);
  return 0;
}
```

The equivalent AArch64 assembly program is as follows:

```
        .data
fmt:    .asciz  "The sum is %d\n"
        .align  2
x:      .word   5
y:      .word   8
        .text
        .type   main, %function
        .global main
main:
        stp     x29, x30, [sp, #-16]!   // Push FP and LR onto the stack
        // sum = x + y
        adr     x14, x                  // Calculate address of x
        adr     x15, y                  // Calculate address of y
        ldr     x4, [x14]               // Load x
        ldr     x5, [x15]               // Load y
        add     x1, x4, x5              // x1 = x4 + x5

        // printf("The sum is %d\n", sum)
        adr     x0, fmt                 // Calculate address of fmt
        bl      printf                  // Call the printf function

        // return 0
        mov     w0, #0
        ldp     x29, x30, [sp], #16     // Pop FP and LR from the stack
        ret                             // Return from main
        .size   main,(. - main)
```

4.2.2 Logical operations

There are seven basic logical operations:

and   Bitwise AND,

bic	Bitwise Bit Clear,
eor	Bitwise Exclusive OR,
eon	Bitwise Exclusive OR NOT,
orr	Bitwise OR,
orn	Bitwise OR NOT, and
mvn	Bitwise NOT.

All of them involve two source operands and a destination register, except for mvn.

4.2.2.1 Syntax

```
<op>{s}     Rd, Rn, <Operand2>
<op2>       Rd, Rn, <Operand2>
mvn         Rd, <Operand2>
```

- <op> is either and or bic.
- <op2> is one of eor, eon, orr, orn, bic, or mvn.
- The optional s specifies whether or not the instruction should affect the nzcv flag bits in PSTATE.
- Recall from Section 4.1 that <Operand2> may use the ror shift and no sign extension.
- <Operand2> may only be used as an immediate in the following cases: and, ands, orr, and eor.
- mvn is an alias of orn where Rn is ZR.

4.2.2.2 Operations

Name	Effect	Description
and	$Rd \leftarrow Rn \wedge Operand2$	Bitwise AND.
bic	$Rd \leftarrow Rn \wedge \neg Operand2$	Bit Clear.
eor	$Rd \leftarrow Rn \oplus Operand2$	Bitwise Exclusive OR.
eon	$Rd \leftarrow Rn \oplus \neg Operand2$	Bitwise Exclusive OR NOT.
orr	$Rd \leftarrow Rn \vee Operand2$	Bitwise OR.
orn	$Rd \leftarrow Rn \vee \neg Operand2$	Bitwise OR NOT.
mvn	$Rd \leftarrow \neg Operand2$	One's complement.

4.2.2.3

```
         ands    x0, x1, #7              // x0 = x1 & 0x111 and set PSTATE

         bic     x3, x3, x4, lsl #5      // x3 = x3 & ~(x4 << 5)

         eor     x4, x5, #0x3fffc000     // x4 = x5 ^ 0x3fffc000
```

4.2.3 Data movement operations

The data movement operations copy data from one register to another:

mov   Move,
movz   Move wide with zero,
movn   Move wide with NOT, and
movk   Move wide with keep.

The mov instruction is an alias, chosen by the assembler, for orr, movz, movn, or add in-structions. It facilitates choosing an immediate value or moving a register value to or from the stack pointer. The wide moves allow the programmer to *widen* the sixteen bit immediate value to fill a 32 or 64 bit destination register.

4.2.3.1 Syntax

```
    mov         Rd|SP, Rn|SP
    mov         Rd, #<imm16>|#<pattern>
    <op>        Rd, #<imm16>{, lsl #<shift>}
```

- <op> is one of the wide moves movz, movn, or movk. It accepts an optional <shift> of either 0, 16, 32, or 48 for a 64-bit destination registers, or a shift of 0 or 16 for 32-bit destination registers.
- Non-pattern immediates are wide (16-bits long).
- Any 64-bit value can be loaded with 4 movk instructions. If it is a pattern, it can be done with just one mov instruction, which is an alias for orr.

4.2.3.2 Operations

Name	Effect	Description
mov	Rd|sp ← Rd|sp	Move general purpose register to or from the sp or to or from another general purpose register.

continued on next page

Name	Effect	Description
mov	Rd ← imm16 or `pattern`	Move an immediate to the destination register.
movz	Rd ← imm16 << `shift`	Set register to immediate shifted.
movn	Rd ← ¬(imm16 << `shift`)	Sets the register to the 1's complement of the shifted immediate.
movk	Rd[`shift` + 15 : `shift`] ← (imm16 << `shift`)	Sets only four bits and keeps the rest of the number.

4.2.3.3 Examples

```
    mov     x0, x1              // x0 = x1
    mov     sp, x3              // sp = x3
    mov     x0, 0xffffffff      // x0 = 0xffffffff
    movz    x0, 0xfedc, lsl #48 // x0 = 0xfedc000000000000
    movk    x0, 0xba09, lsl #32 // x0 = 0xfedcba0900000000
    movk    x0, 0x8765, lsl #16 // x0 = 0xfedcba0987650000
    movk    x0, 0x4321          // x0 = 0xfedcba0987654321
```

4.2.4 Shift operations

The shift operations allow the programmer to shift or rotate the contents of Rn by the number of bits given by the contents of Rm. There are four operations in this group:

asr Arithmetic shift right,
lsr Logical shift right, and
lsl Logical shift left,
ror Rotate right.

4.2.4.1 Syntax

```
    <op>    Rd, Rn, Rm
```

- <op> is one of `asr`, `lsl`, `lsr`, and `ror`.
- When shifting right, the programmer must decide what to do with the most significant bit. An arithmetic right shift means that the sign bit is extended to fill the upper bits, whereas a logical right shift uses zeroes to fill the upper bits.
- The register Rm's value, modulus the register size (32 or 64), is used as the shift amount.
- The `ror` operation rotates n bits to the right. Those bits "wrap around" and are shifted into the upper bits.

4.2.4.2 Operations

Name	Effect	Description
lsl	$Rd \leftarrow Rn \ll Rm$	Shift Left.
lsr	$Rd \leftarrow Rn \gg Rm$	Shift Right.
asr	$Rd \leftarrow signExtend(Rn \gg Rm)$	Shift Right with sign extend.
ror	$Rd \leftarrow Rd[Rn - 1 : 0] :$ $Rd[sizeof(Rd) - 1 : Rn]$	Rotate Right.

4.2.4.3 Examples

```
    lsl     x0, x1, x2      // x0 = x1 << x2
    asr     x3, x3, x0      // x3 = signExtend(x3 >> x0)
    lsr     x4, x5, x5      // x4 = x5 >> x5
    ror     x0, x0, x1      // x0 = x0[x1-1:0]:x0[63:x1]
```

These instructions are redundant because the same results can be achieved with an add or and instruction. For example:

```
    add     x0, xzr, x0, lsl #63    // x0 = x0 << 63
    orr     x0, xzr, x0, ror #47    // x0 = x0[46:0]:x0[63:47]
```

Although the results are identical, use of the lsl, alr, lsr, and ror mnemonics is strongly encouraged, because it results in code that is much easier to read, debug, and maintain.

4.2.5 Multiply operations with overflow

These four instructions perform multiplication using two 32-bit registers to form a 32-bit result, or two 64-bit registers to form a 64-bit result:

mul   Multiply,
madd   Multiply add,
msub   Multiply subtract,
mneg   Multiply negate.

4.2.5.1 Syntax

```
    <op>     Rd, Rn, Rm, Ra
    <op2>    Rd, Rn, Rm
```

- `<op>` is one of `madd`, `msub`, or `mneg`.
- `<op2>` is either `mul` or `mneg`. `mul` is an alias for `madd`, and `mneg` is an alias for `msub`. Both aliases use the zero register ZR for Ra.

4.2.5.2 Operations

Name	Effect	Description
`mul`	Rd ← Rn × Rm	Multiply.
`madd`	Rd ← Ra + Rn × Rm	Multiply and add.
`msub`	Rd ← Ra − Rn × Rm	Multiply and subtract.
`mneg`	Rd ← −(Rn × Rm)	Multiply and negate.

4.2.5.3 Examples

```
    mul     x0, x1, x2      // x0 = x1 * x2
    madd    x0, x1, x2, x3  // x0 = x3 + (x1 * x2)
    msub    x0, x1, x2, x3  // x0 = x3 - (x1 * x2)
    mneg    x0, x1, x2      // x0 = -(x1 * x2)
```

4.2.6 Multiply operations with 64-bit results

These instructions perform multiplication using two 32-bit registers to form a 64-bit result:

smull Signed multiply long,
smaddl
 Signed multiply add long,
smsubl
 Signed multiply subtract long,
smnegl
 Signed multiply negate long,
umull Unsigned multiply long,
umaddl
 Unsigned multiply add long,
umsubl
 Unsigned multiply subtract long, and
umnegl
 Unsigned multiply negate long.

They are the same form as the previous multiplication instructions in Section 4.2.5, except that they only use 32-bit operands, which are sign or unsigned extended, to produce a 64-bit output.

4.2.6.1 Syntax

```
<op>     Xd, Wn, Wm, Xa
<op2>    Xd, Wn, Wm
```

- <op> is any non-alias instruction: `smaddl`, `smsubl`, `umaddl`, or `umsubl`.
- <op2> is any alias instruction: `smull`, `smnegl`, `umull`, or `umnegl`.

4.2.6.2 Operations

Name	Effect	Description
smull	$Xd \leftarrow signExtend(Wn) \times signExtend(Wm)$	Signed multiply long.
smaddl	$Xd \leftarrow Xa + signExtend(Wn) \times signExtend(Wm)$	Signed multiply add long.
smsubl	$Xd \leftarrow Xa - signExtend(Wn) \times signExtend(Wm)$	Signed multiply subtract long.
smnegl	$Xd \leftarrow -(signExtend(Wn) \times signExtend(Wm))$	Signed multiply negate long.
umull	$Xd \leftarrow Wn \times Wm$	Unsigned multiply long.
umaddl	$Xd \leftarrow Ra + Wn \times Wm$	Unsigned multiply add long.
umsubl	$Xd \leftarrow Ra - Wn \times Wm$	Unsigned multiply subtract long.
umnegl	$Xd \leftarrow -(Wn \times Wm)$	Unsigned multiply negate long.

4.2.6.3 Examples

```
smull   x0, w1, w3              // x0 = signExtend(w1) * signExtend(w3)
umsubl  x0, w20, w21, x9        // x0 = x9 - (w20 * w21)
```

4.2.7 Multiply operations with 128-bit results

Signed and unsigned multiply high calculate the upper 64 bits of the 128-bit product of a multiplication between two 64-bit numbers. The lower 64 bits can be obtained using the `mul` instruction.

smulh Signed multiply high and
umulh Unsigned multiply high.

4.2.7.1 Syntax

```
<op>    Xd, Xn, Xm
```

- The <op> is either smulh or umulh.

4.2.7.2 Operations

Name	Effect	Description
smulh	$Xd \leftarrow (signExtend(Xn) \times signExtend(Xm))[127:64]$	Signed multiply high.
umulh	$Xd \leftarrow (Xn \times Xm)[127:64]$	Unsigned multiply high.

4.2.7.3 Examples

```
mul    x0, x1, x2    // x0 = x1 * x2
smulh  x1, x1, x2    // x1 = (signExtend(x1) * signExtend(x2))[127:64]
```

The following example shows how a C program and an AArch64 program could implement a simple command-line tool for multiplying 64-bit unsigned numbers, without rounding or truncating the result. To change the program to work with signed numbers, the only change needed for the AArch64 version would be to replace umulh with smulh.

```c
#include <stdio.h>
#include <stdint.h>

/* longMul takes two unsigned longs as input x and y and returns a
   128-bit product. */
__uint128_t longMul(uint64_t x, uint64_t y)
{
  return (__uint128_t) x * (__uint128_t) y;
}

/* Reads two unsigned longs from stdin and prints the resulting
   product on stdout. */
int main(void)
{
  long x, y, lo, hi;
  printf("Enter a number in hex: ");
  scanf("%lx", &x);
  printf("Enter a number in hex: ");
```

```
19    scanf("%lx", &y);
20    __uint128_t prod = longMul(x, y);
21    lo = (uint64_t) prod;
22    hi = (uint64_t) (prod >> 64);
23    printf("%016lx * %016lx = %016lx%016lx\n", x, y, hi, lo);
24    return 0;
25 }
```

The AArch64 implementation executes the multiplication step with just two instructions; most of the code is devoted to calling **printf** and **scanf** statements, and moving parameters into position:

```
1              .text
2              /* longMul takes two unsigned longs as input on x0 and x1 and
3                 returns a 128-bit product with the lower half "lo" in x0
4                 and the higher half "hi" in x1. */
5              .type    longMul, %function
6  longMul:
7              mul      x9, x0, x1
8              umulh    x1, x0, x1
9              mov      x0, x9
10             ret
11             .type    main, %function
12             .global main
13             /* Reads two unsigned longs from stdin and prints the resulting
14                product on stdout. */
15 main:
16             stp      x29, x30, [sp, #-32]!
17             stp      x19, x20, [sp, #16]     // Callee-saved
18
19             // printf("Enter a number in hex: ")
20             adr      x0, prompt
21             bl       printf
22
23             // scanf("%ld", &x)
24             adr      x0, fmtScan
25             adr      x1, dataIn
26             bl       scanf
27             adr      x1, dataIn
28             ldr      x19, [x1]       // x
29
30             // printf("Enter a number in hex: ")
31             adr      x0, prompt
32             bl       printf
33
34             // scanf("%ld", &y)
35             adr      x0, fmtScan
36             adr      x1, dataIn
```

```
37          bl      scanf
38          adr     x1, dataIn
39          ldr     x20, [x1]      // y
40
41          // longMul(x, y)
42          mov     x0, x19
43          mov     x1, x20
44          bl      longMul
45
46          // printf("%18lx * %18lx = %18lx%18lx\n", x, y, hi, lo)
47          mov     x4, x0         // lo
48          mov     x3, x1         // hi
49          mov     x2, x20        // y
50          mov     x1, x19        // x
51          adr     x0, fmtResult
52          bl      printf
53
54          // return 0
55          mov     w0, #0
56          ldp     x19, x20, [sp, #16]      // Callee-saved
57          ldp     x29, x30, [sp], #32
58          ret
59          .size   main, (. - main)
60
61          .data
62  dataIn:
63          .skip 8
64
65          .section .rodata
66  prompt:
67          .string "Enter a number in hex: "
68  fmtScan:
69          .string "%lx"
70  fmtResult:
71          .string "%016lx * %016lx = %016lx%016lx\n"
```

4.2.8 Division operations

There are only two division instructions in AArch64:

sdiv Signed divide and
udiv Unsigned divide.

4.2.8.1 Syntax

```
<op>    Rd, Rm, Rn
```

- The <op> is either sdiv or udiv.

4.2.8.2 Operations

Name	Effect	Description
sdiv	Rd ← Rm ÷ Rn	Signed divide.
udiv	Rd ← Rm ÷ Rn	Unsigned divide.

4.2.8.3 Examples

```
sdiv  x0, x1, x2   // x0 = x1 / x2 (signed)
udiv  x0, x1, x2   // x0 = x1 / x2 (unsigned)
```

4.2.9 Comparison operations

These three comparison operations update the PSTATE flags, but have no other effect:

cmp   Compare,
cmn   Compare Negative, and
tst   Test Bits.

They each perform an arithmetic operation, but result of the operation is discarded. cmp is an alias of subs, cmn is an alias of adds, and tst is an alias of ands.

4.2.9.1 Syntax

```
<op>    Rn, Operand2
```

- <op> is one of cmp, cmn, or tst.

4.2.9.2 Operations

Name	Effect	Description
cmp	Rn − Operand2	Compare and set PSTATE flags.
cmn	Rn + Operand2	Compare negative and set PSTATE flags.
tst	Rn ∧ Operand2	Test bits and set PSTATE flags.
teq	Rn ⊕ Operand2	Test equivalence and set PSTATE flags.

4.2.9.3 Examples

```
    cmp     w0, 35               // Compare w0 to 35 and set flags
    cmn     x2, w3, sxtw         // Compare negative x2 to sxtw and set flags
    tst     w1, 0xaaaaaaaa       // Test bits in w1 and Set PSTATE flags
```

Example 10. The following listings show how the compare, branch, and add instructions can be used to create a for loop. There are basically three steps for creating a for loop: allocating and initializing the loop variable, testing the loop variable, and modifying the loop variable. In general, any of the registers $x0-x15$, except $x8$ can be used to hold the loop variable. These are the scratch registers shown in Fig. 3.2. Section 5.4 in Chapter 5 introduces some considerations for choosing an appropriate register. For now, it is assumed that $x0$ is available for use as the loop variable in the example.

Suppose we want to implement a loop that is equivalent to the following C code:

```
    ⋮
    for(int i = 0; i < 10; i++)
    {

        /* insert loop body statements here */

    }
    ⋮
```

The loop can be written with the following AArch64 assembly code:

```
         ⋮
         mov     w0, #0          // int i = 0;
loop:    cmp     w0, #10
         bge     endloop         // Exit loop if r0 > 10

         // Insert loop body instructions here

         add     w0, w0, #1      // i++
         b       loop
endloop:
         ⋮
```

Example 11. The following C program waits for the character 'A' to be input on standard input:

```
#include <stdio.h>

int main(void)
{
```

```
5      do
6         {}                            // loop body does nothing
7      while (getchar() != 'A')
8      return 0;
9   }
```

An AArch64 translation of the C program uses the `cmp` instruction to test whether or not the input character is 'A' and a conditional branch instruction to implement the do-while loop:

```
1           .text
2           .type   main, %function
3           .global main
4   main:
5           stp     x29, x30, [sp, #-32]!
6           str     x19, [sp, #16]
7           mov     x19, #'A'
8
9   loop:   // beginning of do-while loop
10
11          // loop body does nothing
12
13          bl      getchar        // call getchar to get next character in x0
14          cmp     x0, x19        // compare it to 'A'
15          bne     loop           // repeat loop if the character is not equal to 'A'
16
17          ldr     x20, [sp, #16]
18          ldp     x29, x30, [sp], #32
19          ret
20          .size   main, (. - main)
```

4.2.10 Conditional operations

These conditional select operations set the destination register to the first operand, Rn if the condition is true, and to the second operand optionally incremented, inverted, or negated if the condition is false:

csel   Conditional Select,
csinc Conditional Select Increment,
csinv Conditional Select Invert,
csneg Conditional Select Negate.

There are five aliases that are derived from the previous instructions:

cinc   Conditional Increment,
cinv   Conditional Invert,

cneg   Conditional Negate,
cset   Conditional Set, and
csetm Conditional Set Mask.

Conditional comparison instructions allow PSTATE to be set to either a comparison or an immediate, depending on the flags in PSTATE:

ccmp   Conditional Compare,
ccmn   Conditional Compare Negative.

The conditional compare and compare negative instructions check the PSTATE flags for the given condition. If it is true, then it sets the PSTATE flags to the comparison of Rn and either Rm or <imm5>. If it is false, it sets the PSTATE flags to an immediate four bit value representing nzcv.

4.2.10.1 Syntax

```
<op>    Rd, Rn, Rm, <cond>
<op2>   Rd, Rn, <cond>
<op3>   Rd, <cond>
ccmp    Rn, Rm|#<imm5>, #<nzcv>, <cond>
ccmn    Rn, Rm|#<imm5>, #<nzcv>, <cond>
```

- <op> is one of csel, csinc, csinv, or csneg.
- <op2> is one of cinc, cinv, or cneg.
- <op3> is cset or csetm.
- The <cond> is any one of the codes from Table 3.2 on page 59.
- The <nzcv> is any number 0-15 (0x0-0xf). It is four bits representing the N, Z, C, and V flags.
- The <imm5> is an unsigned 5-bit immediate.

4.2.10.2 Operations

Name	Effect	Description

csel	**if** < cond > **then**
	    Rd ← Rn
	else
	    Rd ← Rm
	end if
		Set Rd to Rn or Rm.

continued on next page

Name	Effect	Description

csinc	**if** < cond > **then**
	   Rd ← Rn
	else
	   Rd ← Rm + 1
	end if
		Set Rd to Rn or Rm incremented.

csinv	**if** < cond > **then**
	   Rd ← Rn
	else
	   Rd ← ¬Rm
	end if
		Set Rd to Rn or Rm inverted.

csneg	**if** < cond > **then**
	   Rd ← Rn
	else
	   Rd ← 1 + ¬Rm
	end if
		Set Rd to Rn or Rm negated.

cinc	**if** < cond > **then**
	   Rd ← Rn + 1
	else
	   Rd ← Rn
	end if
		Set Rd to Rn incremented or Rn.

cinv	**if** < cond > **then**
	   Rd ← ¬Rn
	else
	   Rd ← Rn
	end if
		Set Rd to Rn inverted or Rn.

cneg	**if** < cond > **then**
	   Rd ← 1 + ¬Rn
	else
	   Rd ← Rn
	end if
		Set Rd to Rn negated or Rn.

cset	**if** < cond > **then**
	   Rd ← 1
	else
	   Rd ← 0
	end if
		Set Rd to 0x1 or 0x0.

continued on next page

Name	Effect	Description

csetm	**if** < cond > **then**
	    Rd ← 0xffffffffffffffff
	else
	    Rd ← 0x0000000000000000
	end if
		Fill Rn with ones or zeroes.

ccmp	**if** < cond > **then**
	    PSTATE ← $flags$(Rn − Rm|imm5)
	else
	    PSTATE ← nzcv
	end if
		Test and set PSTATE flags to a comparison or immediate.

ccmn	**if** < cond > **then**
	    PSTATE ← $flags$(Rn + Rm|imm5)
	else
	    PSTATE ← nzcv
	end if
		Test and set PSTATE flags to a negated comparison or immediate.

4.2.10.3 Examples

```
    csel    x0, x1, x2, ge        // x0 = (ge ? x1 : x2)
    csneg   x2, x3, x4, ne        // x2 = (ne ? x3: -x4)
    cinv    x9, x7, hi            // x9 = (hi ? ~x7 : x7)
    cset    x1, eq               // x1 = (eq ? 1 : 0)
    ccmp    x4, #31, 0b0100, mi   // nzcv = (mi ? flags(x4 - 32) : 4)
```

The following C function will flip the parity of a number and alternate between two values if its output is re-used as the input:

```
int flipFlop(int num) {
  if (num % 2 == 0) {   /* is even */
    return num + 1;
  } else {
    return num - 1;
  }
}
```

The AArch64 translation uses conditional selection instead of branching to implement the if-else statement. The input num is in w0 and is also returned in w0 as the output:

```
        .text
        .type   flipFlop, %function
flipFlop:
        // Tests for parity by setting the PSTATE flags to w0 AND 1
```

```
5     tst     w0, #1              // Z = ~(num & 1)
6     add     w1, w0, #-1         // w1 = w0 - 1
7     csinc   w0, w1, w0, ne      // w0 = (num & 1 ? w1 : w0 + 1)
8     ret                         // return w0
```

This works because instead of testing if the function is even with division and remainders as in the C code, num % 2 == 0, the assembly does a conditional bitwise AND with 0x1 because the least significant bit will always be 0 for an even number and 1 for an odd number in binary. The value w0 - 1 is placed in w1. If num is odd, which means that the Z flag in PSTATE is 0, then w0 is set to the value of w1. Otherwise, if num is even, then the Z flag is 1 and w0 is set to the value of w0 + 1 since it is the conditional select increment instruction.

4.3 Special instructions

There are a few instructions that do not fit into any of the previous categories. They are used to request operating system services and access advanced CPU features.

4.3.1 Count leading zeros

These instructions count the number of leading zeros in the operand register or the number of leading sign bits and stores the result in the destination register:

clz   Count Leading Zeros and
cls   Count Leading Sign Bits.

4.3.1.1 Syntax

```
<op>    Rd, Rn
```

• The <op> is either clz or cls.

4.3.1.2 Operations

Name	Effect	Description
clz	Rd ← $Count\,Leading\,Zeros$(Rn)	Count leading zeros in Rn.
cls	Rd ← $Count\,Leading\,Sign\,Bits$(Rn)	Count leading ones or zeros in Rn.

4.3.1.3 Example

```
1     clz     x8, x0
```

4.3.2 Accessing the PSTATE register

These two instructions allow the programmer to access the status bits of the CPSR and SPSR:

mrs   Move Status to Register, and
msr   Move Register to Status.

4.3.2.1 Syntax

```
mrs    Xt, <field>
msr    <field>, Xt
```

*   The optional `<fields>` is any one of:
    NZCV   Condition Flags
    DAIF   Interrupt Bits
    CurrentEL   Current Exception Level
    PAN   Privileged Access Never (ARMv8.1 only)
    UAO   User Access Override (ARMv8.2-UAO only)
*   The optional `<cond>` can be any of the codes from Table 3.2 specifying conditional execution.

4.3.2.2 Operations

Name	Effect	Description
mrs	Xt ← PSTATE	Move from Process State.
msr	PSTATE ← Xt	Move to Process State.

4.3.2.3 Examples

`nzcv` is the only PSTATE field guaranteed to be accessible at the lowest execution state, EL0, which is unprivileged and where applications are intended to be run:

```
mrs    x0, NZCV      // Read the NZCV flags into r0
and    x0, xzr, x0   // Clear the flags
msr    NZCV, x0      // Write the flags back
```

4.3.3 Supervisor call

The following instruction allows a user program to perform a *system call* to request operating system services:

svc   Supervisor Call.

In Linux, the system calls are documented in section two of the online manual. Each system call has a unique id, which may vary from one computer architecture to the next, or from operating system to another. On Linux, it is generally better to make system calls by using the corresponding C library function, rather than calling them directly from assembly. This is because the C library function may perform additional work before or after making the system call. For instance, the `exit` library function may invoke other functions to cleanly shut down the program before it performs the `exit` system call.

4.3.3.1 Syntax

```
svc       <syscall_number>
```

- The `<syscall_number>` is encoded in the instruction. The operating system may examine it to determine which operating system service is being requested.
- In Linux, `<syscall_number>` should always just be zero. The system call number is passed in `x8` and six other parameters can be passed in on `x0-x5`.

4.3.3.2 Operations

Name	Effect	Description
svc	Request Operating System Service	Perform software interrupt.

4.3.3.3 Example

This example leverages the `write` system call to print a message without using any C standard library functions, like `printf`:

```
     .section .rodata
msg: .string "She Sells Sea Shells By The Sea Shore\n" // 39 bytes
     .text
     .type   main, %function
     .global main
     // the following code asks the operating system
     // to write some characters to standard output
main:
     stp    x29, x30, [sp, #-16]!

     mov  x0, #1    // file descriptor 1 is stdout
     adr  x1, msg   // load address of data to write
     mov  x2, #39   // load number of bytes to write
     mov  x8, #64   // syscall #64 is the write() function
     svc  #0        // invoke syscall
```

```
16
17      // return 0
18      mov    w0, #0
19      ldp    x29, x30, [sp], #16
20      ret
21      .size  main, (. - main)
```

4.3.4 No operation

This instruction does nothing except waste execution time.

nop   No Operation.

4.3.4.1 Syntax

```
nop
```

4.3.4.2 Operations

Name	Effect	Description
nop	No effects	No Operation.

4.3.4.3 Examples

nop's can sometimes be inserted to optimize machine specific code. Other times they are used in computer attacks. They can even be used just to experiment with a debugger. The following example shows how one might make a counter to delay a short period using nop instructions and a loop:

```
1       .text
2       .type   main, %function
3       .global main
4  main:
5       // a programmed delay loop
6       mov    x0, #0x3fffffff // load loop counter
7  loop: nop
8       nop
9       nop
10      nop
11      sub    x0, x0, #1      // decrement counter
12      cmp    w0, #0
13      bgt    loop
14      // return 0
```

```
15    mov    w0, #0
16    ret
17    .size    main, (. - main)
```

4.4 Alphabetized list of AArch64 instructions

This chapter and the previous one introduced the core set of AArch64 instructions. Most of these instructions were introduced with the very first AArch64 processors. There are more architectural extensions that provide special instructions for uses such as cryptography. There are also additional instructions that compute floating point operations, called Vector Floating Point (VFP) and NEON.

The instructions introduced so far are:

Name	Page	Operation
adc	86	Add with Carry
add	86	Add
adr	78	Form PC-Relative Address
adrp	78	Form PC-Relative Address to 4 KB Page
and	88	Bitwise AND
asr	91	Arithmetic shift right
b	70	Branch
bic	89	Bitwise Bit Clear
bl	73	Branch and Link
blr	73	Branch to Register and Link
br	72	Branch to Register
cbnz	76	Compare and Branch if Nonzero
cbz	76	Compare and Branch if Zero
ccmn	101	Conditional Compare Negative
ccmp	101	Conditional Compare
cinc	100	Conditional Increment
cinv	100	Conditional Invert
cls	104	Count Leading Sign Bits
clz	104	Count Leading Zeros
cmn	98	Compare Negative
cmp	98	Compare
cneg	101	Conditional Negate
csel	100	Conditional Select
cset	101	Conditional Set

continued on next page

Name	Page	Operation
csetm	101	Conditional Set Mask
csinc	100	Conditional Select Increment
csinv	100	Conditional Select Invert
csneg	100	Conditional Select Negate
eon	89	Bitwise Exclusive OR NOT
eor	89	Bitwise Exclusive OR
ldp	68	Load Pair
ldr	65	Load Register
ldur	67	Load Register (Unscaled)
lsl	91	Logical shift left
lsr	91	Logical shift right
madd	92	Multiply add
mneg	92	Multiply negate
mov	90	Move
movk	90	Move wide with keep
movn	90	Move wide with NOT
movz	90	Move wide with zero
mrs	105	Move Status to Register
msr	105	Move Register to Status
msub	92	Multiply subtract
mul	92	Multiply
mvn	89	Bitwise NOT
neg	86	Negate
ngc	86	Negate with Carry
nop	107	No Operation
orn	89	Bitwise OR NOT
orr	89	Bitwise OR
ret	72	Return from Subroutine
ror	91	Rotate right
sbc	86	Subtract with Carry
sdiv	97	Signed divide
smaddl	93	Signed multiply add long
smnegl	93	Signed multiply negate long
smsubl	93	Signed multiply subtract long
smulh	95	Signed multiply high
smull	93	Signed multiply long
stp	68	Store Pair

continued on next page

Name	Page	Operation
str	65	Store Register
stur	67	Store Register (Unscaled)
sub	86	Subtract
svc	106	Supervisor Call
tbnz	76	Test Bit and Branch if Nonzero
tbz	76	Test Bit and Branch if Zero
tst	98	Test Bits
udiv	97	Unsigned divide
umaddl	93	Unsigned multiply add long
umnegl	93	Unsigned multiply negate long
umsubl	93	Unsigned multiply subtract long
umulh	95	Unsigned multiply high
umull	93	Unsigned multiply long

4.5 Chapter summary

The AArch64 Instruction Set Architecture includes 31 general-purpose registers and four basic instruction types. This chapter introduced the instructions used for

- moving data from one register to another,
- performing computational operations with two source operands and one destination register,
- multiplication and division,
- performing comparisons, and
- performing special operations.

Most of the data processing instructions are three address instructions, because they involve two source operands and produce one result. For most instructions, the second source operand can be a register, a rotated or shifted register, or an immediate value. This flexibility results in a relatively powerful assembly language.

Exercises

4.1. If x0 initially contains 1, what will it contain after the third instruction in the sequence below?

```
    add    x0, x0, #1
    mov    x1, x0
    add    x0, x1, x0, lsl #1
```

4.2. What will x0 and x1 contain after each of the following instructions? Give your answers in base 10.

```
        mov     x0,#1
        mov     x1,#0x20
        orr     x1,x1,x0
        lsl     x1,x1,#0x2
        orr     x1,x1,x0
        eor     x0,x0,x1
        lsr     x1,x0,#3
```

4.3. What is the difference between lsr and asr?

4.4. Write instructions to load the numbers stored at address num1 and address num2, add them together, and store the result in numsum. Use only x0 and x1.

4.5. Given the following variable definitions:

```
num1:    .word    x
num2:    .word    y
```

where you do not know the values of x and y, write a short sequence of arm64 assembly instructions to load the two numbers, compare them, and move the largest number into register x0.

4.6. Assuming that a is stored in register x0 and b is stored in register x1, show the arm64 assembly code that is equivalent to the following C code:

```
if ( a & 1 )
   a = -a;
else
   b = b+7;
```

4.7. Write a loop to count the number of bits in x0 that are set to 1. Use any other registers that are necessary.

4.8. Without using any mul instruction, give the instructions to multiply x3 by the following constants, leaving the result in x0. You may also use x1 and x2 to hold temporary results, and you do not need to preserve the original contents of x3.
   a. 10
   b. 100
   c. 575
   d. 123

4.9. Assume that x0 holds the least significant 64 bits of a 128-bit integer a, and x1 holds the most significant 64 bits of a. Likewise, x2 holds the least significant 64 bits of a 128-bit integer b, and x3 holds the most significant 64 bits of b. Show the shortest instruction sequences necessary to:
   a. compare a to b, setting the PSTATE flags,

    b.   shift *a* left by one bit, storing the result in *b*,

    c.   add *b* to *a*, and

    d.   subtract *b* from *a*.

4.10.   The C standard library provides the openat() function, which is documented in section two of the Linux manual pages. This function is a very small "wrapper" to allow C programmers to access the openat() system call. Assembly programmers can access the system call directly. In arm64 Linux, the system call number for openat() is 56. The Linux system calls can be found on a Linux system in: /usr/include/asm-generic/unistd.h. Write the arm64 assembly instructions and directives necessary to make a Linux system call to open a file named input.txt for reading, without using the C standard library. In other words, write the assembly equivalent to: openat(-100,"input.txt",O_RDONLY); using the svc instruction.

## CHAPTER 5

Structured programming

Before IBM released FORTRAN in 1957, almost all programming was done in assembly language. Part of the reason for this is that nobody knew how to design a good high-level language, nor did they know how to write a compiler to generate efficient code. Early attempts at high-level languages resulted in languages that were not well structured, difficult to read, and difficult to debug. The first release of FORTRAN was not a particularly elegant language by today's standards, but it *did* generate efficient machine code.

In the 1960s, a new paradigm for designing high-level languages emerged. This new paradigm emphasized grouping program statements into blocks of code that execute from beginning to end. These *basic blocks* have only one entry point and one exit point. Control of which basic blocks are executed, and in what order, is accomplished with highly structured *flow control* statements. The structured program theorem provides the theoretical basis of structured programming. It states that there are three ways of combining basic blocks: sequencing, selection, and iteration. These three mechanisms are sufficient to express any computable function. It has been proven that all programs can be written using only basic blocks, the *pre-test loop*, and *if-then-else* structure. Although most high level languages provide additional flow control statements for the convenience of the programmer, they are just "syntactical sugar." Other structured programming concepts include well-formed functions and procedures, pass-by-reference and pass-by-value, separate compilation, and information hiding.

These structured programming languages enabled programmers to become much more productive. Well-written programs that adhere to structured programming principles are much easier to write, understand, debug, and maintain. Most successful high-level languages are designed to enforce, or at least facilitate, good programming techniques. This is not generally true for assembly language. The burden of writing well-structured code lies with the programmer, and not the language.

The best assembly programmers rely heavily on structured programming concepts. Failure to do so results in code that contains unnecessary branch instructions and, in the worst cases, results in something called *spaghetti code*. Consider a code listing where a line has been drawn from each branch instruction to its destination. If the result looks like someone spilled a plate of spaghetti on the page, then the listing is spaghetti code. If a program is spaghetti code, then the flow of control is difficult to follow. Spaghetti code is much more likely to have bugs, and

Listing 5.1 **if** statement in C.

```
1    :
2    if (x >= y)
3    {
4        :                      // if statement body
5    }
6    :
```

is extremely difficult to debug. If the flow of control is too complex for the programmer to follow, then it cannot be adequately debugged. It is the responsibility of the assembly language programmer to write code that uses a block-structured approach.

Adherence to structured programming principles results in code that has a much higher probability of working correctly. Also, well-written code has fewer branch statements. Therefore, the ratio of data processing statements versus branch statements is higher. High data processing density results in higher throughput of data. In other words, writing code in a structured manner leads to higher efficiency.

5.1 Sequencing

Sequencing simply means executing statements (or instructions) in a linear sequence. When statement n is completed, statement $n + 1$ will be executed next. Uninterrupted sequences of statements form basic blocks. Basic blocks have exactly one entry point and one exit point. Flow control is used to select which basic block should be executed next.

5.2 Selection

The first control structure that we will examine is the basic *selection* construct. It is called selection because it selects one of two (or possibly more) blocks of code to execute, based on some condition. In its most general form, the condition could be computed in a variety of ways, but most commonly it is the result of some comparison operation or the result of evaluating a Boolean expression.

5.2.1 If-then statement

The most important form of selection is the **if** statement. Listing 5.1 demonstrates a simple if statement in C. Listing 5.2 shows the AArch64 assembly translation. The first label is for

Listing 5.2 **if** statement in AArch64 assembly.

```
     :
if:     cmp     x0, x1          // perform test
        blt     endif           // if (x < y) goto endif
     :                  //   if statement body
endif:
     :
```

clarity, to show where the if statement begins, and the second label, endif: is the end of the if statement. The condition in assembly is reversed from the C code. The C code checks if x >= y to know if the if statement should be taken. In AArch64 assembly, the opposite check is used with the instruction blt. If x < y, then the if statement is not executed. Instead, control branches past the if statement.

5.2.2 If-then-else statement

Most high-level languages provide an extension to the basic if statement, which allows the programmer to select one of two possible blocks of code to execute. The same behavior can be achieved by using two sequential if statements, with logically opposite tests. However the if-then-else construct results in code that is more efficient as well as easier to read, understand, and debug.

5.2.2.1 Using branch instructions

Listing 5.3 shows a typical if-then-else statement in C. Listing 5.4 shows the AArch64 code equivalent, using branch instructions. Note that this method requires a conditional branch, an unconditional branch, and two labels. This is the most common method of writing the bodies of the then and else selection constructs.

5.2.2.2 Using conditional selection

Branching is not the only way to write an if statement. AArch64 has selection instructions that are able to choose between two source registers based on a condition code. These instructions can also optionally increment, set, or negate one of the choices.

Listing 5.5 shows the AArch64 code equivalent to Listing 5.3, using conditional operations. Section 4.2.10 covers the eleven conditional selection instructions in AArch64. While they cannot always be used to replace branch-based if statements, in certain examples such as Listing 5.5, they can be used to accomplish the same task without using any branch instructions. In this and similar cases, the code is more efficient than using a branch.

Listing 5.3 Selection in C.

```
       ⋮
static int a,b,x;
       ⋮
       if ( a < b )
          x = 1;
       else
          x = 0;
       ⋮
```

Listing 5.4 Selection in AArch64 assembly using branch instructions.

```
       ⋮
       adr    x0, a        // load pointer to a
       adr    x1, b        // load pointer to b
       ldr    x0, [x0]     // load a
       ldr    x1, [x1]     // load b
if:
       cmp    x0, x1       // compare them
       bge    else         // if a >= b then skip forward
       mov    x0, #1       // THEN section - load 1 into x0
       b      endif        // skip the else section
else:
       mov    x0, #0       // ELSE section - load 0 into x0
endif:
       adr    x1, x        // load pointer to x
       str    x0, [x1]     // store x0 in x
       ⋮
```

Listing 5.5 Selection in AArch64 assembly using a conditional instruction.

```
       ⋮
       adr    x0, a        // load pointer to a
       adr    x1, b        // load pointer to b
       ldr    x0, [x0]     // load a
       ldr    x1, [x1]     // load b
       cmp    x0, x1       // compare them
       cset   x0, lt       // Conditionally set x0 to 1 if a
       adr    x1, x        // load pointer to x
       str    x0, [x1]     // store x0 in x
       ⋮
```

Listing 5.6 Complex selection in C.

```
static long a = 0x57;
static long b = 0x75;
static long c = 0x21;
static long x;
int main(void)
{
  if ((a < b) && (a < c)) {
    x = a;
  } else if ( b < c ) {
    x = b;
  } else {
    x = c;
  }
  return 0;
}
```

5.2.3 Complex selection

More complex selection structures should be written with care. Listing 5.6 shows a fragment of C code which compares the variables *a*, *b*, and *c*, and sets the variable *x* to the least of the three values. In C, Boolean expressions use *short circuit evaluation*. For example, consider the Boolean AND operator in the expression ((a<b)&&(a<c)). If the first sub-expression evaluates to false, then the truth value of the complete expression can be immediately determined to be false, so the second sub-expression is not evaluated. This usually results in the compiler generating very efficient assembly code. Good programmers can take advantage of short-circuiting by checking array bounds early in a Boolean expression, and accessing array elements later in the expression. For example, the expression ((i<15)&&(array[i]<0)) makes sure that the index i is less than 15 before attempting to access the array. If the index is greater than 14, the array access will not take place. This prevents the program from attempting to access the sixteenth element on an array that has only fifteen elements.

Listing 5.7 shows an AArch64 assembly code fragment which is equivalent to Listing 5.6. In this code fragment, x0 is used to store a temporary value for the variable y, and the value is only stored to memory once, at the end of the fragment of code. The first if statement is implemented using branch instructions. Its comparison is performed on line 17 of Listing 5.7. If the comparison evaluates to false, then it immediately branches to the elseif block, but if the first comparison evaluates to true, then it performs the second comparison in the if statement. Again, if that comparison evaluates to false, then it branches to the elseif block. If both comparisons evaluate to true, then it executes x0 = a and then branches to the endif label. In this case, x0 is already set to a, so it just branches to endif.

Listing 5.7 Complex selection in AArch64 assembly.

```
1          .text
2          .type   main, %function
3          .global main
4   main:
5          stp     x29, x30, [sp, #-16]!
6
7          // calculate addresses
8          adr     x0, a           // x0 = &a
9          adr     x1, b           // x1 = &b
10         adr     x2, c           // x2 = &c
11
12         // load from memory
13         ldr     x0, [x0]        // x0 = a
14         ldr     x1, [x1]        // x1 = b
15         ldr     x2, [x2]        // x2 = c
16
17  if:    // if ((a < b) && (a < c))
18         cmp     x0, x1          // compare a and b
19         bge     elseif          // if (a >= b) goto elseif
20         cmp     x0, x2          // compare a and c
21         bge     elseif          // if (a >= c) goto elseif
22         b       endif           // x0 = a
23  elseif:
24         cmp     x1, x2          // compare b and c
25         csel    x0, x1, x2, lt  // x0 = (b < c ? b : c)
26  endif:
27         adr     x9, y           // x9 = &y
28         str     x0, [x9]        // *y = x0
29
30         // return 0
31         mov     w0, #0
32         ldp     x29, x30, [sp], #16
33         ret
34         .size   main, (. - main)
35
36         .data
37  y:     .skip   8
38  a:     .quad   0x57
39  b:     .quad   0x75
40  c:     .quad   0x21
```

The if statement on line 9 of Listing 5.6 is implemented using conditional selection. The comparison is performed on line 23 of Listing 5.7. Line 24 contains the instruction that conditionally selects the result value.

Listing 5.8 Unconditional loop in AArch64 assembly.

```
         .section .rodata
hstr:    .string "Hello World\n"
         .text
         .type   main, %function
         .global main
main:
loop:    adr     x0,hstr          // load pointer to "Hello World\n\0"
         bl      printf           // print "Hello World\n"
         b       loop             // repeat main unconditionally
         .size   main, (. - main)
```

Note that the number of comparisons that execute will always be minimized, and the number of branches has also been minimized. The only way that line 23 can be reached is if one of the first two comparisons evaluates to false. The program fragment will *always* reach line 26 and a value will be stored in *x*. Thus, the AArch64 assembly code fragment in Listing 5.7 can be considered to be a block of code with exactly one entry point and one exit point.

When writing nested selection structures, it is important to maintain a block structure, even if the bodies of the blocks consist of only a single instruction. It is often very helpful to write the algorithm in pseudo-code or a high-level language, such as C or Java, before converting it to assembly. Prolific commenting of the code is also strongly encouraged.

5.3 Iteration

Iteration involves the transfer of control from a statement in a sequence to a *previous* statement in the sequence. The simplest type of iteration is the *unconditional* loop, also known as the *infinite* loop. This type of loop may be used in programs or tasks that should continue running indefinitely. Listing 5.8 shows an AArch64 assembly fragment containing an unconditional loop. Few high-level languages provide a true unconditional loop, but the high-level programmer can achieve a similar effect by using a conditional loop and specifying a condition that always evaluates to true.

5.3.1 Pre-test loop

A pre-test loop is a loop in which a test is performed *before* the block of instructions forming the loop body is executed. If the test evaluates to true, then the loop body is executed. The last instruction in the loop body is a branch back to the beginning of the test. If the test evaluates

Listing 5.9 Pre-test loop in C.

```
    :
while (x >= y)                    // perform loop test
{
    :                             // body of loop
}                                 // done
    :
```

Listing 5.10 Pre-test loop in AArch64 assembly.

```
        :
loop:   cmp     x0, x1            // perform loop test
        blt     done             // exit loop if x0 < x1
        :                         // body of loop
        b       loop             // repeat loop
done:
        :
```

Listing 5.11 Post-test loop in C.

```
        :
do {
        :                         // body of loop
} while (x >= y);                 // perform loop test
        :
```

to false, then execution branches to the first instruction following the loop body. All structured programming languages have a pre-test loop construct. For example, in C, the pre-test loop is called a while loop. In assembly, a pre-test loop is constructed very similarly to an if-then statement. The only difference is that it includes an additional branch instruction at the end of the sequence of instructions that form the body. Listing 5.10 shows a pre-test loop in AArch64 assembly.

5.3.2 Post-test loop

In a post-test loop, the test is performed *after* the loop body is executed. If the test evaluates to true, then execution branches to the first instruction in the loop body. Otherwise, execution continues sequentially. Most structured programming languages have a post-test loop construct. For example, in C, the post-test loop is called a do-while loop. Listing 5.11 shows a

Listing 5.12 Post-test loop in AArch64 assembly.

```
do:                       ⋮
                                        // body of loop
        cmp     x0, x1                  // perform loop test
        bge     do                      // repeat loop if x0 >= x1
                          ⋮
```

Listing 5.13 Guarded loop in AArch64 assembly.

```
                          ⋮
        cmp     x0, x1
        blt     loopEnd                 // if (x0 < x1) goto loopEnd
loop:
                          ⋮             // body of loop
        cmp     x0, x1
        bge     do                      // if (x0 >= x1) goto loop
loopEnd:
                          ⋮
```

do-while loop in C, and Listing 5.12 shows a post-test loop in AArch64 assembly. The body of a post-test loop will always be executed at least once.

A post-test loop is more efficient than a pre-test loop because it will always perform the loop test one time less than a pre-test loop, and has one fewer branch instruction. When the test fails, the post-test loop can exit immediately. A pre-test loop must first branch up to the beginning and then branch back down to the end. When the programmer can guarantee that the loop body must always execute at least once, a post-test loop is obviously the better choice.

Even if the loop body is not guaranteed to execute at least once in all cases, the post-test loop can provide slightly better performance. In order to take advantage of this efficiency, pre-test loops can be turned into post-test loops by adding one check before the pre-test loop begins. This is commonly referred to as a *guarded* loop. The guarded loop is shown in Listing 5.13.

5.3.3 For loop

Many structured programming languages have a **for** loop construct, which is a type of *counting* loop. The **for** loop is not essential, and is only included as a matter of syntactical convenience. In some cases, a **for** loop is easier to write and understand than an equivalent pre-test

Listing 5.14 **for** loop in C.

```
       ⋮
for (int i = 0; i < 10; i++)
{
   printf("Hello World - %d\n", i);
}
       ⋮
```

Listing 5.15 **for** loop re-written as a pre-test loop in C.

```
       ⋮
int i = 0;
while (i < 10)
{
   printf("Hello World - %d\n", i);
   i++;
}
       ⋮
```

or post-test loop. However, with the addition of an if-then construct, *any* loop can be implemented as a pre-test loop. The following sections show how loops can be converted from one form to another.

5.3.3.1 Pre-test conversion

Most assembly languages do not have a **for** loop. However, any **for** loop can be converted to a pre-test loop. Listing 5.14 shows a simple C program with a **for** loop. The program prints "Hello World" ten times, appending an integer to the end of each line.

In order to write an equivalent program in assembly, the programmer must first re-write the **for** loop as a pre-test loop. Listing 5.15 shows the program re-written so that it is easier to translate into assembly. Note that the initialization of the loop variable has been moved to its own line before the **while** statement. Also, the loop variable is modified on the last line of the loop body. This is a straightforward conversion from one type of loop to another type. Listing 5.16 shows a translation of the pre-test loop structure into AArch64 assembly.

5.3.3.2 Post-test conversion

If the programmer can guarantee that the body of a **for** loop will *always* execute at least once, then the **for** loop can be converted to an equivalent post-test loop. This form of loop is more

Listing 5.16 pre-test loop in AArch64 assembly.

```
        .section .rodata
str:    .string "Hello World - %d\n"
        .text
        :
        mov     x4, #0      // x4 = i = 0
loop:
        cmp     x4, #10     // perform comparison
        bge     done        // end loop if i >= 10
        adr     x0, str     // load pointer to format string
        mov     x1, x4      // copy i into x1
        bl      printf      // printf("Hello World - %d\n", i);
        add     x4, x4, #1  // i++
        b       loop        // repeat loop
done:
        :
```

Listing 5.17 **for** loop re-written as a post-test loop in C.

```
        :
    int i = 0;
    do {
      printf("Hello World - %d\n", i);
      i++;
    } while(i < 10);
        :
```

efficient, because the loop control variable is tested one time less than for a pre-test loop. Also, a post-test loop requires only one label and one conditional branch instruction, whereas a pre-test loop requires two labels, a conditional branch, and an unconditional branch.

Since the loop in Listing 5.14 always executes the body exactly ten times, we know that the body will always execute at least once. Therefore, the loop can be converted to a post-test loop. Listing 5.17 shows the program re-written as a post-test loop so that it is easier to translate into assembly. Note that, as in the previous example, the initialization of the loop variable has been moved to its own line before the do-while loop, and the loop variable is modified on the last line of the loop body. This post-test version will produce the same output as the pre-test version. This is a straightforward conversion from one type of loop to an equivalent type. Listing 5.18 shows a straightforward translation of the post-test loop structure into AArch64 assembly.

Listing 5.18 Post-test loop in AArch64 assembly

```
1         .section .rodata
2  str:    .string "Hello World - %d\n"
3         .text
4         :
5         ldr     x4, #0      // x4 = i = 0
6  loop:
7         adr     x0, str     // load pointer to format string
8         mov     x1, x4      // copy i into x1
9         bl      printf      // printf("Hello World - %d\n", i);
10        add     x4, x4, #1  // i++
11        cmp     x4, #10     // perform comparison
12        blt     loop        // continue if i < 10
13        :
```

5.4 Subroutines

A subroutine is a sequence of instructions to perform a specific task, packaged as a single unit. Depending on the particular programming language, a subroutine may be called a procedure, a function, a routine, a method, a subprogram, or some other name. Some languages, such as Pascal, make a distinction between functions and procedures. A function must return a value and must not alter its input arguments or have any other *side effects* (such as producing output or changing static or global variables). A procedure returns no value, but may alter the value of its arguments or have other side effects.

Other languages, such as C, make no distinction between procedures and functions. In these languages, functions may be described as pure or impure. A function is pure if:

1. the function always evaluates the same result value when given the same argument value(s), and
2. evaluation of the result does not cause any semantically observable side effect or output.

The first condition implies that the function result cannot depend on any hidden information or state that may change as program execution proceeds, or between different executions of the program, nor can it depend on any external input from I/O devices. The result value of a pure function does not depend on anything other than the argument values. If the function returns multiple result values, then these two conditions must apply to all returned values. Otherwise the function is impure. Another way to state this is that impure functions have side effects while pure functions have no side effects.

Assembly language does not impose any distinction between procedures and functions, pure or impure. The assembly language will provide a way to call subroutines and return from them. It is up to the programmer to decide how to pass arguments to the subroutines and how to pass return values back to the section of code that called the subroutine. Once again, the expert assembly programmer will use these structured programming concepts to write efficient, readable, debugable, and maintainable code.

5.4.1 Advantages of subroutines

Subroutines help programmers to design reliable programs by decomposing a large problem into a set of smaller problems. It is much easier to write and debug a set of small code pieces than it is to work on one large piece of code. Careful use of subroutines will often substantially reduce the cost of developing and maintaining a large program, while increasing its quality and reliability. The advantages of breaking a program into subroutines include:

- enabling reuse of code across multiple programs,
- reducing duplicate code within a program,
- enabling the programming task to be divided between several programmers or teams,
- decomposing a complex programming task into simpler steps that are easier to write, understand, and maintain,
- enabling the programming task to be divided into stages of development, to match various stages of a project, and
- hiding implementation details from users of the subroutine (a programming principle known as *information hiding*).

5.4.2 Disadvantages of subroutines

There are two minor disadvantages to using subroutines. First, invoking a subroutine (versus using in-line code) imposes overhead. The arguments to the subroutine must be put into some known location where the subroutine can find them. If the subroutine is a function, then the return value must be put into a known location where the caller can find it. Also, a subroutine typically requires some standard entry and exit code to manage the stack and save and restore the return address.

In most languages, the cost of using subroutines is hidden from the programmer. In assembly, however, the programmer is often painfully aware of the cost, since they have to explicitly write the entry and exit code for each subroutine, and must explicitly write the instructions to pass the data into the subroutine. However, the advantages far outweigh the costs. Assembly programs can get very large and failure to modularize the code by using subroutines will result in code that cannot be understood or debugged, much less maintained and extended.

Listing 5.19 Calling **scanf** and **printf** in C.

```
#include <stdio.h>

static char str1[] = "%d";
static char str2[] = "You entered %d\n";
static int n = 0;

int main()
{
    scanf(str1, &n);
    printf(str2, n);
    return 0;
}
```

5.4.3 Standard C library functions

Subroutines may be defined within a program, or a set of subroutines may be packaged together in a library. Libraries of subroutines may be used by multiple programs, and most
languages provide some built-in library functions. The C language has a very large set of
functions in the C standard library. All of the functions in the C standard library are available to any program that has been *linked* with the C standard library. Even assembly programs
can make use of this library. Linking is done automatically when gcc is used to assemble the
program source. All that the programmer needs to know is the name of the function and how
to pass parameters to it.

5.4.4 Passing parameters

Listing 5.19 shows a very simple C program which reads an integer from standard input using
scanf and prints the integer to standard output using printf. An equivalent program written
in AArch64 assembly is shown in Listing 5.20. These examples show how parameters can be
passed to subroutines in C and equivalently in assembly language.

All processor families have their own standard methods, or *function calling conventions*,
which specify how arguments are passed to subroutines and how function values are returned.
The function call standard allows programmers to write subroutines and libraries of subroutines that can be called by other programmers. In most cases, the function calling standards
are not enforced by hardware, but assembly programmers and compiler writers conform to the
standards in order to make their code accessible to other programmers. The basic subroutine
calling rules for the AArch64 processor are simple:

Listing 5.20 Calling **scanf** and **printf** in AArch64 assembly.

```
     .data
str1:    .asciz  "%d"
str2:    .asciz  "You entered %d\n"
n:       .word   0
         .text
         .type   main, %function
         .global main
main:
         stp     x29, x30, [sp, #-16]!    // push FP, LR onto stack
         adr     x0, str1                 // x0 = &str1
         adr     x1, n                    // x1 = &n
         bl      scanf                    // scanf("%d",&n)
         adr     x0, str2                 // x0 = &str2
         adr     x1, n                    // x1 = &n
         ldr     w1, [x1]                 // w1 = n
         bl      printf                   // print message
         mov     w0, #0                   // load return value
         ldp     x29, x30, [sp], #16      // pop FP, LR from stack
         ret                              // return from main
         .size   main, (. - main)
```

- The first eight parameters go in registers **x0-x7**.
- Any remaining parameters are pushed to the stack (in reverse order).

If the subroutine returns a value, then the value is stored in **x0** before the function returns to its caller. Calling a subroutine in AArch64 assembly usually requires several lines of code. The number of lines required depends on how many arguments the subroutine requires, and where the data for those arguments are stored. Some variables may already be in the correct register. Others may need to be moved from one register to another. Still others may need to be loaded from memory into a register or loaded and then stored to the stack. Careful programming is required to minimize the amount of work that must be done just to move the subroutine arguments into their required locations.

The AArch64 register set was introduced in Chapter 3. Some registers have special purposes that are dictated by the hardware design. Others have special purposes that are dictated by *programming conventions*. Programmers follow these conventions so that their subroutines are compatible with each other. These conventions are simply a set of rules for how registers should be used. In AArch64 assembly, all of the registers have alternate names that can be used to help remember the rules for using them. Fig. 5.1 shows an expanded view of the AArch64 registers, including their alternate names and conventional use.

63	32	31	0	
X0		W0		Argument Registers (volatile). Used to pass argument values into a subroutine and to return a result value from a function. They may also be used to hold intermediate values within a routine. Caller assumes they will be modified.
X1		W1		
X2		W2		
X3		W3		
X4		W4		
X5		W5		
X6		W6		
X7		W7		
X8 (XR)		W8		Indirect Result Register. (volatile) Linux system call number.
X9		W9		Caller-Saved Registers (volatile)
X10		W10		
X11		W11		
X12		W12		
X13		W13		
X14		W14		
X15		W15		
X16 (IP0)		W16		Intra-procedure-call Registers (volatile)
X17 (IP1)		W17		
X18 (PR)		W18		Platform Register. (volatile) For portability, avoid using.
X19		W19		Callee-Saved Registers (non-volatile). A subroutine must preserve (or save and restore) the contents of these registers. If they are used, they must be pushed to the stack at the beginning of the subroutine/function, and restored before returning.
X20		W20		
X21		W21		
X22		W22		
X23		W23		
X24		W24		
X25		W25		
X26		W26		
X27		W27		
X28		W28		
X29 (FP)		W29		Frame Pointer (non-volatile)
X30 (LR)		W30		Procedure Link Register
XZR (31)		WZR (31)		Zero Register (hardware special)
SP (31)				Stack Pointer (hardware special)
ELR				Exception Link Register (hardware special)
PC				Program Counter (hardware special)
PSTATE				Processor State (hardware special)

Figure 5.1: AArch64 User Program Registers.

Registers x0-x7 are used for passing *arguments* to subroutines. When calling a subroutine, the first argument must be placed in x0, the second argument in x1, and so on. Conversely, when writing a function, it is safe to assume that the arguments have been placed in these registers by the caller. If the function accepts more than eight arguments, then the additional arguments are passed using the stack, as will be explained in Section 5.4.5.2. The argument registers, x0-x7, are considered to be *volatile*, because their contents can change whenever a subroutine is called. If the contents are needed after the subroutine call, then they must be saved either to a non-volatile register or to the stack before the subroutine is called.

Registers x8-x17 are used for holding local *variables* in a subroutine. These registers are also considered to be *volatile* Some of these registers are used for special purposes by the operating system and/or compiler. From the perspective of the programmer who is writing a user-level program, the special purposes are not important, but we will explain them briefly nonetheless. Register x8 is used when using the svc instruction on the Linux operating system, to specify which system call is being invoked. The intra-procedure scratch registers, x16 and x17, are used by the C library when calling dynamically linked functions. If a subroutine does not call any C library functions, then it can use x16 and x17 as extra registers to store local variables. If a C library function is called, it may change the contents of x16 and x17. Therefore, if x16 or x17 are being used to store a local variable, then their contents should be saved to another register or to the stack before a C library function is called. This is called caller-saved because the caller is responsible for saving the contents of the registers and restoring them after a subroutine.

Registers x19-x28 can also be used for holding local variables. However, before using them, the subroutine must save their contents (usually on the stack) and their contents must be restored before the subroutine exits. These registers are considered *non-volatile* because their contents will not be changed by a subroutine call. More precisely, the subroutine may use them, but it will restore their contents before it returns.

As mentioned in Section 3.2.2, register x29 can also be referred to colloquially as the *frame pointer* FP because it is used by the C compiler to track the *stack frame*, unless the code is compiled using the --omit-frame-pointer command line option. This register must be called x29 in GNU AArch64 code.

5.4.5 Calling subroutines

The stack pointer (sp), link register (x30), and program counter (pc), along with the argument registers, are all involved in performing subroutine calls. The calling subroutine must place arguments in the argument registers, and possibly on the stack as well. Placing the arguments in their proper locations is known as *marshaling* the arguments. After marshaling the arguments, the calling subroutine executes the bl instruction, which will modify the program counter and link register. The bl instruction copies the contents of the program counter to the link register, then loads the program counter with the address of the first instruction in the subroutine that is being called. The CPU will then fetch and execute its next instruction from the address in the program counter, which is the first instruction of the subroutine that is being called.

Listing 5.21 Simple function call in C.

```
printf("Hello World");
```

Listing 5.22 Simple function call in AArch64 assembly.

```
// load first param (pointer to  format string) in x0
adr     x0, hellostr
// call printf
bl      printf
```

Our first examples of calling a function will involve the printf function from the C standard library. The printf function can be a bit confusing at first, but it is an extremely useful and flexible function for printing formatted output. The first argument is the address of a *format string*, which is a null-terminated ASCII string. The printf function examines the format string to determine how many other arguments have been passed to it. The format string may include conversion specifiers, which start with the % character.

For each conversion specifier, printf assumes that an argument has been passed in the correct register or location on the stack. The argument is retrieved, converted according to the specified format, and printed. The %d format specifier causes the matching argument to be printed as a signed decimal number. Other specifiers include %X to print the matching argument as an integer in hexadecimal, %c to print the matching argument as an ASCII character, and %s to print a null-terminated string. The integer specifiers can include an optional width and zero-padding specification. For example %8X will print an integer in hexadecimal, using eight characters. Any leading zeros will be printed as spaces. The format string %08X will also print an integer in hexadecimal, using eight characters, but in this case, any leading zeros will be printed as zeros. Similarly, %15d can be used to print an integer in base ten using spaces to pad the number up to fifteen characters, while %015d will print an integer in base ten using zeros to pad up to fifteen characters.

Listing 5.21 shows a call to printf in C. The printf function requires one argument, but can accept more than one. In this case, there is only one argument, the format string. Since the format string contains no conversion specifiers, printf does not require any additional arguments. Listing 5.22 shows an equivalent call made in AArch64 assembly language. The single argument (the address of the format string) is loaded into r0 in conformance with the AArch64 subroutine calling convention.

Listing 5.23 A larger function call in C.

```
#include <stdio.h>

static int i = 120;
static int k = 122;

int main(void)
{
  int j = 121;
  printf("The results are: %d %d %d\n",i,j,k);
  return 0;
}
```

5.4.5.1 Passing arguments in registers

Listing 5.23 shows a call to `printf` in C having four arguments. The format string is the first argument. The format string contains three conversion specifiers, and is followed by three more arguments. Arguments are matched to conversion specifiers according to their positions. The type of each argument matches the type indicated in the conversion specifier. The first conversion specifier is applied to the second argument, the second conversion specifier is applied to the third argument, and the third conversion specifier is applied to the fourth argument. The %d conversion specifiers indicate that the arguments are to be interpreted as integers and printed in base ten. Listing 5.24 shows an equivalent call made in AArch64 assembly language. The arguments are loaded into x0, w1, w2, and w3 in conformance with the AArch64 subroutine calling convention.

As long as there are eight or fewer arguments that must be passed, they can all fit in registers x0-x7, but when there are more arguments, things become a little more complicated. Any remaining arguments must be passed on the program stack. Care must be taken to ensure that the arguments are pushed to the stack in the proper order. Also, after the function call, the arguments must be removed from the stack, so that the stack pointer is restored to its original value. The tricky part is that, on AArch64 processors, the stack is only allowed to grow or shrink in 16-byte increments.

5.4.5.2 Passing arguments on the stack

Listing 5.25 shows a call to `printf` in C having more than eight arguments. The format string is the first argument. The format string contains ten conversion specifiers, which implies that the format string must be followed by ten additional arguments. Arguments are matched to conversion specifiers according to their positions. The type of each argument must match

Listing 5.24 A larger function call in AArch64 assembly.

```
        .section .rodata
format:
        .asciz  "The results are: %d %d %d\n"
i:      .word   120
k:      .word   122
        .text
        .type   main, %function
        .global main
main:
        stp     x29, x30, [sp, #-16]!    // push FP & LR

        mov     w6, #121                // j is in w6

        // printf("The results are: %d %d %d\n", i, j, k)
        adr     x0, format   // x0 = &format
        adr     x1, i        // x1 = &i
        ldr     w1, [x1]     // w1 = i
        mov     w2, w6       // w2 = j
        adr     x3, k        // x3 = &k
        ldr     x3, [x3]     // w3 = k
        bl      printf       // call printf

        // return 0
        mov     w0, #0
        ldp     x29, x30, [sp], #16      // pop FP & LR
        ret
        .size   main, (. - main)
```

Listing 5.25 A function call using the stack in C.

```
#include <stdio.h>

static int i = -1, j = 2, k = -3, l = 4, m = -5, n = 6, o = -7,
        p = 8, q = -9, r = 10;


int main(void)
{
  printf("Results: %d, %d, %d, %d, %d, %d, %d, %d, %d, %d\n",
        i, j, k, l, m, n, o, p, q, r);
  return 0;
}
```

the type indicated in the conversion specifier. The first conversion specifier is applied to the second argument, the second conversion specifier is applied to the third argument, the third conversion specifier is applied to the fourth argument, and so on. The %d conversion specifiers indicate that the arguments are to be interpreted as integers and printed in base ten.

Listing 5.26 shows an equivalent call made in AArch64 assembly language. Since there are eleven arguments, the last three must be pushed to the program stack. The arguments are loaded into w9-w11 and then the arguments are stored on the stack in reverse order. Even though the parameters are 4 bytes each, they are pushed onto the stack using eight bytes each. The top four bytes are padding. Note that the eleventh argument is placed in the stack at a higher location than the ninth and tenth arguments. Line 10 decrements the stack pointer by 32 bytes, providing room for up to four 8-byte double-words. This is necessary because the AArch64 stack pointer must *always* be on a 16-byte boundary. Since we have three 8-byte arguments to push, we must allocate 32 bytes, rounding up to the next 16-byte boundary. It then stores x9 at address of the stack pointer, and x10 eight bytes above that. Line 11 stores x11 16 bytes above the (decremented) stack pointer. The remaining arguments are loaded in x0-x7. Note that we assume that format has previously been defined to be "The results are: %d %d %d %d %d\n" using an .asciz or .string assembler directive.

Listing 5.27 shows how the eleventh, tenth, and ninth arguments can be pushed to the stack using a stp instruction and str instruction. This would be slightly more efficient than the code shown in Listing 5.26. The eleventh argument is loaded into w11, the tenth argument is loaded into w10, and the ninth argument is loaded into w9, then the stp instruction is used to store w9 and w10 on the stack and adjust the stack pointer. After w11 is stored on the stack, note that they have been placed in reverse order.

A little care must be taken to ensure that the arguments are stored in the correct order on the stack. Remember that the stp instruction will always push the first register to the lower address, and the stack grows downward. Therefore, x9, which is the ninth argument will be pushed onto the stack first. Because pre-indexing is used, x9 is actually at the highest address. The 64-bit notation is used because otherwise stp would store both registers in just 8 bytes, and 16 bytes are needed. Additionally, the printf instruction does not care what the rest of the bits are for each int, they are just treated as padding on the stack. Moreover, the x9 register has zeroes in the upper 32-bits because 32-bit operations, such as ldr zero the upper bits unless they are explicitly sign-extended using an instruction like ldrsw. Finally, the stack must always be 16-byte aligned, which is why it is pushed down 32 bytes instead of just 24.

After the printf function is called, the ninth, tenth, and eleventh arguments must be popped from the stack. If those values are no longer needed, then there is no need to load them into registers. The quickest way to pop them from the stack is to simply adjust the stack pointer back to its original value. In this case, we pushed three arguments onto the stack, using a total

Listing 5.26 A function call using the stack in AArch64 assembly.

```
       .section .rodata
format: .asciz  "Results: %d, %d, %d, %d, %d, %d, %d, %d, %d, %d\n"
       // give variable i a label, and let j, k, l, m, n, o, p,
       // q, and r follow it.
i:      .word   -1, 2, -3, 4, -5, 6, -7, 8, -9, 10


       .text
       .type   main, %function
       .global main
main:   stp     x29, x30, [sp, #-32]!     // push FP & LR
       str     x19, [sp, #16]           // save x19
       adr     x19, i                   // x19 = &i

       // load the eight register parameters
       adr     x0, format               // x0 = &format
       ldr     w1, [x19, #(4*0)]        // w1 = i
       ldr     w2, [x19, #(4*1)]        // w2 = j
       ldr     w3, [x19, #(4*2)]        // w3 = k
       ldr     w4, [x19, #(4*3)]        // w4 = l
       ldr     w5, [x19, #(4*4)]        // w5 = m
       ldr     w6, [x19, #(4*5)]        // w6 = n
       ldr     w7, [x19, #(4*6)]        // w7 = o

       // marshal params on stack in reverse order - p at bottom
       ldr     w9, [x19, #(4*7)]        // w9 = p
       ldr     w10, [x19, #(4*8)]       // w10 = q
       ldr     w11, [x19, #(4*9)]       // w11 = r
       add     sp, sp, #-32
       str     w11, [sp, #16]           // 8 bytes per parameter
       str     w10, [sp, #8]
       str     w9, [sp]

       // call printf
       bl      printf

       // pop p, q, and r from the stack
       add     sp, sp, #32

       //      return 0
       mov     w0, #0
       ldr     x19, [sp, #16]      // restore x19
       ldp     x29, x30, [sp], #32      // pop FP & LR
       ret
       .size   main, (. - main)
```

Listing 5.27 A function call using **stp** to push arguments on the stack.

```
      :
   // marshaling onto the stack.
   ldr    w9, [x19, #(4*7)]      // w9 = p
   ldr    w10, [x19, #(4*8)]     // w10 = q
   ldr    w11, [x19, #(4*9)]     // w11 = r
   stp    x9, x10, [sp, #-32]!   // 8 bytes per parameter
   str    x11, [sp, #16]

   // call printf
   bl     printf

   // pop p, q, and r from the stack
   add    sp, sp, #32
      :
```

of thirty two bytes (each argument required 8 bytes, but the stack is only allowed to grow or shrink in multiples of 16 bytes). Therefore, all we need to do is add thirty two to the stack pointer, thereby restoring its original value.

5.4.6 Writing subroutines

We have looked at the conventions that programmers use to pass arguments when calling functions. Now we will examine these same conventions from the point of view of the function being called. Because of the calling conventions, the programmer writing a function can assume that

- the first eight parameters are in x0–x7,
- any additional parameters can be accessed with ldr Rd, [sp, #offset],
- the *calling* function will remove parameters from the stack, if necessary,
- if the function return type is not void, then they must ensure that the return value is in x0 (and possibly x1, x2, and x3, if the function returns multiple values), and
- the return address will be in x30.

Also because of the conventions, there are certain registers that can be used freely, while others must be preserved or restored so that the calling function can continue operating correctly. Registers which can be used freely are referred to as *volatile*, and registers which must be preserved or restored before returning are referred to as *non-volatile*. When writing a subroutine (function),

- registers x0-x18 are *volatile*,

Listing 5.28 A small C function with a register variable.

```
int doit(void)
{
  int x[20];
  register int i;          /* try to keep i in a register */
  for(i = 0; i < 20; i++)
    x[i] = i;
  return i;
}
```

- registers **x19-x29** are *non-volatile* (they can be used, but their contents *must* be restored to their original value before the function returns),
- register **x30** can be used by the function, but its contents must be saved so that they can be loaded into the program counter, which will cause the function to return to its caller.
- Although the platform register, **x18**, has no special use on Linux systems, it should be avoided because it may have a special use on some platforms, making user-level assembly code that uses it less portable. If portability to another operating system is not a concern, then it can be used as another volatile register.

5.4.7 Automatic variables

In block-structured high-level languages, an automatic variable is a variable that is local to a block of code and not declared with static duration. It has a lifetime that lasts only as long as its block is executing. Automatic variables can be stored in one of two ways:

1. the stack is temporarily adjusted to hold the variable, or
2. the variable is held in a register during its entire life.

When writing a subroutine in assembly, it is the responsibility of the programmer to decide what automatic variables are required, and also where they will be stored. In high-level languages this decision is usually made by the compiler. In some languages, including C, it is possible to request that an automatic variable be held in a register. The compiler will attempt to comply with the request, but it is not guaranteed. Listing 5.28 shows a small function which requests that one of its variables be kept in a register instead of on the stack.

Listing 5.29 shows how the function could be implemented in assembly. Note that the array of integers consumes 80 bytes of storage on the stack, and that the loop control variable can easily be stored in one of the registers for the duration of the function. Also notice that on line 4 the storage for the array is allocated simply by adjusting the stack pointer, and on line

Listing 5.29 Automatic variables in AArch64 assembly.

```
        .type   doit, %function
        .global doit
doit:
        sub     sp, sp, #80         // int x[20];
        mov     w2, #0              // w2 is i
loop:
        cmp     w2, #20             // pre-test loop
        bge     done                // if (i >= 20) goto done
        str     w2, [sp, x2, lsl #2] //   x[i] = i;
        add     w2, w2, #1          //   i++
        b       loop                // go back to loop test
done:
        mov     w0, w2              // return i
        add     sp, sp, #80         // destroy automatic variable
        ret                         // return from function
        .size   doit,(. - doit)
```

14 the storage is released by restoring the stack pointer to its original contents. It is *critical* that the stack pointer be restored, no matter how the function returns. Otherwise, the program will mysteriously fail at some point when it attempts to use data from the stack. For this reason, each function should have exactly one block of instructions for returning. If the function needs to return from some location other than the end, then it should branch to the return block rather than returning directly in the middle of the function.

5.4.8 Recursive functions

A function that calls itself is said to be *recursive*. Certain problems are easy to implement recursively, but are more difficult to solve iteratively. A problem exhibits recursive behavior when it can be defined by two properties:

1. a simple base case (or cases), and
2. a set of rules that reduce all other cases toward the base case.

For example, we can define person's ancestors recursively as follows:

1. one's parents are one's ancestors (base case),
2. the ancestors of one's ancestors are also one's ancestors (recursion step).

Recursion is a very powerful concept in programming. Many functions are naturally recursive, and can be expressed very concisely in a recursive way. Numerous mathematical axioms are based upon recursive rules. For example, the formal definition of the natural numbers by the Peano axioms can be formulated as:

1. 0 is a natural number, and
2. each natural number has a successor, which is also a natural number.

Using one base case and one recursive rule, it is possible to generate the set of all natural numbers. Other recursively defined mathematical objects include functions and sets.

Listing 5.30 shows the C code for a small program which uses recursion to reverse the order of characters in a string. The base case where recursion ends is when there are less than two characters remaining to be swapped. The recursive rule is that the reverse of a string can be created by swapping the first and last characters and then reversing the string between them. In short, a string is reversed if:

1. the string has a length of zero or one character, or
2. the first and last characters have been swapped and the remaining characters have been reversed.

In Listing 5.30, line 6 checks for the base case. If the string has not been reversed according to the first rule, then the second rule is applied. Lines 8–10 swap the first and last characters, and line 11 recursively reverses the characters between them.

Listing 5.30 A C program that uses recursion to reverse a string.

```
1   #include <stdio.h>
2   #include <string.h>
3
4   void reverse(char *a, int left, int right)
5   {
6     if(left < right)
7     {
8       char tmp = a[left];
9       a[left] = a[right];
10      a[right] = tmp;
11      reverse(a, left+1, right-1);
12    }
13  }
14
15  int main()
16  {
17    char str[] = "\nThis is the string to reverse\n";
18    printf(str);
19    reverse(str, 0, strlen(str)-1);
20    printf(str);
21    return 0;
22  }
```

Listing 5.31 shows how the reverse function can be implemented using recursion in AArch64 assembly. Line 8 saves the link register to the stack and decrements the stack pointer. Lines 10 and 11 test for the base case. If the current case is the base case, then the function simply returns (restoring the stack as it goes). Otherwise, the first and last characters are swapped in lines 12 through 15 and a recursive call is made in lines 17 through 19.

Listing 5.31 AArch64 assembly implementation of the reverse function.

```
1          .data    // Declare static data
2  str:     .string "\nThis is the string to reverse\n"
3
4          .text
5          .type    reverse, %function
6          .global reverse
7  reverse:
8          stp      x29, x30, [sp, #-16]!   // push FP and LR
9          // if( left >= right) goto endif
10         cmp      x1, x2
11         bge      endif
12         ldrb     w3, [x0, x1]            // w3 = a[left]
13         ldrb     w4, [x0, x2]            // w4 = a[right]
14         strb     w4, [x0, x1]            // a[left] = w4
15         strb     w3, [x0, x2]            // a[right] = w3
16         // reverse(a, left+1, right-1)
17         add      w1, w1, #1              // left += 1
18         sub      w2, w2, #1              // right -= 1
19         bl       reverse
20 endif:  ldp      x29, x30, [sp], #16     // pop FP and LR
21         ret
22         .size    reverse, (. - reverse)
23
24         .type    main, %function
25         .global main
26 main:   stp      x29, x30, [sp, #-16]!
27         // printf(str)
28         adr      x0, str
29         bl       printf
30         // reverse(str, 0, strlen(str)-1)
31         adr      x0, str
32         bl       strlen
33         add      w0, w0, #-1
34         mov      w2, w0
35         mov      w1, wzr
36         adr      x0, str
37         bl       reverse
38         // printf(str)
39         adr      x0, str
40         bl       printf
```

```
41        // return 0
42        mov     w0, #0
43        ldp     x29, x30, [sp], #16
44        ret
45        .size   main, (. - main)
```

The code in Listing 5.31 is slightly more efficient than a literal translation of the C code shown in Listing 5.30. First, the string is stored in the data section instead of actually on the stack. This saves instructions that would be required to load the string onto the stack and possibly remove it. Also, the local variable `tmp` is not needed in the assembly implementation. Instead, both characters are loaded, then stored to their new destination. This means that the function can use half as much stack space and will run much faster than an implementation that stores the local variable on the stack.

The previous examples used the concept of an array of characters to access the string that is being reversed. Listing 5.32 shows how this problem can be solved in C using pointers to the first and last characters in the string, rather than using array indices. This version only has two parameters in the reverse function, and uses pointer dereferencing rather than array indexing to access each character. Other than that difference, it works the same as the original version. Listing 5.33 shows how the reverse function can be implemented efficiently in AArch64 assembly. This reverse function implementation has the same number of instructions as the previous version, but lines 11 through 14 use a different addressing mode. On an AArch64 processor, the pointer method and the array index method are equally efficient. However, many processors do not have the rich set of addressing modes available on the AArch64. On those processors, the pointer method may be significantly more efficient.

Listing 5.32 String reversing in C using pointers.

```
1   #include <stdio.h>
2   #include <string.h>
3
4   void reverse(char *left, char *right)
5   {
6     char tmp;
7     if(left <= right)
8     {
9       tmp = *left;
10      *left = *right;
11      *right = tmp;
12      reverse(left+1, right-1);
13    }
14  }
15
16  int main()
17  {
```

```
18    char str[] = "\nThis is the string to reverse\n";
19    printf(str);
20    reverse(str, str + strlen(str)-1);
21    printf(str);
22    return 0;
23 }
```

Listing 5.33 String reversing in assembly using pointers.

```
1          .data    // Storing in the data section instead of on the stack
2  str:    .string "\nThis is the string to reverse\n"
3
4          .text
5          .type   reverse, %function
6          .global reverse
7  reverse:stp     x29, x30, [sp, #-16]!    // push FP and LR
8          // if( left >= right) goto endif
9          cmp     x0, x1
10         bge     endif
11         ldrb    w2, [x0]                 // w2 = *left
12         ldrb    w3, [x1]                 // w3 = *right
13         strb    w3, [x0]                 // a[left] = w3
14         strb    w2, [x1]                 // a[right] = w2
15         // reverse(a, left+1, right-1)
16         add     w0, w0, #1               // left += 1
17         sub     w1, w1, #1               // right -= 1
18         bl      reverse
19 endif:  ldp     x29, x30, [sp], #16      // pop FP and LR
20         ret
21         .size   reverse, (. - reverse)
22
23         .type   main, %function
24         .global main
25 main:   stp     x29, x30, [sp, #-16]!
26         // printf(str)
27         adr     x0, str
28         bl      printf
29         // reverse(str, str + strlen(str)-1)
30         adr     x0, str
31         bl      strlen
32         add     x1, x0, #-1
33         adr     x0, str
34         add     x1, x1, x0
35         bl      reverse
36         // printf(str)
37         adr     x0, str
38         bl      printf
39         // return 0
40         mov     w0, #0
```

```
41      ldp     x29, x30, [sp], #16
42      ret
43      .size   main, (. - main)
```

5.5 Aggregate data types

An aggregate data item can be referenced as a single entity, and yet consists of more than one piece of data. Aggregate data types are used to keep related data together, so that the programmer's job becomes easier. Some examples of aggregate data are arrays, structures or records, and objects. In most programming languages, aggregate data types can be defined to create higher-level structures. Most high-level languages allow aggregates to be composed of basic types as well as other aggregates. Proper use of structured data helps to make programs less complicated and easier to understand and maintain.

In high-level languages, there are several benefits to using aggregates. Aggregates make the relationships between data clear, and allow the programmer to perform operations on blocks of data. Aggregates also make passing parameters to functions simpler and easier to read.

5.5.1 Arrays

The most common aggregate data type is an array. An array contains zero or more values of the same data type, such as characters, integers, floating point numbers, or fixed point numbers. An array may also be contain values of another aggregate data type. Every element in an array must have the same type. Each data item in an array can be accessed by its *array index*.

Listing 5.34 shows how an array can be allocated and initialized in C. Listing 5.35 shows the equivalent code in AArch64 assembly. Note that in this case, the scaled register offset addressing mode was used to access each element in the array. This mode is often convenient when the size of each element in the array is an integer power of 2. If that is not the case, then it may be necessary to use a different addressing mode. An example of this will be given later in Section 5.5.3.

5.5.2 Structured data

The second common aggregate data type is implemented as the struct in C or the record in Pascal. It is commonly referred to as a structured data type or a record. This data type can contain multiple *fields*. The individual fields in the structured data are referred to as structured data elements or simply elements. In most high-level languages, each element of a structured data type may be one of the base types, an array type, or another structured data type.

Listing 5.34 Initializing an array of integers in C.

```
int x[100];
...
for(int i = 0; i < 100; i++)
    x[i] = i;
```

Listing 5.35 Initializing an array of integers in assembly.

```
       ...
sub    sp, sp, #400           // int x[100]
mov    x0, #0                 // x0 = i = 0
loop:
str    w0, [sp, x0, lsl #2]   // x[i] = i
add    x0, x0, #1             // i++
cmp    x0, #100               // loop test
blt    loop                   // if (i < 100) goto loop
       ...
```

Listing 5.36 shows how a struct can be declared, allocated, and initialized in C. Listing 5.37 shows the equivalent code in AArch64 assembly.

Care must be taken using assembly to access data structures that were declared in higher level languages such as C and C++. The compiler will typically pad a data structure to ensure that the data fields are aligned for efficiency. On most systems, it is more efficient for the processor to access word-sized data if the data is aligned to a word boundary. Some processors simply cannot load or store a word from an address that is not on a word boundary, and attempting to do so will result in an exception. The assembly programmer must somehow determine the relative address of each field within the higher-level language structure. One way that this can be accomplished in C is by writing a small function which prints out the offsets to each field in the C structure. The offsets can then be used to access the fields of the structure from assembly language. Another method for finding the offsets is to run the program under a debugger and examine the data structure.

5.5.3 Arrays of structured data

It is often useful to create arrays of structured data. For example, a color image may be represented as a two dimensional array of pixels, where each pixel consists of three integers which specify the amount of red, green, and blue that are present in the pixel. Typically, each of the

Listing 5.36 Initializing a structured data type in C.

```
1  #include <string.h>
2
3  struct student {
4      char first_name[30];
5      char last_name[30];
6      unsigned char class;
7      int grade;
8  };
9
10 int main(void)
11 {
12     struct student newstudent;    /* allocate struct on the stack */
13     strcpy(newstudent.first_name, "Sam");
14     strcpy(newstudent.last_name, "Smith");
15     newstudent.class = 2;
16     newstudent.grade = 88;
          :
18     return 0;
19 }
```

three values is represented using an unsigned eight bit integer. Image processing software often adds a fourth value, α, specifying the transparency of each pixel.

Listing 5.38 shows how an array of pixels can be allocated and initialized in C. The listing uses the malloc() function from the C standard library to allocate storage for the pixels from the heap (See Section 1.4). Note that the code uses the sizeof() function to determine how many bytes of memory are consumed by a single pixel, then multiplies that by the width and height of the image. Listing 5.39 shows the equivalent code in AArch64 assembly.

Note that the code in Listing 5.39 is far from optimal. It can be greatly improved by combining the two loops into one loop. This will remove the need for the multiply on line 31 and the addition and will simplify the code structure. An additional improvement would be to increment the single loop counter by three on each loop iteration, which will make it very easy to calculate the pointer for each pixel. Listing 5.40 shows the AArch64 assembly implementation with these optimizations.

Although the implementation shown in Listing 5.40 is more efficient than the previous version, there are several more improvements that can be made. If we consider that the goal of the code is to allocate some number of bytes and initialize them all to zero, then the code can be written more efficiently. Rather than using three separate store instructions to set 3 bytes

```
1        .data
2   sam:    .asciz  "Sam"
3   smith:  .asciz  "Smith"
4        .text
5        .equ    s_first_name, 0
6        .equ    s_last_name, 30
7        .equ    s_class, 60
8        .equ    s_grade, 64
9        .equ    s_size, 68
10       .equ    s_size_aligned, 80
11
12       .type   main, %function
13       .global main
14  main:
15       stp     x29, x30, [sp, #32]!
16       str     x19, [sp, #16]            // Callee-saved save
17
18       // struct student newstudent
19       sub     sp, sp, #s_size_aligned
20       mov     x19, sp                   // x19=pointer to struct
21       add     x0, x19, #s_first_name    // offset to first name
22       adr     x1, sam                   // Load pointer to "Sam"
23       bl      strcpy                    // copy the string
24       add     x0, x19, #s_last_name     // offset to last name
25       adr     x1, smith                 // Load pointer to "Smith"
26       bl      strcpy                    // copy the string
27       mov     w1, #2                    // load constant value of 2
28       strb    w1, [x19, #s_class]       // store with offset
29       mov     w1, #88                   // load value of 88
30       str     w1, [x19, #s_grade]       // store with offset
31
32       // use the struct
33       :
34       // delete the struct
35       add     sp, sp, #s_size_aligned
36
37       // return 0
38       mov     w0, #0
39       ldr     x19, [sp, #16]            // Callee-saved restore
40       ldp     x29, x30, [sp], #32
41       ret
42       .size   main,(. - main)
```

Listing 5.37 Initializing a structured data type in AArch64 assembly.

Listing 5.38 Initializing an array of structured data in C.

```
 1  #include <stdio.h>
 2  #include <stdlib.h>
 3
 4  /* image size */
 5  const int width = 100;
 6  const int height = 100;
 7
 8  /* define structure for a pixel */
 9  struct pixel {
10      unsigned char red;
11      unsigned char green;
12      unsigned char blue;
13  };
14
15  int main(void)
16  {
17      // allocate image
18      struct pixel *image = malloc(width * height * sizeof(struct pixel));
19      if (image == NULL) {
20          fprintf(stderr, "Error: Out of memory.\n");
21          return 1;
22      }
23
24      // initialize all pixels in the image to black
25      for (int j = 0; j < height; j++) {
26          for (int i = 0; i < width; i++) {
27              image[j*width+i].red = 0;
28              image[j*width+i].blue = 0;
29              image[j*width+i].green = 0;
30          }
31      }
32
33      // delete the image
34      free(image);
35
36      return 0;
37  }
```

to zero on each iteration of the loop, why not use a single store instruction to set sixteen bytes to zero on each iteration? The only problem with this approach is that we must consider the possibility that the array may end in the middle of the quad-word. However this can be dealt with by using two consecutive loops. The first loop sets one *quad-word* of the array to zero on each iteration, and the second loop finishes off any remaining bytes. Listing 5.41 shows the

Listing 5.39 Initializing an array of structured data in assembly.

```
1       .equ    i_red,      0
2       .equ    i_green,    1
3       .equ    i_blue,     2
4       .equ    i_size,     3
5       .equ    width,      100
6       .equ    height,     100
7
8       .text
9       .type   main, %function
10      .global main
11  main:       stp     x29, x30, [sp, #-16]!
12      // Call malloc to allocate the array of pixels
13              mov     x0, #(100*100*3)        // precompute w*h*size
14              bl      malloc                  // allocate storage.
15              cmp     x0, xzr                 // if (image != NULL)
16              bne     endif                   // goto endif
17              mov     w0, #1                  // else return 1
18              b       return
19  endif:      mov     w3, #100                // w3 = width
20              mov     w4, #100                // w4 = height
21              mov     w5, #0                  // w5 = 0
22              mov     w1, #0                  // int j = 0
23  loop1:      mov     w2, #0                  // int i = 0
24  loop2:      madd    x9, x1, x3, x2          // x9 = j * width + i
25              add     x9, x9, x0              // x9 += image address
26              strb    w5, [x9, #i_red]        // image[j*w+i].red = 0
27              strb    w5, [x9, #i_green]      // image[j*w+i].green = 0
28              strb    w5, [x9, #i_blue]       // image[j*w+i].blue = 0
29              add     w2, w2, #1              // i++
30              cmp     w2, w3                  // if (i < width) goto loop2
31              ble     loop2                   // end of loop2
32
33              add     w1, w1, #1              // j++
34              cmp     w1, w4                  //if (j < height) goto loop1
35              ble     loop1                   // end of loop1
36
37              bl      free                    // free(image)
38              mov     w0, #0                  // return 0
39  return:     ldp     x29, x30, [sp], #16
40              ret
41      .size   main,(. - main)
```

Listing 5.40 Improved initialization in assembly.

```
 1        .equ   i_red, 0
 2        .equ   i_green, 1
 3        .equ   i_blue, 2
 4        .equ   i_size, 3
 5        .equ   width, 100
 6        .equ   height, 100
 7
 8        .text
 9        .type  main, %function
10        .global main
11 main:  stp    x29, x30, [sp, #-16]!
12        // Call malloc to allocate the array of pixels
13        mov    x0, #(width*height*i_size)  // calculate space needed
14        bl     malloc                      // allocate storage.
15        cmp    x0, #0                       // if (image != NULL)
16        bne    endif                        // goto endif
17        mov    w0, #1                       // else return 1
18        b      return
19 endif: mov    w2, #(width*height*i_size)  // w2 = w * h * size
20        mov    w3, #0                       // w3 = 0
21        mov    w1, #0                       // int i = 0
22 loop:  add    x9, x1, x0                   // x9 = image + i
23        strb   w3, [x9, #i_red]             // image[i].red = 0
24        strb   w3, [x9, #i_green]           // image[i].green = 0
25        strb   w3, [x9, #i_blue]            // image[i].blue = 0
26        add    w1, w1, #3                   // i += 3
27        cmp    w1, w2                       // if(i<width*height*3)
28        blt    loop                         // repeat loop
29
30        bl     free                         // free(image)
31        mov    w0, #0                       // return 0
32 return: ldp   x29, x30, [sp], #16
33        ret
34        .size  main, (. - main)
```

results of these additional improvements. This third implementation will run *much* faster than the previous implementations on a very large array, or image.

5.6 Chapter summary

Spaghetti code is the bane of assembly programming, but it can easily be avoided. Although assembly language does not enforce structured programming, it does provide the low-level mechanisms required to write structured programs. The assembly programmer must be aware of, and assiduously practice, proper structured programming techniques. The burden of writing properly structured code, with selection structures and iteration structures, lies with

Listing 5.41 Very efficient initialization in assembly.

```
        .equ    i_red,      0
        .equ    i_green,    1
        .equ    i_blue,     2
        .equ    i_size,     3
        .equ    width,      100
        .equ    height,     100

        .text
        .type   main, %function
        .global main
main:   stp     x29, x30, [sp, #-16]!
        // Call malloc to allocate the array of pixels
        mov     x0, #(width*height*i_size) // calculate space needed
        bl      malloc          // allocate storage.
        cmp     x0, #0          // if (image != NULL)
        bne     endif           // goto endif
        mov     x0, #1          // else return 1
        b       return
endif:  mov     x3, #0
        mov     x4, #0
        mov     x2, #(width*height*i_size) // w2=w * h *size
        add     x2, x2, x0      // get pointer to end
        // move end addr back to quad boundary
        orr     x1, x2, #0xF    // get # of leftover bytes
        sub     x2, x2, x1
        mov     x9, x0          // x9 = i = image
loop:   stp     x3, x4, [x9], #16   // clear quadword and i+=16
        cmp     x9, x2          // while (i < end-1)
        blt     loop            //    goto loop
        // clear leftover bytes
loop2:  strb    x3, [x9], #1    // clear byte; i += 1
        subs    x1,x1,#1
        blt     loop2
done:   bl      free            // free(image)
        mov     x0, #0          // return 0
return: ldp     x29, x30, [sp], #16
        ret
        .size   main, (. - main)
```

the programmer, and failure to apply structured programming techniques will result in code that is difficult to understand, debug, and maintain.

Subroutines provide a way to split programs into smaller parts, each of which can be written and debugged individually. This allows large projects to be divided among team members. In assembly language, defining and using subroutines is not as easy as in higher level languages. However, the benefits far outweigh the costs. The C library provides a large number of functions. These can be accessed by an assembly program as long as it is linked with the C standard library.

Assembly provides the mechanisms to access aggregate data types. Arrays can be accessed using various addressing modes on the AArch64 processor. The Pre-indexing, and post-indexing modes allow array elements to be accessed using pointers, with the pointers being incremented after each element access. Fields in structured data records can be accessed using immediate offset addressing mode. The rich set of addressing modes available on the AArch64 processor allow the programmer to use aggregate data types more efficiently than on most processors.

Exercises

5.1. What does it mean for a register to be volatile? Which AArch64 registers are considered volatile according to the AArch64 function calling convention?

5.2. Fully explain the differences between static variables and automatic variables.

5.3. In AArch64 Assembly language, write a function that is equivalent to the following C function.

```
1  int max(int a, int b)
2  {
3      if( a > b )
4          return a;
5      return b;
6  }
```

5.4. What are the two places where an automatic variable can be stored?

5.5. You are writing a function and you decided to use registers x19 and x20 within the function. Your function will not call any other functions. It is self-contained. Modify the following skeleton structure to ensure that x19 and x20 can be used within the function and are restored to comply with the AArch64 standards, but without unnecessary memory accesses.

```
1      .text
2      .type    myfunc, %function
3  myfunc:
```

```
4      stp      x29, x30, [sp, #-16]!
5               ⋮
6      // function statements
7               ⋮
8      ldp      x29, x30, [sp], #16
9      ret
```

5.6. Convert the following C program to AArch64 assembly, using a post-test loop:

```
1  int main()
2  {
3    for(i = 0; i < 10; i++)
4      printf("Hi!\n");
5    return 0;
6  }
```

5.7. Write a complete AArch64 function to shift a 128 bit value left by any amount between 0 and 127 bits. The function should expect its arguments to be in registers x0, x1, and w2. The lower 64 bits of the value are passed in x0, the upper 64 bits of the value are passed in x1, and the shift amount is passed in w2.

5.8. Write a complete subroutine in AArch64 assembly that is equivalent to the following C subroutine.

```
1   /* This function copies 'count' bytes from 'src' to 'dest'. */
2   void bytecopy(char dest[], char src[], int count)
3   {
4     count = count - 1;
5     while(count >= 0)
6     {
7       dest[count] = src[count];
8       count = count - 1;
9     }
10  }
```

5.9. Write a complete function in AArch64 assembly that is equivalent to the following C function.

```
1    /* This function returns the minimum of six values. */
2    int minsix(int a, int b, int c, int d, int e, int f)
3    {
4      if(b < a)
5        a = b;
6      if(d < c)
7        c = d;
8      if(c < a)
9        a = c;
10     if(f < e)
11       e = f;
```

```
12   if(e < a)
13       a = e;
14   return a;
15 }
```

5.10. Write an AArch64 assembly function to calculate the average of an array of integers, given a pointer to the array and the number of items in the array. Your assembly function must implement the following C function prototype:

`int average(int *array, int number_of_items);`

5.11. Write a complete function in AArch64 assembly that is equivalent to the following C function. Note that a and b must be allocated on the stack, and their addresses must be passed to scanf so that it can place their values into memory.

```
1 int read_and_add()
2 {
3   int a, b, sum;
4   scanf("%d",&a);
5   scanf("%d",&b);
6   sum = a + b;
7   return sum;
8 }
```

5.12. The factorial function can be defined as:

$$x! = \begin{cases} 1, & \text{if } x \le 1, \\ x \times (x-1)! & \text{otherwise.} \end{cases}$$

The following C program repeatedly reads x from the user and calculates $x!$ It quits when it reads end-of-file or when the user enters a negative number or something that is not an integer.

```
1  #include <stdio.h>
2
3  /* The factorial function calculates x! */
4  unsigned long factorial(int x)
5  {
6      if (x < 2)
7          return 1;
8      return x * factorial(x-1);
9  }
10
11 /* main repeatedly asks for x, and prints x! */
12 int main()
13 {
14     int x, goodval;
15     do {
```

```
16    printf("Enter x: ");
17    goodval = scanf("%d", &x);
18    if(goodval == 1)
19      printf("%d! = %lu\n", x, factorial(x));
20  } while(goodval == 1);
21  return 0;
22 }
```

Write this program in AArch64 assembly.

5.13.   For large x, the factorial function is slow. However, a lookup table can be added to the function to avoid performing deep recursion x every time. This technique is commonly known as memoization or tabling, but is sometimes called dynamic programming. The following C implementation of the factorial function uses memoization. Modify your AArch64 assembly program from the previous problem to include memoization.

```
 1 #define TABSIZE 50
 2
 3 /* The factorial function calculates x! */
 4 unsigned long factorial(int x)
 5 {
 6   /* declare table and initialize to all zero */
 7   static unsigned long table[TABSIZE] = {0};
 8
 9   /* handle base case */
10   if(x<2)
11     return 1;
12
13   /* if x! is not in the table and x is small enough,
14      then compute x! and put it in the table */
15   if((x < TABSIZE) && (table[x] == 0))
16     table[x] =  x * factorial(x-1);
17
18   /* if x is small enough, then
19       return the value from the table */
20   if(x < TABSIZE)
21     return table[x];
22
23   /* if x is too large to be in the table, use
24      a recursive call */
25   return x * factorial(x-1);
26 }
```

## CHAPTER 6

Abstract data types

An Abstract Data Type (ADT) is composed of data and the operations that work on that data. The ADT is one of the cornerstones of structured programming. Proper use of ADTs has many benefits. Most importantly, abstract data types help to support information hiding. An ADT hides information by encapsulating the information into a module or other construct, and providing a well-defined *interface*. The interface typically consists of the names of data types provided by the ADT and a set of subroutine definitions, or function prototypes, for operating on the data types. The *implementation* of the ADT is hidden from the client code that uses the ADT.

A common use of information hiding is to hide the physical storage layout for data so that if it is changed, the change is restricted to a small subset of the total program. For example, if a three-dimensional point (x, y, z) is represented in a program with three floating point scalar variables and later, the representation is changed to a single array variable of size three, a module designed with information hiding in mind would protect the remainder of the program from such a change.

Information hiding reduces software development risk by shifting the code's dependency on an uncertain implementation onto a well-defined interface. Clients of the interface perform operations purely through the interface, which does not change. If the implementation changes, the client code does not have to change.

Encapsulating software and data structures behind an interface allows the construction of objects that mimic the behavior and interactions of objects in the real world. For example, a simple digital alarm clock is a real-world object that most people can use and understand. They can understand what the alarm clock does, and how to use it through the provided interface (buttons and display) without having to understand every part inside of the clock. If the internal circuitry of the clock were to be replaced with a different implementation, people could continue to use it in the same way, provided that the interface did not change.

6.1 ADTs in assembly language

As with all other structured programming concepts, ADTs can be implemented in assembly language. In fact, most high level compilers convert structured programming code into assembly during compilation. All that is required is that the programmer define the data structure(s),

ARM 64-Bit Assembly Language
https://doi.org/10.1016/B978-0-12-819221-4.00013-4

and the set of operations that can be used on the data. Listing 6.1 gives an example of an ADT *interface* in C. The type Image is not fully defined in the interface. This prevents client software from accessing the internal structure of the image data type. Therefore, programmers using the ADT can modify images only by using the provided functions. Other structured programming and object-oriented programming languages such as C++, Java, Pascal, and Modula 2 provide similar protection for data structures so that client code can access the data structure only through the provided interface. Note that only the pval definition is exposed, indicating to client programs that the red, green, and blue components of a pixel must be a number between 0 and 255. In C, as with other structured programming languages, the *implementation* of the subroutines can also be hidden by placing them in separate compilation modules. Only the ADT implementation code will have access to the internal structure of the Image data type.

Assembly language does not have the ability to define a data structure as such, but it does provide the mechanisms needed to specify the location of each field with respect to the beginning of a data structure, as well as the overall size of the data structure. With a little thought and effort, it is possible to implement ADTs in Assembly language.

Listing 6.1 Definition of an Abstract Data Type in a C header file.

```
1   #ifndef IMAGE_H
2   #define IMAGE_H
3   #include <stdio.h>
4
5   typedef unsigned char pval;
6
7   struct imageStruct;
8
9   typedef struct imageStruct Image;
10
11  Image *allocateImage();
12  void freeImage(Image *image);
13
14  int readImage(FILE *f, Image *image);
15  int writeImage(FILE *f, Image *image);
16
17  int setPixelRGB(Image *image, int row, int col, pval r, pval g, pval b);
18  int setPixelGray(Image *image, int row, int col, pval v);
19
20  pixel getPixelRGB(Image *image, int row, int col);
21  pval getPixelGray(Image *image, int row, int col);
22
23  #endif
```

Listing 6.2 Definition of the image structure is hidden in a separate header file, and not available to *client* code that uses the ADT.

```
1   #ifndef IMAGE_PRIVATE_H
2   #define IMAGE_PRIVATE_H
3   #include <image.h>
4
5   typedef struct {
6     pval r,g,b;
7   } Pixel;
8
9   struct imageStruct {
10    int rows;      // number of rows in the image
11    int cols;      // number of columns in the image
12    Pixel *pixels; // array of pixel data
13  };
14  #endif
```

Listing 6.3 Definition of an ADT in Assembly.

```
1   ### Definitions for pixel and image data structures
2
3   ## pixel
4   .equ   p_red,   #0   // offset to red value
5   .equ   p_green, #1   // offset to green value
6   .equ   p_blue,  #2   // offset to blue value
7   .equ   p_size,  #3   // size of the pixel data structure
8
9   ## image
10  .equ   i_rows,  #0   // offset to number of rows
11  .equ   i_cols,  #4   // offset to number of columns
12  .equ   i_pixels,#8   // offset to pointer to image data
13  .equ   i_size,  #12  // size of the image data structure
```

Listing 6.2 shows the private implementation of the Image data type, which is included by the C files which implement the Image abstract data type. A pixel is defined as a C struct with one byte each for the red, blue, and green color components. Since the largest item in the struct is a byte, the C compiler will create a structure that is exactly three bytes long, without any extra bytes for alignment. The image struct contains three integer values, so it will occupy twelve bytes. Listing 6.3 shows how the data structures from the previous listings can be defined in assembly language. With those definitions, any of the functions declared in Listing 6.1 can be written in assembly language.

6.2 Word Frequency counts

Counting the frequency of words in written text has several uses. In digital forensics, it can be used to provide evidence as to the author of written communications. Different people have different vocabularies, and use words with differing frequency. Word counts can also be used to classify documents by type. Scientific articles from different fields contain words specific to the field, and historical novels will differ from western novels in word frequency.

Listing 6.4 shows the main function for a simple C program which reads a text file and creates a list of all of the words contained in a file, along with their frequency of occurrence. The program has been divided into two parts: the main program, and an ADT. The ADT is used to keep track the words and their frequencies, and to print a table of word frequencies.

Listing 6.4 C program to compute word frequencies.

```
1    #include <stdlib.h>
2    #include <string.h>
3    #include <stdio.h>
4    #include <ctype.h>
5    #include <list.h>
6    /**********************************************************/
7    /* remove_punctuation copies the input string to a new    */
8    /* string, but omits any punctuation characters           */
9    char *remove_punctuation(char *word)
10   { char* newword = (char*)malloc(strlen(word)+1);
11       char* curdst = newword;
12       char* cursrc = word;
13       while( *cursrc != 0 )
14       {
15           if(strchr(".,\'\"!$()::{}[]\\/", *cursrc) == NULL)
16           { /* Current character is not punctuation */
17               *curdst = tolower(*cursrc);
18               curdst++;
19           }
20           cursrc++;
21       }
22       *curdst=0;
23       return newword;
24   }
25
26   /**********************************************************/
27   /* The main function reads whitespace separated words     */
28   /* from stdin, removes punctuation, and generates a word  */
29   /* frequency list.                                        */
30   int main()
```

```
{ int MaxWordLength = 1024;
  char *nextword, *cleanword;
  wordlist *list;
  nextword = (char*)malloc(MaxWordLength*sizeof(char));
  list = wl_alloc();
  while(scanf("%s",nextword) == 1)
    {
      cleanword = remove_punctuation(nextword);
      if(strlen(cleanword)>0)
        wl_increment(list,cleanword);
      free(cleanword);
    }
  printf("Alphabetical List\n");
  wl_print(list);
  printf("\nNumerical List\n");
  wl_print_numerical(list);
  wl_free(list);
  return 0;
}
```

The interface for the ADT is shown in Listing 6.5. There are several ways that the ADT could be implemented. Note that the *interface* given in the header file does not show the internal fields of the word list data type. Thus, any file which includes this header are allowed to declare pointers to wordlist data types, but cannot access or modify any internal fields. The list of words could be stored in an array, a linked list, a binary tree, or some other data structure. Also, the subroutines could be implemented in C or in some other language, including assembly. Listing 6.6 shows an implementation in C using a linked list. Note that the function for printing the word frequency list in numerical order has not been implemented. It will be written in assembly language. Since the program is split into multiple files, it is a good idea to use the make utility to build the executable program. A basic makefile is shown in Listing 6.7.

Listing 6.5 C header for the wordlist ADT.

```
#ifndef LIST_H
#define LIST_H

/********************************************************/
/* Define an opaque type, named wordlist                */
typedef struct wlist wordlist;

/********************************************************/
/* wl_alloc allocates and initializes a new word list.  */
wordlist* wl_alloc();

/********************************************************/
```

```
13  /* wl_free frees all the storage used by a wordlist      */
14  void wl_free(wordlist* wl);
15
16  /***********************************************************/
17  /* wl_increment adds one to the count of the given word.  */
18  /* If the word is not in the list, then it is added with  */
19  /* a count of one.                                        */
20  void wl_increment(wordlist *list, char *word);
21
22  /***********************************************************/
23  /* wl_print prints a table showing the number             */
24  /* of occurrences for each word, followed by the word.    */
25  void wl_print(wordlist *list);
26
27  /***********************************************************/
28  /* wl_print_numerical prints a table showing the number   */
29  /* of occurrences for each word, followed by the word,    */
30  /* sorted in reverse order of occurence.                  */
31  void wl_print_numerical(wordlist *list);
32
33  #endif
```

Listing 6.6 C implementation of the wordlist ADT.

```
1   #include <stdlib.h>
2   #include <string.h>
3   #include <stdio.h>
4   #include <list.h>
5
6   /***********************************************************/
7   /* The wordlistnode type is a linked list of words and    */
8   /* the number of times each word has occurred.            */
9   typedef struct wlist_node{
10    char *word;
11    int count;
12    struct wlist_node *next;
13  }wordlistnode;
14
15  /***********************************************************/
16  /* The wordlist type holds a pointer to the linked list   */
17  /* and keeps track of the number of nodes in the list     */
18  typedef struct wlist{
19    int nwords;
20    wordlistnode *head;
21  }wordlist;
22
23  /***********************************************************/
24  /* wl_alloc allocates and initializes a new word list.    */
25  wordlist* wl_alloc()
```

```c
26    { wordlist* tmp;
27      tmp = (wordlist*)malloc(sizeof(wordlist));
28      if(tmp == NULL)
29      {
30        printf("Unable to allocate wordlist\n");
31        exit(1);
32      }
33      tmp->nwords = 0;
34      tmp->head = NULL;
35      return tmp;
36    }

38    /**********************************************************/
39    /*  wl_free frees all the storage used by a wordlist    */
40    void wl_free(wordlist* wl)
41    {
42      wordlistnode* tmpa, *tmpb;
43      tmpa = wl->head;
44      while(tmpa != NULL)
45      {
46        tmpb = tmpa;
47        tmpa = tmpa->next;
48        free(tmpb->word);
49        free(tmpb);
50      }
51      free(wl);
52    }

54    /**********************************************************/
55    /*  wln_lookup is used internally to search the list of  */
56    /*  words. It returns a pointer to the wordlistnode.  If */
57    /*  the word is not in the list, then it returns a pointer*/
58    /*  to the place where the word should be inserted.  If  */
59    /*  the insertion point is the head of the list, then it */
60    /*  returns NULL.                                        */
61    wordlistnode* wln_lookup(wordlistnode* lst, char *word)
62    {
63      wordlistnode *prev = NULL;
64      while((lst != NULL)&&(strcmp(lst->word, word)<0))
65      {
66        prev = lst;
67        lst = lst->next;
68      }
69      if((lst != NULL)&&(strcmp(lst->word, word) == 0))
70        return lst;
71      else
72        return prev;
73    }
74
```

```
75   /****************************************************************/
76   /* wl_increment adds one to the count of the given word.  */
77   /* If the word is not in the list, then it is added with  */
78   /* a count of one.                                        */
79   void wl_increment(wordlist *list, char *word)
80   {
81     wordlistnode *newword;
82     wordlistnode *wlst = wln_lookup(list->head,word);
83     if((wlst == NULL)||(strcmp(wlst->word, word) != 0))
84       {
85         list->nwords++;
86         newword = (wordlistnode*)malloc(sizeof(wordlistnode));
87         if(newword == NULL)
88           {
89             printf("Unable to allocate wordlistnode\n");
90             exit(1);
91           }
92         newword->word = strdup(word);
93         newword->count = 1;
94         if(wlst == NULL)
95           {
96             newword->next = list->head;
97             list->head = newword;
98           }
99         else
100          {
101            newword->next = wlst->next;
102            wlst->next = newword;
103          }
104      }
105    else
106      wlst->count++;
107  }
108
109  /****************************************************************/
110  /* wl_print prints a table showing the number           */
111  /* of occurrences for each word, followed by the word.  */
112  void wl_print(wordlist *list)
113  {
114    wordlistnode *wlist = list->head;
115    while(wlist != NULL) {
116      printf("%10d %s\n",wlist->count,wlist->word);
117      wlist=wlist->next;
118    }
119    printf("There are %d unique words in the document\n",
120            list->nwords);
121  }
122
123  /****************************************************************/
```

```
124  /* wl_print_numerical prints a table showing the number    */
125  /* of occurrences for each word, followed by the word,     */
126  /* sorted in reverse order of occurence.                   */
127  void wl_print_numerical(wordlist *list)
128  {
129      printf("wl_print_numerical has not been implemented");
130  }
```

Listing 6.7 Makefile for the wordfreq program.

```
1   C_OBJS=wordfreq.o list.o
2   ASM_OBJS=
3   OBJS=$(C_OBJS) $(ASM_OBJS)
4
5   LFLAGS=-O2 -g
6   CFLAGS=-I. -O2 -g -Wall
7   SFLAGS=-I. -O2 -g -Wall
8   DEPENDFLAGS=-I. -M
9
10  wordfreq: $(OBJS)
11      gcc $(LFLAGS) -o wordfreq $(OBJS)
12
13  .S.o:
14      gcc $(SFLAGS) -c $<
15
16  .c.o:
17      gcc $(CFLAGS) -c $<
18
19  clean:
20      rm -f *.o *~ wordfreq
21
22  # make depend will create a file ".depend" with all the dependencies
23  depend:
24      rm -f .depend
25      $(CC) $(DEPENDFLAGS) $(C_OBJS:.o=.c) > .depend
26
27  # if we have a .depend file, include it
28  ifeq (.depend,$(wildcard .depend))
29  include .depend
30  endif
```

Suppose we wish to implement one of the functions from Listing 6.6 in AArch64 assembly language. We would

1. delete the function from the C file,
2. create a new file with the assembly version of the function, and
3. modify the makefile so that the new file is included in the program.

The header file and the main program file would not require any changes. The header file provides function prototypes that the C compiler uses to determine how parameters should be passed to the functions. As long as our new assembly function conforms to its C header definition, the program will work correctly.

6.2.1 Sorting by word frequency

The linked list is created in alphabetical order, but the wl_print_numerical function is required to print it sorted in reverse order of number of occurrences. There are several ways in which this could be accomplished, with varying levels of efficiency. Possible approaches include, but are not limited to:

- Re-ordering the linked list using an insertion sort. This approach creates a complete new list by removing each item, one at a time, from the original list, and inserting it into a new list sorted by the number of occurrences rather than the words themselves. The time complexity for this approach would be $O(N^2)$, but would require no additional storage. However, if the list were later needed in alphabetical order, or any more words were to be added, then it would need to be re-sorted in the original order.

- Sorting the linked list, using a merge sort algorithm. Merge sort is one of the most efficient sorting algorithms known, and can be efficiently applied to data in files and linked lists. The merge sort works as follows:

  1. The sub-list size, i, is set to 1.
  2. The list is divided into sub-lists, each containing i elements. Each sub-list is assumed to be sorted. (A sub-list of length one is sorted by definition.)
  3. The sub-lists are merged together to create a list of sub-lists of size $2i$, where each sub-list is sorted.
  4. The sub-list size, i is set to $2i$.
  5. The process is repeated from step 2 until $i \geq N$, where N is the number of items to be sorted.

  The time complexity for the merge sort algorithm is $O(N \log N)$, which is far more efficient than the insertion sort. This approach would also require no additional storage. However, if the list were later needed in alphabetical order, or any more words were to be added, then it would need to be re-sorted in the original alphabetical order.

- Create an index, and sort the index rather than rebuilding the list. Since the number of elements in the list is known, we can allocate an array of pointers. Each pointer in the array is then initialized to point to one element in the linked list. The array forms an index, and the pointers in the array can be re-sorted in any desired order, using any common sorting method such as bubble sort $(O(N^2))$, in-place insertion sort $(O(N^2))$, quick sort $(O(N \log N))$, or others. This approach requires additional storage, but has the advantage that it does not modify the original linked list.

There are many other possibilities for re-ordering the list. Regardless of which method is chosen, the main program and the interface (header file) need not be changed. Different implementations of the sorting function can be substituted without affecting any other code, because the implementation of the ADT is hidden from the code that uses it. This separation of interface and implementation provides many benefits.

The wl_print_numerical function can be implemented in assembly as shown in Listing 6.8. The function operates by re-ordering the linked list using an insertion sort as described above. Listing 6.9 shows the change that must be made to the makefile. Now, when make is run, it compiles the two C files and the assembly file into object files, then links them all together. The C implementation of wl_print_numerical in list.c must be deleted or commented out or the linker will emit an error indicating that it found two versions of wl_print_numerical.

Listing 6.8 AArch64 assembly implementation of wl_print_numerical.

```
1  ### Definitions for the wordlistnode type
2  .equ    wln_word,   0   // word pointer field
3  .equ    wln_count,  8   // count field and padding
4  .equ    wln_next,   16  // next pointer field
5  .equ    wln_size,   24  // sizeof(wordlistnode)
6  ### Definitions for the wordlist type
7  .equ    wl_nwords,  0   // number of words in list and padding
8  .equ    wl_head,    8   // head of linked list
9  .equ    wl_size,    16  // sizeof(wordlist)
10 ### Define NULL
11 .equ    NULL,0
12
13 ### --------------------------------------------------------
14 ### The sort_numerical function sorts the list of words in
15 ### reverse by number of occurrences, and returns a
16 ### pointer to the head of the re-ordered list.
17 ### Records with identical counts will maintain their
18 ### original ordering with respect to each other.
19 ### x0 holds head of source list (head)
20 ### x1 holds destination list (dest)
21 ### x2 holds pointer to node currently being moved (node)
22 ### x3 holds pointer to current node in destination list (curr)
23 ### x4 holds pointer to previous node in destination list (prev)
24 ### x5 holds count of current node in destination list
25 ### w6 holds count of node currently being moved
26 .type   sort_numerical, %function
27 sort_numerical:
28 stp     x29, x30, [sp, #-16]!
29 mov     x1, #NULL       // initialize new list to NULL
30 ## Loop until source list is empty
31 loopa:  cmp     x0, #NULL
```

```
32          beq     endloopa
33                  ## detatch first node from source list
34          mov     x2, x0                  // node <- head
35          ldr     w5, [x2, #wln_count]    // load count for node
36          ldr     x0, [x0, #wln_next]     // head <- head->next
37                  ## find location to insert into destination list
38          mov     x3, x1                  // curr <- dest
39          mov     x4, #NULL               // prev <- NULL
40  loopb:  cmp     x3, #NULL               // Reached end of list?
41          beq     found
42          ldr     w6, [x3, #wln_count]    // load count for curr
43          cmp     w5, w6                  // compare with count for node
44          bgt     found
45          mov     x4, x3                  // previous <- current
46          ldr     x3, [x3, #wln_next]     // current <- current->next
47          b       loopb
48                  ## insert into destination list at current location
49  found:  str     x3, [x2, #wln_next]     // node-> next <- current
50          cmp     x4, #NULL               // if prev == NULL
51                  ## repeat with next list item
52          csel    x1, x2, x1, eq          // dest <- node
53          beq     loopa
54          str     x2, [x4, #wln_next]     // else prev->next <- node
55          b       loopa
56  endloopa:
57          mov     x0, x1                  // return dest (sorted list)
58          ldp     x29, x30, [sp], #16
59          ret
60          .size   sort_numerical, (. - sort_numerical)
61  ### ----------------------------------------------------------
62  ### wl_print_numerical prints a table showing the number
63  ### of occurrences for each word, followed by the word,
64  ### sorted in reverse order of occurence.
65          .type   wl_print_numerical, %function
66          .global wl_print_numerical
67  wl_print_numerical:
68          stp     x29, x30, [sp, #-32]!
69          str     x0, [sp, #16]           // save pointer to wordlist
70          ldr     x0, [x0, #wl_head]      // load pointer to the linked list
71          bl      sort_numerical          // re-sort the list
72          ldr     x1, [sp, #16]           // load pointer to wordlist
73          str     x0, [x1, #wl_head]      // update list pointer
74          mov     x0, x1                  // prepare to print
75          bl      wl_print                // print the sorted list
76          ldp     x29, x30, [sp], #32     // restore FP & LR
77          ret                             // return
78          .size   wl_print_numerical, (. - wl_print_numerical)
```

```
 1    C_OBJS=wordfreq.o list.o
 2    ASM_OBJS=wl_print_numerical.o
 3    OBJS=$(C_OBJS) $(ASM_OBJS)
 4
 5    CC=aarch64-linux-gnu-gcc
 6    LFLAGS=-O2 -g
 7    CFLAGS=-I. -O2 -g -Wall
 8    SFLAGS=-I. -O2 -g -Wall
 9    DEPENDFLAGS=-I. -M
10
11    wordfreq: $(OBJS)
12        $(CC) $(LFLAGS) -o wordfreq $(OBJS)
13
14    .S.o:
15        $(CC) $(SFLAGS) -c $<
16
17    .c.o:
18        $(CC) $(CFLAGS) -c $<
19
20    clean:
21        rm -f *.o *~ wordfreq
22
23    # make depend will create a file ".depend" with all the dependencies
24    depend:
25        rm -f .depend
26        $(CC) $(DEPENDFLAGS) $(C_OBJS:.o=.c) > .depend
27
28    # if we have a .depend file, include it
29    ifeq (.depend,$(wildcard .depend))
30    include .depend
31    endif
```

Listing 6.9 Revised makefile for the wordfreq program.

6.2.2 Better performance

The word frequency counter, as previously implemented, takes several minutes to count the frequency of words in this textbook on a Raspberry Pi. Most of the time is spent building the list of words, and re-sorting the list in order of word frequency. Most of the time for both of these operations is spent in searching for the word in the list before incrementing its count or inserting it in the list. There are much more efficient ways to build ordered lists of data.

Since the code is well modularized using an ADT, the internal mechanism of the list can be modified without affecting the main program. A major improvement can be made by changing the data structure from a linked list to a binary tree. Fig. 6.1 shows an example binary tree

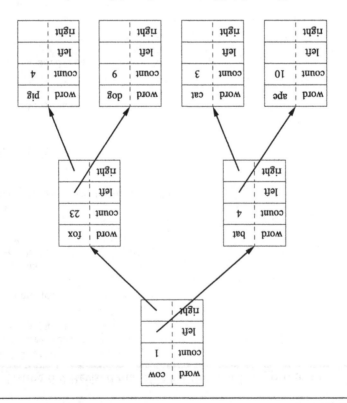

Figure 6.1: Binary tree of word frequencies.

storing word frequency counts. The time required to insert into a linked list is $O(N)$, but the time required to insert into a binary tree is $O(\log_2 N)$. To give some perspective, the author's manuscript for this textbook contains about 125,000 words. Since $\log_2(125,000) < 17$, we would expect the linked list implementation to require about $\frac{125000}{17} \approx 7353$ times as long as a binary tree implementation to process the author's manuscript for this textbook. In reality, there is some overhead to the binary tree implementation. Even with the extra overhead, we should see a significant speedup. Listing 6.10 shows the C implementation using a balanced binary tree instead of a linked list.

Listing 6.10 C implementation of the wordlist ADT using a tree.

```
1  #include <stdlib.h>
2  #include <string.h>
3  #include <stdio.h>
4  #include <list.h>
5
6  #define MAX(x,y) (x<y?y:x)
7
```

```c
/***********************************************************/
/* The wordlistnode type is a binary tree of words and     */
/* the number of times each word has occurred.             */
typedef struct wlist_node{
    char *word;
    int count;
    struct wlist_node *left, *right;
    int height;
}wordlistnode;

/***********************************************************/
/* wln_alloc allocates and initializes a wordlistnode      */
wordlistnode *wln_alloc(char *word)
{
    wordlistnode *newword;
    newword = (wordlistnode*)malloc(sizeof(wordlistnode));
    if(newword == NULL)
    {
        printf("Unable to allocate wordlistnode\n");
        exit(1);
    }
    newword->word = strdup(word);
    newword->count = 1;
    newword->left = NULL;
    newword->right = NULL;
    newword->height = 1;
    return newword;
}

/***********************************************************/
/* wln_free frees the storage of a wordlistnode            */
void wln_free(wordlistnode *root)
{
    if(root == NULL)
        return;
    free(root->word);
    wln_free(root->left);
    wln_free(root->right);
    free(root);
}

/***********************************************************/
/* wln_lookup is used to search the tree of words. It      */
/* returns a pointer to the wordlistnode. If the word is   */
/* not in the list, then it returns NULL.                  */
wordlistnode* wln_lookup(wordlistnode* root, char *word)
{
    int cmp;
    if(root != NULL)
```

```
57       {
58         cmp = strcmp(word,root->word);
59         if(cmp < 0)
60           root = wln_lookup(root->left,word);
61         else
62           if(cmp>0)
63             root = wln_lookup(root->right,word);
64       }
65   return root;
66 }
67
68 /***************************************************************/
69 /* wln_height finds the height of a node and returns     */
70 /* zero if the pointer is NULL.                          */
71 int wln_height(wordlistnode *node)
72 {
73   if(node == NULL)
74     return 0;
75   return node->height;
76 }
77
78 /***************************************************************/
79 /* wln_balance finds the balance factor of a node and    */
80 /* returns zero if the pointer is NULL.                  */
81 int wln_balance(wordlistnode *node)
82 {
83     if (node == NULL)
84         return 0;
85     return wln_height(node->left) - wln_height(node->right);
86 }
87
88 /***************************************************************/
89 /* wln_rotate_left rotates counterclockwise              */
90 wordlistnode* wln_rotate_left(wordlistnode* rt)
91 {
92   wordlistnode* nrt = rt->right;
93   rt->right = nrt->left;
94   nrt->left = rt;
95   rt->height =
96     MAX(wln_height(rt->left),wln_height(rt->right)) + 1;
97   nrt->height =
98     MAX(wln_height(nrt->left),wln_height(nrt->right)) + 1;
99   return nrt;
100 }
101
102 /***************************************************************/
103 /* wln_rotate_left rotates clockwise                     */
104 wordlistnode* wln_rotate_right(wordlistnode* rt)
105 {
```

```
wordlistnode* nrt = rt->left;
rt->left = nrt->right;
nrt->right = rt;
rt->height =
  MAX(wln_height(rt->left),wln_height(rt->right)) + 1;
nrt->height =
  MAX(wln_height(nrt->left),wln_height(nrt->right)) + 1;
return nrt;
}

/*******************************************************************/
/* wln_insert performs a tree insertion, and re-balances  */
wordlistnode * wln_insert(wordlistnode *root, wordlistnode *node)
{
  int balance;
  if (root == NULL)
    /* handle case where tree is empty, or we reached a leaf */
    root = node;
  else
    {
      /* Recursively search for insertion point, and perform the
         insertion. */
      if (strcmp(node->word,root->word) < 0)
        root->left  = wln_insert(root->left, node);
      else
        root->right = wln_insert(root->right, node);

      /* As we return from the recursive calls, recalculate the heights
         and perform rotations as necessary to re-balance the tree */
      root->height = MAX(wln_height(root->left),
                         wln_height(root->right)) + 1;

      /* Calculate the balance factor */
      balance = wln_balance(root);
      if (balance > 1)
        {
          /* the tree is deeper on the left than on the right) */
          if(strcmp(node->word,root->left->word) <= 0)
            root = wln_rotate_right(root);
          else
            {
              root->left = wln_rotate_left(root->left);
              root = wln_rotate_right(root);
            };
        }
      else
        if(balance < -1)
          {
            /* the tree is deeper on the right than on the left) */
```

```
155              if(strcmp(node->word,root->right->word) >= 0)
156                root = wln_rotate_left(root);
157              else
158                {
159                  root->right = wln_rotate_right(root->right);
160                  root = wln_rotate_left(root);
161                }
162          }
163      }
164    return root;
165 }
166
167 /*****************************************************************/
168 /* The wordlist type holds a pointer to the binary tree    */
169 /* and keeps track of the number of nodes in the list      */
170 typedef struct wlist{
171   int nwords;
172   wordlistnode *root;
173 }wordlist;
174
175 /*****************************************************************/
176 /* wl_alloc allocates and initializes a new word list.     */
177 wordlist* wl_alloc()
178 { wordlist* tmp;
179   tmp = (wordlist*)malloc(sizeof(wordlist));
180   if(tmp == NULL)
181     {
182       printf("Unable to allocate wordlist\n");
183       exit(1);
184     }
185   tmp->nwords = 0;
186   tmp->root = NULL;
187   return tmp;
188 }
189
190 /*****************************************************************/
191 /* wl_free frees all the storage used by a wordlist        */
192 void wl_free(wordlist* wl)
193 {
194   wln_free(wl->root);
195   free(wl);
196 }
197
198 /*****************************************************************/
199 /* wl_increment adds one to the count of the given word.   */
200 /* If the word is not in the list, then it is added with   */
201 /* a count of one.                                         */
202 void wl_increment(wordlist *list, char *word)
203 {
```

```c
204      wordlistnode *newword;
205      wordlistnode *wlst = wln_lookup(list->root,word);
206      if((wlst == NULL)||(strcmp(wlst->word, word) != 0))
207      { /* create a new node */
208          list->nwords++;
209          newword = wln_alloc(word);
210          list->root = wln_insert(list->root,newword);
211      }
212      else
213          wlst->count++;
214  }
215
216  /*******************************************************
217   /* wln_print is an interal functino to print a table  */
218   /* showing the number of occurrences for each word,   */
219   /* followed by the word.                              */
220  void wln_print(wordlistnode *list)
221  {
222      if(list != NULL)
223      {
224          wln_print(list->left);
225          printf("%10d '%s'\n",list->count,list->word);
226          wln_print(list->right);
227      }
228  }
229
230  /*******************************************************
231   /* wl_print_alphabetical prints a table showing the number*/
232   /* of occurrences for each word, followed by the word. */
233  void wl_print(wordlist * list)
234  {
235      wln_print(list->root);
236      printf("There are %d unique words in the document\n",
237          list->nwords);
238  }
239
240  #ifndef USE_ASM
241  /*******************************************************
242   /* wl_print_numerical prints a table showing the number */
243   /* of occurrences for each word, followed by the word,  */
244   /* sorted in reverse order of occurence.                */
245  void wl_print_numerical(wordlist *list)
246  {
247      printf("wl_print_numerical has not been implemented\n");
248  }
249  #endif
```

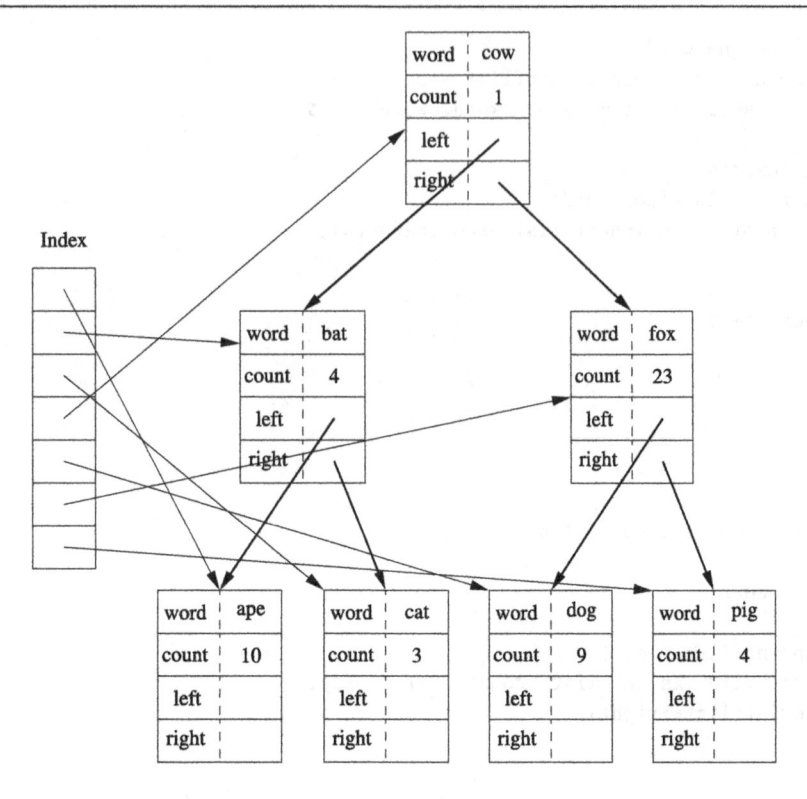

Figure 6.2: Binary tree of word frequencies with index added.

6.2.3 Indexing and sorting by frequency

With the tree implementation, wl_print_numerical could build a new tree, sorted on the word frequency counts. However, it may be more efficient to build a separate index, and sort the index by word frequency counts. The assembly code will allocate an array of pointers, and set each pointer to one of the nodes in the tree, as shown in Fig. 6.2. Then, it will use a quicksort to rearrange the pointers into descending order by word frequency count, as shown in Fig. 6.3. This implementation is shown in Listing 6.11.

Listing 6.11 AArch64 assembly implementation of wl_print_numerical with a tree.

```
1   ### Definitions for the wordlistnode type
2           .equ    wln_word,       0       // word field
3           .equ    wln_count,      8       // count filed
4           .equ    wln_left,       16      // left field
5           .equ    wln_right,      24      // right field
6           .equ    wln_height,     32      // height of this node
7           .equ    wln_size,       40      // sizeof(wordlistnode)
8   ### Definitions for the wordlist type
```

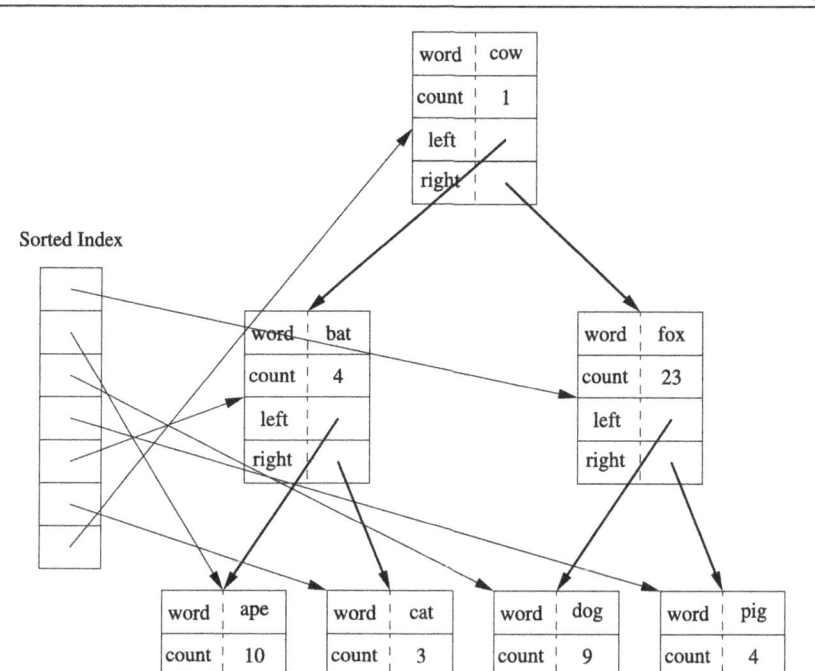

Figure 6.3: Binary tree of word frequencies with sorted index.

```
 9          .equ    wl_nwords,      0           // number of words in list
10          .equ    wl_head,        8           // head of linked list
11          .equ    wl_size,        16          // sizeof(wordlist)
12  ### Define NULL
13          .equ    NULL,           0
14          .data
15  failstr:.asciz  "Unable to allocate index\n"
16  fmtstr: .asciz  "%10d '%s'\n"
17          .text
18  ### -----------------------------------------------------------
19  ### Recursively traverses the tree, filling in the array of
20  ### pointers.
21  ### wordlistnode **getptrs(wordlistnode *ptrs[], wordlistnode *node)
22          .type   getptrs, %function
23  getptrs:
24          cmp     x1, #0                      // if (node == NULL)
25          beq     getptrs_exit                // return ptrs;
26          stp     x29, x30, [sp, #-32]!
27          str     x19, [sp, #16]
```

```
28          mov     x19, x1                 // save address of node
29
30          // ptrs = getptrs(ptrs, node->left)
31          ldr     x1, [x19, #wln_left]
32          bl      getptrs
33
34          // *(ptrs++)=node
35          str     x19, [x0], #8
36          // return getptrs(ptrs, node->right)
37
38          ldr     x1, [x19, #wln_right]
39          bl      getptrs
40          ldr     x19, [sp, #16]
41          ldp     x29, x30, [sp], #32
42  getptrs_exit:
43          ret
44          .size   getptrs,(. - getptrs)
45
46  ### -----------------------------------------------------------
47  ### wl_print_numerical prints a table showing the number
48  ### of occurrences for each word, followed by the word,
49  ### sorted in reverse order of occurence.
50  ### void wl_print_numerical(wordlist *list)
51          .type   wl_print_numerical, %function
52          .global wl_print_numerical
53  wl_print_numerical:
54          stp     x29, x30, [sp, #-48]!
55          stp     x19, x20, [sp, #16]
56          str     x21, [sp, #32]
57
58          // wordlistnode **ptrs=malloc(list->nwords*sizeof(wordlistnode*));
59          mov     x19, x0                 // x19 = list
60          ldr     w20, [x0, #wl_nwords]   // w20 = list->nwords
61          lsl     x0, x20, #3             // x0 = list->nwords*8
62          bl      malloc                  // allocate storage
63
64          // check return value
65          cmp     x0, #0
66          bne     malloc_ok
67          adr     x0, failstr             // load pointer to string
68          bl      printf
69          mov     w0, #1
70          ldr     x21, [sp, #32]
71          ldp     x19, x20, [sp, #16]
72          ldp     x29, x30, [sp], #48
73          bl      exit                    // exit(1)
74  malloc_ok:
75
76          // getptrs(ptrs, list->root)
```

```
77                mov     x21, x0                    // x21 = ptrs
78                ldr     x1, [x19, #wl_head]        // x1 = list->root
79                bl      getptrs                    // fill in the pointers
80
81        // wl_quicksort(**start, **end)
82                mov     x0, x21                    // get pointer to array
83                add     x1, x0, x20, lsl #3        // get pointer to end of array
84                sub     x1, x1, #8
85                bl      wl_quicksort               // re-sort the array of pointers
86
87        // Print the word frequency list.
88                mov     x19, #0                    // do a for loop
89        // for(i = 0; i < list->nwords; i++)
90        loop:   cmp     x19, x20
91                bge     done
92
93        // printf("%10d '%s'\n", ptrs[i]->count, ptrs[i]->word)
94                adr     x0, fmtstr
95                ldr     x3, [x21, x19, lsl #3]     // get next pointer
96                add     x19, x19, #1
97                ldr     w1, [x3, #wl_count]        // load count
98                ldr     x2, [x3, #wl_word]         // load ptr to word
99                bl      printf
100               b       loop
101       done:
102               ldr     x21, [sp, #32]
103               ldp     x19, x20, [sp, #16]
104               ldp     x29, x30, [sp], #48
105               ret
106               .size   wl_print_numerical,(. - wl_print_numerical)
107
108       ### -------------------------------------------------------
109       ### function wl_quicksort(wln **left,wln **right) quicksorts
110       ### the array of pointers in order of the word counts
111               .type   wl_quicksort, %function
112       wl_quicksort:
113               cmp     x0, x1
114               bge     wl_quicksort_exit          // return if length<=1
115               stp     x29, x30, [sp, #-48]!
116               stp     x19, x20, [sp, #16]
117               stp     x21, x22, [sp, #32]
118
119               ldr     x12, [x0]                  // use count of first item as
120               ldr     w12, [x12, #wl_count]      // pivot value in x12
121               mov     x19, x0                    // current left
122               mov     x20, x1                    // current right
123               mov     x21, x0                    // original left(first)
124               mov     x22, x1                    // original right(last)
125       loopa:  cmp     x19, x20                   // while left <= right &&
```

```
126          bgt     finale
127          ldr     x0, [x19]           // (*left)->count > pivot
128          ldr     w1, [x0, #wln_count]
129          cmp     x1, x12
130          ble     loopb
131          add     x19, x19, #8        // increment left
132          b       loopa
133 loopb:   cmp     x19, x20            // while left < right &&
134          bgt     finale
135          ldr     x2, [x20]           // (*right)->count < pivot
136          ldr     w3, [x2, #wln_count]
137          cmp     x3, x12
138          bge     cmp
139          sub     x20, x20, #8        // decrement right
140          b       loopb
141 cmp:     cmp     x19, x20            // if( left <= right )
142          bgt     finale
143          str     x0, [x20], #-8      // swap pointers and
144          str     x2, [x19], #8       // change indices
145          b       loopa
146 finale:  mov     x0, x21             // quicksort array from
147          mov     x1, x20             // first to current right
148          bl      wl_quicksort
149          mov     x0, x19             // quicksort array from
150          mov     x1, x22             // current left to last)
151          bl      wl_quicksort
152          ldp     x19, x20, [sp, #16]
153          ldp     x21, x22, [sp, #32]
154          ldp     x29, x30, [sp], #48
155 wl_quicksort_exit:
156          ret
157          .size   wl_quicksort,(. - wl_quicksort)
```

The tree-based implementation gets most of its speed improvement through using two $O(N \log N)$ algorithms to replace $O(N^2)$ algorithms. These examples show how a small part of a program can be implemented in assembly language, and how to access C data structures from assembly. The functions could just as easily have been written in C rather than assembly, without greatly affecting performance. Later chapters will show examples where the assembly implementation *does* have significantly better performance than the C implementation.

6.3 Ethics case study: Therac-25

The Therac-25 was a device designed for radiation treatment of cancer. It was produced by Atomic Energy of Canada Limited (AECL), which had previously produced the Therac-6 and

Therac-20 units in partnership with CGR of France. It was capable of treating tumors close to the skin surface using electron beam therapy, but could also be configured for Megavolt X-ray therapy to treat deeper tumors. The X-ray therapy required the use of a tungsten radiation shield to limit the area of the body that was exposed to the potentially lethal radiation produced by the device.

The Therac-25 used a double pass accelerator, which provided more power, in a smaller space, at less cost, compared to its predecessors. The second major innovation was that computer control was a central part of the design, rather than an add-on component as in its predecessors. Most of the hardware safety interlocks that were integral to the designs of the Therac-6 and Therac-20, were seen as unnecessary, because to software would perform those functions. Computer control was intended to allow operators to set up the machine more quickly, allowing them to spend more time communicating with patients and to treat more patients per day. It was also seen as a way to reduce production costs by relying on software, rather than hardware, safety interlocks.

There were design issues with both the software and the hardware. Although this machine was built with the goal of saving lives, between 1985 and 1986, three deaths and other injuries were attributed to the hardware and software design of this machine. Death due to radiation exposure is usually slow and painful, and the problem was not identified until the damage had been done.

6.3.1 History of the Therac-25

AECL was required to obtain US Food and Drug Administration (FDA) approval before releasing the Therac-25 to the US market. They obtained approval quickly by declaring ''pre-market equivalence,'' effectively claiming that the new machine was not significantly different from its predecessors. This practice was common in 1984, but was overly optimistic, considering that most of the safety features had been changed from hardware to software implementations. With FDA approval, AECL made the Therac-25 commercially available and performed a Fault Tree Analysis to evaluate the safety of the device.

Fault Tree Analysis, as its name implies, requires building a tree to describe every possible fault, and assign probabilities to those faults. After building the tree, the probabilities of hazards, such as overdose, can be calculated. Unfortunately, the engineers assumed that the software (much of which was re-used from the previous Therac models) would operate correctly. This turned out not to be the case, because the hardware interlocks present in the previous models had hidden some of the software faults. The analysts did consider some possible computer faults, such as an error being caused by cosmic rays, but assigned extremely low probabilities to those faults. As a result, the assessment was very inaccurate.

When the first report of an overdose was reported to AECL in 1985, they sent an engineer to the site to investigate. They also filed a report with the FDA and the Canadian Radiation Protection Board (CRPB). AECL also notified all users of the fact that there had been a report and recommended that operators should visually confirm hardware settings before each treatment. The AECL engineers were unable to reproduce the fault, but suspected that it was due to the design and placement of a microswitch. They redesigned the microswitch and modified all of the machines that had been deployed. They also retracted their recommendation that operators should visually confirm hardware settings before each treatment.

Later that year, a second incident occurred. In this case, there is no evidence that AECL took any action. In January of 1986, AECL received another incident report. An employee at AECL responded by denying that the Therac-25 was at fault, and stated that no other similar incidents had been reported. Another incident occurred in March of that year. AECL sent an engineer to investigate. The engineer was unable to determine the cause, and suggested that it was due to an electrical problem, which may have caused an electrical shock. An independent engineering firm was called to examine the machine and reported that it was very unlikely that the machine could have delivered an electrical shock to the patient. In April of 1986, another incident was reported. In this case, the AECL engineers, working with the medical physicist at the hospital, were able to reproduce the sequence of events that lead to the overdose.

As required by law, AECL filed a report with the FDA. The FDA responded by declaring the Therac-25 defective. AECL was ordered to notify all of the sites where the Therac-25 was in use, investigate the problem, and file a corrective action plan. AECL notified all sites, and recommended removing certain keys from the keyboard on the machines. The FDA responded by requiring them to send another notification with more information about the defect and the consequent hazards. Later in 1986, AECL filed a revised corrective action plan.

Another overdose occurred in January 1987, and was attributed to a different software fault. In February, the FDA and CRPB, both ordered that all Therac-25 unites be shut down, pending effective and permanent modifications. AECL spent six months developing a new corrective action plan, which included a major overhaul of the software, the addition of mechanical safety interlocks, and other safety-related modifications.

6.3.2 Overview of design flaws

The Therac-25 was controlled by a DEC PDP-11 computer. The PDP-11 was the most popular minicomputer ever produced. Around 600,000 were produced between 1970 and 1990 and used for a variety of purposes, including embedded systems, education, and general data processing. It was a 16-bit computer and was far less powerful than a Raspberry Pi. The Therac-25 computer was programmed in assembly language by one programmer and the source code was not documented. Documentation for the hardware components was written in French. After the faults were discovered, a commission concluded that the primary problems with the Therac-25 were attributable to poor software design practices, and not due to any one of several specific coding errors. This is probably the best known case where a poor overall software design, and insufficient testing, led to loss of life.

The worst problems in the design and engineering of the machine were:

- The code was not subjected to independent review.
- The software design was not considered during the assessment of how the machine could fail or malfunction.
- The operator could ignore malfunctions and cause the machine to proceed with treatment.
- The hardware and software were designed separately and not tested as a complete system until the unit was assembled at the hospitals where it was to be used.
- The design of the earlier Therac-6 and Therac-20 machines included hardware interlocks which would ensure that the X-ray mode could not be activated unless the tungsten radiation shield was in place. The hardware interlock was replaced with a software interlock in the Therac-25.
- Errors were displayed as numeric codes, and there was no indication of the severity of the error condition.

The operator interface consisted of a keyboard and text-mode monitor, which was common in the early 1980s. The interface had a data entry area in the middle of the screen and a command line at the bottom. The operator was required to enter parameters in the data entry area, then move the cursor to the command line to initiate treatment. When the operator moved the cursor to the command line, internal variables were updated and a flag variable was set to indicate that data entry was complete. That flag was cleared when a command was entered on the command line. If the operator moved the cursor back to the data entry area without entering a command, then the flag was not cleared, and any subsequent changes to the data entry area did not affect the internal variables.

A global variable was used to indicate that the magnets were currently being adjusted. This variable was modified by two functions, and did not always contain the correct value. Adjusting the magnets required about eight seconds, and the flag was correct for only a small period at the beginning of this time period.

Due to the errors in the design and implementation of the software, the following sequence of events could result in the machine causing injury to, or even death of, the patient:

1. The operator mistakenly specified high power mode during data entry.
2. The operator moved the cursor to the command line area.
3. The operator noticed the mistake, and moved the cursor back to the data entry area without entering a command.
4. The operator corrected the mistake and moved the cursor back to the command line.
5. The operator entered the command line area, left it, made the correction, and returned within the eight-second window required for adjusting the magnets.

If the above sequence occurred in less than eight seconds, then the operator screen could indicate that the machine was in low power mode, although it was actually set in high power mode. During a final check before initiating the beam, the software would find that the magnets were set for high power mode but that the operator setting was for low power mode. It displayed a numeric error code and prevented the machine from starting. The operator could clear the error code by resetting the computer (which only required one key to be pressed on the keyboard). This caused the tungsten shield to withdraw but left the machine in X-ray mode. When the operator entered the command to start the beam, the machine could be in high power mode without having the tungsten shield in place. X-rays were applied to the unprotected patient.

It took some time for this critical flaw to appear. The failure only occurred when the operator initially made a one-keystroke mistake in entering the prescription data, moved to the command area, and then corrected the mistake within eight seconds. Initially, operators were slow to enter data, and spent a lot of time making sure that the prescription was correct before initiating treatment. As they became more familiar with the machine, they were able to enter data, and correct mistakes more quickly. Eventually, operators became familiar enough with the machine that they could enter data, make a correction, and return to the command area within the critical eight-second window. Also, the operators became familiar with the machine reporting numeric error codes, without any indication of the severity of the code. The operators were given a table of codes and their meanings. The code reported was "no dose" and indicated "treatment pause." There is no reason why the operator should consider that to be a serious problem; they had become accustomed to frequent malfunctions that did not have any consequences to the patient.

Although the code was written in assembly language, that fact was not cited as an important factor. The fundamental problems were poor software design and over-confidence. The re-use of code in an application for which it was not initially designed also may have contributed to the system flaws. A proper design using established software design principles, including structured programming and abstract data types, would almost certainly have avoided these fatalities.

6.4 Chapter summary

The abstract data type is a structured programming concept which contributes to software reliability, eases maintenance, and allows for major revisions to be performed in as safe way. Many high-level languages enforce, or at least facilitate, the use of ADTs. Assembly language does not. However, the ethical assembly language programmer will give the extra effort to write code that conforms to the standards of structured programming, using abstract data types to help ensure safety, reliability, and maintainability.

ADTs also facilitate the implementation of software modules in more than one language. The interface specifies the components of the ADT, but not the implementation. The implementation can be in any language. As long as assembly programmers and compiler authors generate code that conforms to a well-known standard, their code can be linked to code written in other languages.

Poor coding practices and poor design can lead to dire consequences, including loss of life. It is the responsibility of the programmer, regardless of the language used, to make ethical decisions in the design and implementation of software. Above all, the programmer must be aware of the possible consequences of the decisions they make.

Exercises

6.1. What are the advantages of designing software using abstract data types?

6.2. Why is the `Pixel` data type hidden from client code in Listing 6.2?

6.3. High level languages provide mechanisms for information hiding, but assembly does not. Why should the assembly programmer not simply bypass all information hiding and access the internal data structures directly?

6.4. The assembly code in `wl_print_numerical` accesses the internal structure of the `wordlistnode` data type. Why is it allowed to do so? *Should* it be allowed to do so?

6.5. Given the following definitions for a stack ADT:

```
1    /* File: stack.h */
2
3    typedef struct IntStackStruct *IntStack;
4
5    /* create an empty stack and return a pointer to it */
6    IntStack InitStack ();
7
8    /* Push value onto the stack.
9       Return 1 for success. Return 0 if stack is full. */
10   int Push (IntStack stack, int k);
11
12   /* Remove value from top of stack.
```

```
13          Return 1 for success. Return 0 if stack was empty. */
14     int Pop (IntStack stack);
15
16     /* Return the value that is at the top of the stack. */
17     int Top (IntStack stack);
18
19     /* Print the elements of the stack. */
20     extern PrintStack (IntStack stack);
```

```
1      /* File: stack.c */
2
3      #define STACKSIZE 100
4
5      /* The stack is implemented as  an array of items and
6         the index of the item at the top */
7      struct IntStackStruct {
8        int stackItems [STACKSIZE];
9        int top;
10     };
11
12     typedef struct IntStackStruct *IntStack;
```

Write the `InitStack` function in AArch64 assembly language.

6.6. Referring to the previous question, write the `Push` function in AArch64 assembly language.

6.7. Referring to the previous two questions, write the `Pop` function in AArch64 assembly language.

6.8. Referring to the previous three questions, write the `Top` function in AArch64 assembly language.

6.9. Referring to the previous three questions, write the `PrintStack` function in AArch64 assembly language.

6.10. Re-implement all of the previous stack functions using a linked list rather than a static array.

6.11. The "Software Engineering Code of Ethics And Professional Practice" states that a responsible software engineer should "Approve software only if they have well-founded belief that it is safe, meets specifications, passes appropriate tests..." (sub-principle 1.03) and "Ensure adequate testing, debugging, and review of software...on which they work." (sub-principle 3.10). Unfortunately, defects did make their way into the system. The software engineering code of ethics also states that a responsible software engineer should "Treat all forms of software maintenance with the same professionalism as new development."

a. Explain how the Software Engineering Code of Ethics And Professional Practice were violated by the Therac 25 developers.

b. How should the engineers and managers at AECL have responded when problems were reported?

c. What other ethical and non-ethical considerations may have contributed to the deaths and injuries?

ARM 64-Bit Assembly Language
https://doi.org/10.1016/B978-0-12-819221-4.00014-6

CHAPTER 7

Integer mathematics

There are some differences between the way calculations are performed in a computer versus the way most of us were taught as children. The first difference is that calculations are performed in binary instead of base ten. Another difference is that the computer is limited to a fixed number of binary digits, which raises the possibility of having a result that is too large to fit in the number of bits available. This occurrence is referred to as *overflow*. The third difference is that subtraction is performed using complement addition.

Addition in base b is very similar to base ten addition, except that the result of each column is limited to $b - 1$. For example, binary addition works exactly the same as decimal addition, except that the result of each column is limited to 0 or 1. The following figure shows an addition in base ten and the equivalent addition in base two.

$$
\begin{array}{rcr}
& & 1 \\
7\,5 & & 0\,1\,0\,0\,1\,0\,1\,1 \\
+\ 1\,9 & = & +\ 0\,0\,0\,1\,0\,0\,1\,1 \\
\hline
9\,4 & & 0\,1\,0\,1\,1\,1\,1\,0 \\
\end{array}
$$

$$
\begin{array}{r}
1\ 1 \\
\end{array}
$$

The carry from one column to the next is shown as a small number above the column that it is being carried into. Note that carries from one column to the next are done the same way in either base. The only difference is that there are more columns in the base two addition because it takes more binary digits to represent a number than it does in decimal.

7.1 Subtraction by addition

Finding the complement was explained in Section 1.3.3. Subtraction can be computed by adding the radix complement of the subtrahend to the menuend. Example 11 shows a complement subtraction with positive results. When the menuend is less than the subtrahend, the result will be negative. In the complement method, we deal with this possibility by adding a *sign digit* at the beginning of the numbers. Example 12 shows complement subtraction with negative results. Example 13 shows several more signed addition and subtraction operations in base ten and binary.

Example 11. Ten's complement subtraction.

Suppose we wish to calculate $384_{10} - 56_{10}$ using the complements method. We first pad the numbers, so that they have leading zeros (the sign digit), and are the same length, resulting in $0384_{10} - 0056_{10}$ From Chapter 1, the ten's complement of 0056_{10} is 9944_{10}. Adding them together gives us $0384_{10} + 9944_{10} = 10328_{10}$. Since we are using four digits for the operation, we expect a four digit result. After discarding the leading '1', we have 328, which is the correct result. Note that if the carry into the most significant column does not match the final carry out, then overflow has occurred. Both methods of subtraction are shown below:

$$
\begin{array}{r}
{\scriptstyle 7\ \ 14} \\
3\ \cancel{8}\ \cancel{4} \\
-\ \ \ 5\ 6 \\
\hline
3\ 2\ 8
\end{array}
\quad = \quad
\begin{array}{r}
{\scriptstyle 1\ \ 1} \\
0\ 3\ 8\ 4 \\
+9\ 9\ 4\ 4 \\
\hline
\cancel{1}0\ 3\ 2\ 8
\end{array}
$$

Example 12. Ten's complement subtraction with a negative result.

Suppose we want to calculate $284 - 481$. First, we pad with leading zeros (sign digits), to get $0284 - 0481$. Then, we take the tens complement of 0481, which is 9519. Adding the ten's complement of the subtrahend to the menuend gives $0284 + 9519 = 9803$. The First digit is greater than 4, indicating that the result is negative. Therefore, we take the ten's complement of the result to get -0197.

Example 13. Signed addition and subtraction in decimal and binary.

$$
\begin{array}{r}
0\ 2\ 3 \\
+\ \ 0\ 1\ 5 \\
\hline
0\ 3\ 8
\end{array}
\quad = \quad
\begin{array}{r}
+\ \ 0\ 0\ 0\ 1\ 0\ 1\ 1\ 1 \\
+\ \ 0\ 0\ 0\ 0\ 1\ 1\ 1\ 1 \\
\hline
0\ 0\ 1\ 0\ 0\ 1\ 1\ 0
\end{array}
$$

$$
\begin{array}{r}
0\ 2\ 3 \\
-\ \ 0\ 1\ 5 \\
\hline
0\ 0\ 8
\end{array}
=
\begin{array}{r}
0\ 2\ 3 \\
+\ \ 9\ 8\ 5 \\
\hline
\cancel{1}\ 0\ 0\ 8
\end{array}
=
\begin{array}{r}
0\ 0\ 0\ 1\ 0\ 1\ 1\ 1 \\
+\ \ 1\ 1\ 1\ 1\ 0\ 0\ 0\ 1 \\
\hline
\cancel{1}\ 0\ 0\ 0\ 0\ 1\ 0\ 0\ 0
\end{array}
$$

$$
\begin{array}{r}
-\ \ 2\ 3 \\
+\ \ 1\ 5 \\
\hline
-\ \ 8
\end{array}
=
\begin{array}{r}
9\ 7\ 7 \\
+\ \ 0\ 1\ 5 \\
\hline
9\ 9\ 2
\end{array}
=
\begin{array}{r}
1\ 1\ 1\ 0\ 1\ 0\ 0\ 1 \\
+\ \ 0\ 0\ 0\ 0\ 1\ 1\ 1\ 1 \\
\hline
1\ 1\ 1\ 1\ 1\ 0\ 0\ 0
\end{array}
$$

$$
\begin{array}{rcrcr}
-\ 0\,2\,3 & & 9\,7\,7 & & 1\,1\,1\,0\,1\,0\,0\,1 \\
-\ 0\,1\,5 & = & -\ 9\,8\,5 & = & +\ 1\,1\,1\,1\,0\,0\,0\,1 \\
\hline
-\ 0\,3\,8 & & -\,\cancel{1}\,9\,6\,2 & & \cancel{1}\,1\,1\,0\,1\,1\,0\,1\,0
\end{array}
$$

7.2 Binary multiplication

Many processors have hardware multiply instructions. However hardware multipliers require a large number of transistors, and consume significant power. Processors designed for extremely low power consumption or very small size usually do not implement a multiply instruction, or only provide multiply instructions that are limited to a small number of bits. On these systems, the programmer must implement multiplication using basic data processing instructions. Even when a hardware multiplier is available, there are some techniques available which may achieve higher performance.

7.2.1 Multiplication by a power of two

If the multiplier is a power of two, then multiplication can be accomplished with a shift to the left. Consider the four bit binary number $x = x_3 \times 2^3 + x_2 \times 2^2 + x_1 \times 2^1 + x_0 \times 2^0$, where x_n denotes bit n of x. If x is shifted left by one bit, introducing a zero into the least significant bit, then it becomes $x_3 \times 2^4 + x_2 \times 2^3 + x_1 \times 2^2 + x_0 \times 2^1 + 0 \times 2^0 = 2\left(x_3 \times 2^3 + x_2 \times 2^2 + x_1 \times 2^1 + x_0 \times 2^0 + 0 \times 2^{-1}\right)$. Therefore, a shift of one bit to the left is equivalent to multiplication by two. This argument can be extended to prove that a shift left by n bits is equivalent to multiplication by 2^n.

7.2.2 Multiplication of two variables

Most techniques for binary multiplication involve computing a set of partial products and then summing the partial products together. This process is similar to the method taught to primary schoolchildren for conducting long multiplication on base ten integers, but has been modified here for application to binary. The method typically taught in school for multiplying decimal numbers is based on calculating partial products, shifting them to the left and then adding them together. The most difficult part is to obtain the partial products, as that involves multiplying a long number by one base ten digit. The following example shows how the partial products are formed.

```
  1 2 3
× 4 5 6
-------
  7 3 8        (this is 123 x 6)
 6 1 5 0       (this is 123 x 5, shifted one position to the left)
+4 9 2 0 0     (this is 123 x 4, shifted two positions to the left)
---------
5 6 0 8 8
```

The first partial product can be written as $123 \times 6 \times 10^0 = 738$. The second is $123 \times 5 \times 10^1 = 6150$, and the third is $123 \times 4 \times 10^2 = 49200$. In practice, we usually leave out the trailing zeros. The procedure is the same in binary, but is simpler because the partial product involves multiplying a long number by a single base 2 digit. Since the multiplier is always either zero or one, the partial product is very easy to compute. The product of multiplying any binary number x by a single binary digit is always either 0 or x. Therefore, the multiplication of two binary numbers comes down to shifting the multiplicand left appropriately for each non-zero bit in the multiplier, and then adding the shifted numbers together.

Suppose we wish to multiply two four-bit numbers, 1011 and 1010:

```
        1 0 1 1      this is 11₁₀
      × 1 0 1 0      this is 10₁₀
      ---------
        0 0 0 0      0 x 1011
        1 0 1 1      1011 x 1, shifted one position to the left
      0 0 0 0        1011 x 0, shifted two positions to the left
   +  1 0 1 1        1011 x 1, shifted three positions to the left
   ---------------
   0 1 1 0 1 1 1 0   this is 110₁₀
```

Notice in the previous example that each partial sum is either zero or x shifted by some amount. A slightly quicker way to perform the multiplication is to leave out any partial sum which is zero. Example 14 shows the results of multiplying 101_{10} by 89_{10} in decimal and binary using this shorter method. For implementation in hardware and software, it is easier to *accumulate* the partial products, by adding each to a running sum, rather than building a circuit to add multiple binary numbers at once. This results in an algorithm with $O(n)$ complexity, where n is the number of bits in the multiplier. There are multiplication algorithms that give even better performance, an some processors use combinatorial multiplication circuits, which are very fast, but require a large number of transistors and consume a great deal of power.

Binary multiplication can be implemented as a sequence of shift and add instructions. Given two registers, x and y, and an *accumulator register* a, the product of x and y can be computed using Algorithm 1. When applying the algorithm, it is important to remember that, in the general case, the result of multiplying an n bit number by an m bit number is (at most) an $n + m$ bit number. For instance $11_2 \times 11_2 = 1001_2$. Therefore, when applying Algorithm 1, it is necessary to know the number of bits in x and y. Since x is shifted left on each iteration of the loop, the registers used to store x and a must both be at least as large as the number of bits in x plus the number of bits in y.

To multiply two n bit numbers, you must be able to add two $2n$ bit numbers. Adding 128 bit numbers requires two add instructions and the carry from the least-significant 64 bits must be added to the sum of the most-significant 64 bits. AArch64 provides a convenient way to perform the add with carry. Assume we have two 128 bit numbers, x and y. We have x in x0, x1 and y in x2, x3, where the high order words of each number are in the higher-numbered registers, and we want to calculate $x = x + y$. Listing 7.1 shows a two instruction sequence for an AArch64 processor. The first instruction adds the two least-significant words together and sets (or clears) the carry bit and other flags in the PSTATE register. The second instruction

Example 14. Equivalent multiplication in decimal and binary.

```
        1000100                      101
     ×  1100101                   ×   68
     ----------                   ------
        1000100                      808
       0000000                      606
      0000000                     ------
     1000100                       6868
    0000000
   1000100
   1000100
   -------------
   1101011010100
```

Algorithm 1: Algorithm for binary multiplication.

```
a → 0
while y ≠ 0 do
    if LSB(y) = 1 then
        a → a + x
    end if
    x → x shifted left 1 bit
    y → y shifted right 1 bit
end while
```

Listing 7.1 AArch64 assembly code for adding two 128 bit numbers.

```
adds    x0, x0, x2    // add the low-order double-words,
                      // ash set flags in PSTATE
adc     x1, x1, x3    // add the high-order double-words
                      // plus the carry flag
```

adds the two most significant words along with the carry bit. This technique can be extended to add integers with any number of bits.

Example 15. Binary multiplication using Algorithm 1.

Assume we wish to multiply two numbers, $x = 01101001$ and $y = 01011010$. Applying Algorithm 1 results in the following sequence:

a	x	y	Next operation
0000000000000000	0000000001101001	01011010	shift only
0000000000000000	0000000011010010	00101101	add, then shift
0000000011010010	0000000110100100	00010110	shift only
0000000011010010	0000001101001000	00001011	add, then shift
0000010000011010	0000011010010000	00000101	add, then shift
0000101010101010	0000110100100000	00000010	shift only
0000101010101010	0001101001000000	00000001	add, then shift
0010010011101010	0011010010000000	00000000	return result

$$105 \times 90 = 9450$$

On an AArch64 processor, the algorithm to multiply two 64-bit unsigned integers is very efficient. Listing 7.2 shows one possible algorithm for multiplying two 64-bit numbers to obtain a 128-bit result. The code is a straightforward implementation of the algorithm, and some modifications can be made to improve efficiency. For example, if we only want a 64-bit result, we do not need to perform 128-bit addition. This significantly simplifies the code, as shown in Listing 7.3.

Listing 7.2 AArch64 assembly code for multiplication with a 128 bit result without umulh or smulh.

```
        .section .rodata
x:      .8byte   0x57
y:      .8byte   0x75
msg:    .asciz   "%lx * %lx = %016lx%016lx\n"
```

```
5          .text
6          .type   main %function
7          .global main
8   /* Multiplies x and y by adding a series of partial products.
9      Shifts x (multiplicand) left and y (multiplier) right. Only
10     adds if the least significant bit of multiplier is 1. */
11  main:  stp   x29, x30, [sp, #-16]!
12         mov   x0, #0      // resultLow = 0
13         mov   x1, #0      // resultHigh = 0
14         adr   x2, x       // x2 = &x
15         ldr   x2, [x2]    // x2 = multiplicandLow
16         mov   x3, #0      // x3 = multiplicandHigh
17         adr   x4, y       // x4 = &y
18         ldr   x4, [x4]    // x4 = multiplier
19  loop:  tst   x4, #1      // Only add if lowest bit of
20         beq   endif       // multiplier is 1:
21         adds  x0, x0, x2  // resultLow += multiplicandLow.
22         adc   x1, x1, x3  // resultHigh +=
23                           //   multiplicandHigh + Carry
24  endif: // Shift 128-bit multiplicand left. Carry bit from lo to hi.
25         ands  xzr, x2, 0x8000000000000000
26         lsl   x2, x2, #1
27         lsl   x3, x3, #1
28         cinc  x3, x3, ne
29         // shift multiplier right
30         lsr   x4, x4, #1
31         // if (y != 0) repeat loop
32         cmp   x4, xzr
33         bne   loop
34         // print results
35         mov   x4, x0
36         mov   x3, x1
37         adr   x2, y
38         ldr   x2, [x2]
39         adr   x1, x
40         ldr   x1, [x1]
41         adr   x0, msg
42         bl    printf
43         mov   w0, #0      // return 0
44         ldp   x29, x30, [sp], #16
45         ret
46         .size main,(. - main)
```

Listing 7.3 AArch64 assembly code for multiplication with a 64 bit result without using mul.

```
1          .section .rodata
2   x:     .8byte 0x57
```

```
 3   y:          .8byte  0x75
 4   msg:        .asciz  "%lx * %lx = %lx\n"
 5               .text
 6               .type   main %function
 7               .global main
 8   main:       stp     x29, x30, [sp, #-16]!
 9               mov     x0, #0          // x0 is result
10               ldr     x1, x           // x1 is multiplicand
11               ldr     x2, y           // x2 is multiplier
12   loop:       tst     x2, 0x1         // check LSB
13               lsr     x2, x2, #1      // shift multiplier right
14               add     x3, x0, x1
15               csel    x0, x3, x0, ne  // Add if LSB was 1
16               lsl     x1, x1, #1      // shift multiplicand left
17               cmp     x2, xzr
18               bne     loop            // if (multiplier == 0), we are done
19   // printf("%lx * %lx = %lx\n")
20               mov     x3, x0
21               ldr     x2, x
22               ldr     x1, y
23               adr     x0, msg
24               bl      printf
25               mov     w0, #0          // return 0
26               ldp     x29, x30, [sp], #16
27               ret
28               .size   main,(. - main)
```

7.2.3 Signed multiplication

Consider the two multiplication problems shown in Fig. 7.1 and Fig. 7.2. Note that the result of a multiply depends on whether the numbers are interpreted as unsigned numbers or signed numbers. For this reason, most computer CPU's have two different multiply operations for signed and unsigned numbers.

$$
\begin{array}{r}
-39 \\
\times \quad 73 \\
\hline
657 \\
219 \\
\hline
-2847
\end{array}
\qquad = \qquad
\begin{array}{r}
1\,1\,0\,1\,1\,0\,0\,1 \\
\times \quad 0\,1\,0\,0\,1\,0\,0\,1 \\
\hline
1\,1\,1\,1\,1\,1\,1\,1\,1\,1\,0\,1\,1\,0\,0\,1 \\
1\,1\,1\,1\,1\,1\,0\,1\,1\,0\,0\,1 \\
1\,1\,1\,1\,0\,1\,1\,0\,0\,1 \\
\hline
1\,1\,1\,1\,0\,1\,0\,0\,1\,1\,1\,0\,0\,0\,0\,1
\end{array}
$$

Figure 7.1: In signed 8-bit math, 11011001_2 is -39_{10}.

$$
\begin{array}{r}
2\ 1\ 7 \\
\times\quad 7\ 3 \\
\hline
5\ 1\ 1 \\
7\ 3 \\
1\ 4\ 6 \\
\hline
1\ 5\ 8\ 4\ 1
\end{array}
\qquad
\begin{array}{r}
1\ 1\ 0\ 1\ 1\ 0\ 0\ 1 \\
\times\quad 0\ 1\ 0\ 0\ 1\ 0\ 0\ 1 \\
\hline
0\ 0\ 0\ 0\ 0\ 0\ 0\ 0\ 1\ 1\ 0\ 1\ 1\ 0\ 0\ 1 \\
0\ 0\ 0\ 0\ 0\ 1\ 1\ 0\ 1\ 1\ 0\ 0\ 1 \\
0\ 0\ 1\ 1\ 0\ 1\ 1\ 0\ 0\ 1 \\
\hline
0\ 0\ 1\ 1\ 1\ 1\ 0\ 1\ 1\ 1\ 1\ 0\ 0\ 0\ 0\ 1
\end{array}
$$

Figure 7.2: In unsigned 8-bit math, 11011001_2 is 217_{10}.

If the CPU provides only an unsigned multiply, then a signed multiply can be accomplished by using the unsigned multiply operation along with a conditional complement. The following procedure can be used to implement signed multiplication.

1. if the multiplier is negative, take the two's complement,
2. if the multiplicand is negative, take the two's complement,
3. perform unsigned multiply, and
4. if the multiplier or multiplicand was negative (but not both), then take two's complement of the result.

Example 16. Signed multiplication using unsigned math.

$$73 \times -39 = 73 \times 39 \times -1$$

$$
\begin{array}{r}
7\ 3 \\
\times\quad 3\ 9 \\
\hline
6\ 5\ 7 \\
2\ 1\ 9 \\
\hline
2\ 8\ 4\ 7 \\
\times\quad -1 \\
\hline
-2\ 8\ 4\ 7
\end{array}
$$

$$
\begin{array}{r}
0\ 0\ 1\ 0\ 0\ 1\ 1\ 1 \\
\times\quad 0\ 1\ 0\ 0\ 1\ 0\ 0\ 1 \\
\hline
0\ 0\ 1\ 0\ 0\ 1\ 1\ 1 \\
0\ 0\ 1\ 0\ 0\ 1\ 1\ 1 \\
0\ 0\ 1\ 0\ 0\ 1\ 1\ 1 \\
\hline
0\ 0\ 0\ 0\ 1\ 0\ 1\ 1\ 0\ 0\ 0\ 1\ 1\ 1\ 1\ 1
\end{array}
$$

one's complement:

$$1\ 1\ 1\ 1\ 0\ 1\ 0\ 0\ 1\ 1\ 1\ 0\ 0\ 0\ 0\ 0$$

two's complement:

$$1\ 1\ 1\ 1\ 0\ 1\ 0\ 0\ 1\ 1\ 1\ 0\ 0\ 0\ 0\ 1$$

7.2.4 Multiplication of a variable by a constant

If x or y is a small constant, then we don't need the loop. We can directly translate the multiplication into a sequence of shift and add operations. This will result in much more efficient code than the general algorithm. For small constants, this method is often faster than using a hardware multiply instruction. If we inspect the constant multiplier, we can usually find a pattern to exploit that will save a few instructions. For example, suppose we want to multiply a variable x by 10_{10}. The multiplier $10_{10} = 1010_2$, so we only need to add x shifted left 1 bit to x shifted left 3 bits as shown below:

```
1    adr    x0, x
2    ldr    x0, [x0]          // load x (x0 = x)
3    lsl    x0, x0, #1        // shift x left (x0=2x)
4    add    x0, x0, x0, lsl #2   // x0 = 2x + 8x
```

Now suppose we want to multiply a number x by 11_{10}. The multiplier $11_{10} = 1011_2$, so we will add x to x shifted left one bit plus x shifted left 3 bits as in the following:

```
1    adr    x1, x
2    ldr    x1, [x1]          // load x (x1 <= x)
3    add    x0, x1, x1, lsl #1   // shift and add (x0=x+2x)
4    add    x0, x0, x1, lsl #3   // x0 = 3x + 8x
```

If we wish to multiply a number x by 1000_{10}, we note that $1000_{10} = 1111101000_2$. It looks like we need 1 shift plus 5 add/shift operations, or 6 add/shift operations. With a little thought, we can reduce the number of operations, as shown below:

```
1    adr    x1, x
2    ldr    x1, [x1]          // load x
3    add    x0, x1, x1, lsl #1   // shift and add: x0=3x
4    add    x0, x0, x0, lsl #2   // x0 = 3x + 3x*4 = 15x
5    add    x0, x1, x0, lsl #1   // x0 = 15x*2 + x = 31x
6    lsl    x0, x0, #5        // x0 = 31x * 32 = 992x
7    add    x0, x0, x1, lsl #3   // x0 = 992x + x*8 = 1000x
```

Applying the basic multiplication algorithm to multiply a number x by 255_{10} would result in seven add/shift operations, but we can do it with only three operations and use only one register, as shown below:

```
1    adr    x0, x
2    ldr    x0, [x0]          // load x
3    add    x0, x0, x0, lsl #1   // shift and add: x0=3x
4    add    x0, x0, x0, lsl #2   // x0 = 3x + x*3*4 = 15x
5    add    x0, x0, x0, lsl #4   // x0 = 15x + x*15*16 = 255x
```

Most modern systems have assembly language instructions for multiplication. However, on most processors, it is often more efficient to use the shift, add, and subtract operations when multiplying by a small constant. The AArch64 processors have a particularly powerful hardware multiplier. They can typically perform multiplication with a 64-bit result in one clock cycle. The long multiply instructions take between three and five clock cycles, depending on the size of the operands. Using the multiply instruction on an AArch64 processor to multiply by a constant requires loading the constant into a register before performing the multiply. Therefore, if the multiplication can be performed using two or fewer shift, add, and/or subtract instructions, then it will be equal to or better than using the multiply instruction.

7.2.5 Multiplying large numbers

Consider the method used for multiplying two digit numbers in base ten, using only the one-digit multiplication tables. Fig. 7.3 shows how a two digit number $a = a_1 \times 10^1 + a_0 \times 10^0$ is multiplied by another two digit number $b = b_1 \times 10^1 + b_0 \times 10^0$ to produce a four digit result using basic multiplication operations which only take one digit from a and one digit from b at each step.

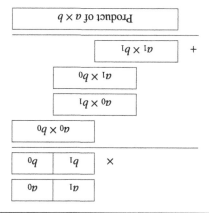

Figure 7.3: Multiplication of large numbers.

This technique can be used for any base and for any number of digits. Recall that one hexadecimal digit is equivalent to exactly four binary digits. If a and b are both eight-bit numbers, then they are also two-digit hexadecimal numbers. In other words eight bit numbers can be divided into groups of four bits, each representing one digit in base sixteen. Given a multiply operation that is capable of producing an eight-bit result from two four-bit inputs, the technique shown above can then be used to multiply two eight-bit numbers using only four-bit multiplication operations.

Carrying this one step further, suppose we are given two sixteen-bit numbers, but our computer only supports multiplying eight bits at a time and producing a sixteen bit result. We can consider each sixteen bit number to be a two digit number in base 256, and use the

$$
\begin{array}{r}
1110110101 \\
\hline
1101\,)\,11000000111001 \\
1101000000000 \\
\hline
1011000111001 \\
110100000000 \\
\hline
100100111001 \\
11010000000 \\
\hline
1010111001 \\
110100000 \\
\hline
100011001 \\
11010000 \\
\hline
1001001 \\
110100 \\
\hline
10101 \\
1101 \\
\hline
1000
\end{array}
$$

$$
\begin{array}{r}
949 \\
\hline
13\,)\,12345 \\
11700 \\
\hline
645 \\
520 \\
\hline
125 \\
117 \\
\hline
8
\end{array}
$$

Figure 7.4: Longhand division in decimal and binary.

above technique to perform four eight bit multiplies with 16-bit results, then shift and add the 16-bit results to obtain the final 32-bit result. This technique can be extended to perform multiplication with any number of bits, even with limited hardware multiplication instructions.

7.3 Binary division

Binary division can be implemented as a sequence of shift and subtract operations. When performing binary division by hand, it is convenient to perform the operation in a manner very similar to the way that decimal division is performed. As shown in Fig. 7.4, the operation is identical, but takes more steps in binary.

7.3.1 Division by a power of two

If the divisor is a power of two, then division can be accomplished with a shift to the right. Using the same approach as was used in Section 7.2.1, it can be shown that a shift right by n bits is equivalent to division by 2^n. However, care must be taken to ensure that an arithmetic shift is used if the dividend is a signed two's complement number, and a logical shift is used if the dividend is unsigned.

7.3.2 Division by a variable

The algorithm for dividing binary numbers is somewhat more complicated than the algorithm for multiplication. The algorithm consists of two main phases: The first phase is to shift the divisor left until it is greater than the dividend, and count the number of shifts. The second phase is to repeatedly shift the divisor back to the right and subtract whenever possible. Fig. 7.5 shows the algorithm in more detail. Before we introduce the AArch64 code, we will

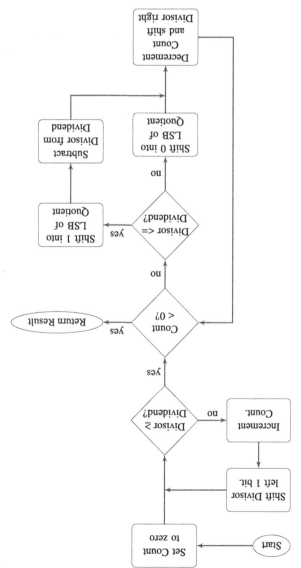

Figure 7.5: Flowchart for binary division.

take some time to step through the algorithm using an example. Let us begin by dividing 94 by 7. The result is shown below:

$$94 \div 7 =$$

$$
\begin{array}{r}
1101 \\
111 \overline{)\,1011110} \\
111000 \\
\hline
100110 \\
11100 \\
\hline
1010 \\
111 \\
\hline
11
\end{array}
$$

To implement the algorithm, we need three registers, one for the dividend, one for the divisor, and one for a counter. The dividend and divisor are loaded into their registers and the counter is initialized to zero as shown below:

Dividend:	0	1	0	1	1	1	1	0
Divisor:	0	0	0	0	0	1	1	1
Counter:	0	0	0	0	0	0	0	0

Next, the divisor is shifted left and the counter incremented repeatedly until the divisor is greater than the dividend. This is shown in the following sequence:

Dividend:	0	1	0	1	1	1	1	0
Divisor:	0	0	0	0	1	1	1	0
Counter:	0	0	0	0	0	0	0	1

Dividend:	0	1	0	1	1	1	1	0
Divisor:	0	0	0	1	1	1	0	0
Counter:	0	0	0	0	0	0	1	0

Dividend:	0	1	0	1	1	1	1	0
Divisor:	0	0	1	1	1	0	0	0
Counter:	0	0	0	0	0	0	1	1

Dividend:	0	1	0	1	1	1	1	0
Divisor:	0	1	1	1	0	0	0	0
Counter:	0	0	0	0	0	1	0	0

Next, we allocate a register for the quotient and initialize it to zero. Then, according to the algorithm, we repeatedly subtract if possible, shift to the right, and decrement the counter. This sequence continues until the counter becomes negative. For our example this results in the following sequence:

Divisor > Dividend: No subtract, shift 0 into Quotient, decrement Counter, shift Divisor right

Quotient:	0	0	0	0	0	0	0	0
Dividend:	0	1	0	1	1	1	1	0
Divisor:	0	1	1	1	0	0	0	0
Counter:	0	0	0	0	0	1	0	0

Divisor <= Dividend: Subtract, shift 1 into Quotient, decrement Counter, shift Divisor right

Quotient:	0	0	0	0	0	0	0	0
Dividend:	0	1	0	1	1	1	1	0
Divisor:	0	0	1	1	1	0	0	0
Counter:	0	0	0	0	0	0	1	1

Divisor <= Dividend: Subtract, shift 1 into Quotient, decrement Counter, shift Divisor right

Quotient:	0	0	0	0	0	0	0	1
Dividend:	0	0	1	0	0	1	1	0
Divisor:	0	0	0	1	1	1	0	0
Counter:	0	0	0	0	0	0	1	0

Divisor > Dividend: No subtract, shift 0 into Quotient, decrement Counter, shift Divisor right

Quotient:	0	0	0	0	0	0	1	1
Dividend:	0	0	0	0	1	0	1	0
Divisor:	0	0	0	0	1	1	1	0
Counter:	0	0	0	0	0	0	0	1

Divisor <= Dividend: Subtract, shift 1 into Quotient, decrement Counter, shift Divisor right

Quotient:	0	0	0	0	0	1	1	0
Dividend:	0	0	0	0	1	0	1	0
Divisor:	0	0	0	0	0	1	1	1
Counter:	0	0	0	0	0	0	0	0

Counter < 0: We are finished

Quotient:	0	0	0	0	1	1	0	1
Dividend:	0	0	0	0	0	0	1	1
Divisor:	0	0	0	0	0	0	1	1
Counter:	1	1	1	1	1	1	1	1

When the algorithm terminates, the quotient register contains the result of the division, and the modulus (remainder) is in the dividend register. Thus, one algorithm is used to compute both the quotient and the modulus at the same time. There are variations on this algorithm. For example, on variation is to shift a single bit left in a register, rather than incrementing a count. This variation has the same two phases as the previous algorithm, but counts in powers of two rather than by ones. The following sequence shows what occurs after each iteration of the first loop in the algorithm.

Dividend:	0	1	0	1	1	1	1	0
Divisor:	0	0	0	0	0	1	1	1
Power:	0	0	0	0	0	0	0	1

Dividend:	0	1	0	1	1	1	1	0
Divisor:	0	0	0	0	1	1	1	0
Power:	0	0	0	0	0	0	1	0

Dividend:	0	1	0	1	1	1	1	0
Divisor:	0	0	0	1	1	1	0	0
Power:	0	0	0	0	0	1	0	0

Dividend:	0	1	0	1	1	1	1	0
Divisor:	0	0	1	1	1	0	0	0
Power:	0	0	0	0	1	0	0	0

Dividend:	0	1	0	1	1	1	1	0
Divisor:	0	1	1	1	0	0	0	0
Power:	0	0	0	1	0	0	0	0

The divisor is greater than the dividend, so the algorithm proceeds to the second phase. In this phase, if the divisor is less than or equal to the dividend, then the power register is added to the quotient and the divisor is subtracted from the dividend. Then, the power and Divisor registers are shifted to the right. The process is repeated until the power register is zero. The following sequence shows what will be in the registers at the end of each iteration of the second loop.

Quotient:	0	0	0	0	0	0	0	0
Dividend:	0	1	0	1	1	1	1	0
Divisor:	0	1	1	1	0	0	0	0
Power:	0	0	0	1	0	0	0	0

Divisor > Dividend: shift Power right, shift Divisor right

Quotient:	0	0	0	0	0	0	0	0
Dividend:	0	1	0	1	1	1	1	0
Divisor:	0	0	1	1	1	0	0	0
Power:	0	0	0	0	1	0	0	0

Divisor ≤ Dividend: Dividend -= Divisor, Quotient += Power, shift Power right, shift Divisor right

Quotient:	0	0	0	0	1	0	0	0
Dividend:	0	0	1	0	0	1	1	0
Divisor:	0	0	0	1	1	1	0	0
Power:	0	0	0	0	0	1	0	0

Divisor ≤ Dividend: Dividend -= Divisor, Quotient += Power, shift Power right, shift Divisor right

Quotient:	0	0	0	0	1	1	0	0
Dividend:	0	0	0	0	1	0	1	0
Divisor:	0	0	0	0	1	1	1	0
Power:	0	0	0	0	0	0	1	0

Divisor > Dividend: shift Power right, shift Divisor right

Quotient:	0	0	0	0	1	1	0	0
Dividend:	0	0	0	0	1	0	1	0
Divisor:	0	0	0	0	0	1	1	1
Power:	0	0	0	0	0	0	0	1

Divisor ≤ Dividend: Dividend -= Divisor, Quotient += Power, shift Power right, shift Divisor right

Quotient:	0	0	0	0	1	1	0	1
Dividend:	0	0	0	0	0	0	1	1
Divisor:	0	0	0	0	0	0	1	1
Power:	0	0	0	0	0	0	0	0

Power = 0: We are finished

As with the previous version, when the algorithm terminates, the quotient register contains the result of the division, and the modulus (remainder) is in the dividend register. Listing 7.4 shows the AArch64 assembly code to implement this version of the division algorithm for 64-bit numbers, and the counting method for 128-bit numbers.

Listing 7.4 AArch64 assembly implementation of signed and unsigned 64-bit and 128-bit division functions

```
1   ### --------------------------------------------------------------
2   ### divide.S
3   ### Author: Larry Pyeatt with William Ughetta
4   ### Date: 10/16/2014. Revised: 07/31/2018.
5   ###
6   ### Division functions in AArch64 assembly language
7   ### --------------------------------------------------------------
8
9           .text
10          .align  2
11          .global udiv64
12          .type   udiv64, %function
13   /* udiv64 divides the dividend in x0 by the divisor in x1. It
14    * returns the quotient in x0 and the remainder in x1. */
15   udiv64:
16          cbnz    x1, endif1      // if (divisor == 0)
17          mov     x0, #0          // return 0
18          mov     x1, #0
19          ret
20   endif1:
21          clz     x2, x1          // x2 = count
22          lsl     x1, x1, x2      // divisor <<= count
23          mov     x3, #0          // x3 = quotient
24          add     x2, x2, #1      // x2 = count+1
25   divloop:
26          lsl     x3, x3, #1      // Shift 0 into quotient LSB
27          cmp     x0, x1
28          blo     endif2          // if (dividend >= divisor)
29          orr     x3, x3, #1      // Set LSB of quotient
30          sub     x0, x0, x1      // dividend -= divisor
31   endif2:
32          sub     x2, x2, #1      // Decrement count
33          lsr     x1, x1, #1      // Shift divisor right
34          cbnz    x2, divloop     // while (count+1 != 0)
35          mov     x1, x0          // remainder is the dividend
36          mov     x0, x3          // return quotient
37          ret
38          .size   udiv64, (. - udiv64)
39   ### --------------------------------------------------------------
40          .global sdiv64
41          .type   sdiv64, %function
42   /* sdiv64 divides a signed 64-bit dividend in x0 with a signed
43    * 64-bit divisor in x1. Returns the quotient in x0 and the
44    * remainder in x1. Uses udiv64 to do the real work. */
45   sdiv64:
46          stp     x29, x30, [sp, #-32]!
```

```
         str     x19, [sp, #16]           // Push FP,LR,& Callee-saved

         cmp     x0, #0                   // If dividend is negative
         cneg    x0, x0, lt               // Complement
         cset    x19, lt                  // Set sign bit for result

         cmp     x1, #0                   // If divisor is negative
         cneg    x1, x1, lt               // Complement
         eor     x9, x19, #1
         csel    x19, x9, x19, lt         // Complement sign bit

         bl  udiv64

         cmp     x19, #0                  // Complement remainder if
         cneg    x0, x0, ne               // sign bit is set
         cneg    x1, x1, ne

         ldr     x19, [sp, #16]
         ldp     x29, x30, [sp], #32      // Pop FP,LR,& Callee-saved
         ret
         .size   sdiv64,(. - sdiv64)
### ------------------------------------------------------------------
         .global udiv128
         .type   udiv128, %function
/* udiv128 divides the dividend in x0:x1 by the divisor in
 * x2:x3. It returns the 128 bit result in x0:x1 and the
 * remainder in x2:x3. x2 is low order bits (i.e. lo:hi). */
udiv128:
         orr     x9, x2, x3
         cbnz    x9, endif3               // if (divisor == 0)
         mov     x0, #0                   //    return 0
         mov     x1, #0
         ret
endif3:
         mov     x4, #0                   // x4:x5 = quotient
         mov     x5, #0

         ## Count leading zeroes
         clz     x6, x3                   // count high order bits
         cbnz    x6, endif4               // if (divisorHigh == 0)
         clz     x6, x4                   //    count low order bits
         add     x6, x6, #64
endif4:
         neg     x7, x6                   // x7 = -count
         add     x8, x6, #128             // x8 = 128 - count
         add     x9, x7, #64              // x9 = 64 - count
         cmp     x9, #0
         csel    x7, x8, x9, lt           // x7 = (128 - count) % 64
```

```
 96            ## shift divisor x2:x3 << count
 97            lsl     x3, x3, x6              // divisorHigh <<= count
 98            lsr     x9, x2, x7              // tmp=Low >> (128-count)%64
 99            orr     x3, x3, x9              // divisorHigh |= tmp
100            lsl     x2, x2, x6              // divisorLow <<= count
101
102            add     x6, x6, #1              // x6 = count+1
103    divloop128:
104            lsl     x5, x5, #1              // Shift 0 into quotient LSB
105            lsr     x9, x4, #63             // tmp = Low >> 63
106            orr     x5, x5, x9
107            lsl     x4, x4, #1
108
109            cmp     x1, x3
110            bne     endif5
111            cmp     x0, x2
112    endif5:
113            blo     endif6                  // if (dividend >= divisor)
114            orr     x4, x4, #1              // Set LSB of quotient
115            sub     x0, x0, x2              // dividend -= divisor
116            sub     x1, x1, x3
117    endif6:
118            sub     x6, x6, #1              // Decrement count
119            lsr     x2, x2, #1              // Shift divisor right
120            lsl     x9, x3, #63
121            orr     x2, x2, x9
122            lsr     x3, x3, #1
123            cbnz    x6, divloop128          // while (count+1 != 0)
124            mov     x2, x0                  // remainder is the dividend
125            mov     x3, x1
126            mov     x0, x4                  // return quotient
127            mov     x1, x5
128            ret
129            .size   udiv128,(. - udiv128)
130    ### ----------------------------------------------------------------
131
132    /* sdiv128 divides the signed dividend in x0:x1 by the signed
133     * divisor in x2:x3. It returns the 128 bit result in x0:x1
134     * and the remainder in x2:x3. Uses udiv128. */
135            .global sdiv128
136            .type   sdiv128, %function
137    sdiv128:
138            stp     x29, x30, [sp, #-32]!
139            str     x19, [sp, #16]
140            cmp     x1, #0                  // If dividendHigh < 0
141            bge     endif7
142            mvn     x0, x0                  // Bitwise NOT
143            mvn     x1, x1
144            adds    x0, x0, #1              // Add 1 for 2's complement
```

```
145    adc    x1, x1, xzr
146    eor    x19, x19, #1     // Keep track of sign
147 endif7:
148    cmp    x3, #0           // If divisorHigh < 0
149    bge    endif8
150    mvn    x2, x2           // Bitwise NOT
151    mvn    x3, x3
152    adds   x2, x2, #1       // Add 1 for 2's complement
153    adc    x3, x3, xzr
154    eor    x19, x19, #1     // Keep track of sign
155 endif8:
156    bl     udiv128
157    // If sign bit is set, then complement the result
158    cmp    x19, #0
159    bge    endif8
160    mvn    x0, x0           // Bitwise NOT
161    mvn    x1, x1
162    adds   x0, x0, #1       // Add 1 for 2's complement
163    adc    x1, x1, xzr
164    // return
165    ldr    x19, [sp, #16]
166    ldp    x29, x30, [sp], #32
167    ret
168    .size  sdiv128, (. - sdiv128)
```

7.3.3 Division by a constant

In general, division is slow. AArch64 processors provide a hardware divide instruction which requires multiple clock cycles to produce a result, depending on the size of the operands. Some older ARMv7 processors must perform division using software, as previously described. In either case, division is by far the slowest of the basic mathematical operations. However, division by a constant c can be converted to a multiply by the reciprocal of c. It is obviously much more efficient to use a multiply instead of a divide wherever possible. Efficient division of a variable by a constant is achieved by applying the following equality:

$$x \div c = x \times \frac{1}{c}. \tag{7.1}$$

The only difficulty is that we have to do it in binary, using only integers. If we modify the right-hand side by multiplying and dividing by some power of two (2^n), we can rewrite Eq. (7.1) as follows:

$$x \div c = x \times \frac{2^n}{c} \times 2^{-n}. \tag{7.2}$$

Recall that, in binary, multiplying by 2^n is the same as shifting left by n bits, while multiplying by 2^{-n} is done by shifting right by n bits. Therefore, Eq. (7.1) is just Eq. (7.1) with two shift operations added. The two shift operations cancel each other out. Now, let

$$m = \frac{2^n}{c}.$$ (7.3)

We can rewrite Eq. (7.2) as:

$$x \div c = x \times m \times 2^{-n}.$$ (7.4)

We now have a method for dividing by a constant c which involves multiplying by a different constant, m, and shifting the result. In order to achieve the best precision, we want to choose n such that m is as large as possible with the number of bits we have available.

Suppose we want efficient code to calculate $x \div 23$ using 8-bit *signed* integer multiplication. Our first task is to find $m = \frac{2^n}{c}$ such that $01111111_2 \geq m \geq 01000000_2$. In other words, we want to find the value of n where the most significant bit of m is zero, and the next most significant bit of m is one. If we choose $n = 11$, then

$$m = \frac{2^{11}}{23} \approx 89.043478260 9.$$

Rounding to the nearest integer gives $m = 89$. In 8 bits, m is 01011001_2 or 59_{16}. We now have values for m and n, and therefore we can apply Eq. (7.4) to divide any number x by 23. The procedure is simple: calculate $y = x \times m$, then shift y right by 11 bits.

However, there are two more considerations. First, when the divisor is positive, the result for some values of x may be incorrect due to rounding error. It is usually sufficient to increment the reciprocal value by one in order to avoid these errors. In the previous example, the number would be changed from 59_{16} to $5A_{16}$. When implementing this technique for finding the reciprocal, the programmer should always verify that the results are correct for all input values. The second consideration is when the dividend is negative. In that case it is necessary to subtract one from the final result.

For example, to calculate $101_{10} \div 23_{10}$ in binary, with eight bits of precision, we first perform the multiplication as follows:

```
        0 1 1 0 0 1 0 1
      × 0 1 0 1 1 0 1 0
      ─────────────────
        0 1 1 0 0 1 0 1
    0 1 1 0 0 1 0 1
  0 1 1 0 0 1 0 1
0 1 1 0 0 1 0 1
─────────────────────────
1 0 0 0 1 1 0 0 0 1 1 1 1 0
```

Then shift the result right by 11 bits. 10001100011101_2 shifted right 11_{10} bits is: $100_2 = 4_{10}$. If the modulus is required, it can be calculated as $101 \mod 23 = 101 - (4 \times 23) = 9$, which once again requires multiplication by a constant.

In the previous example the shift amount of 11 bits provided the best precision possible. But how was that number chosen? The shift amount, n, can be directly computed as

$$n = p + \lceil \log_2 c \rceil - 1, \tag{7.5}$$

where p is the desired number of bits of precision. The value of m can then be computed as

$$m = \begin{cases} \dfrac{2^n}{c} + 1, & c > 0, \\[2mm] \dfrac{2^n}{c} & \text{otherwise.} \end{cases} \tag{7.6}$$

For example, to divide by the constant 33, with 16 bits of precision, we compute n as

$$n = 16 + \lceil \log_2 33 \rceil - 1 = 16 + \lceil 5.044394 \rceil - 1 = 16 + 5 - 1 = 20.$$

and then we compute m as

$$m = \frac{2^{20}}{33} + 1 = 31776.030303 \approx 31776 = 7C20_{16}.$$

Therefore, multiplying a 16 bit number by $7C20_{16}$ and then shifting right 20 bits is equivalent to dividing by 33.

Example 17. Division by constant 193.

To divide by a constant 193, with 32 bits of precision, the multiplier is computed using Eq. (7.5) and Eq. (7.6) with $p = 32$ as follows:

$$m = \frac{2^{32+7-1}}{193} + 1 = \frac{2^{38}}{193} + 1 = 1424237860.81 \approx 1424237860 = 54EA2524_{16}.$$

The shift amount, n, is 38 bits.

Listing 7.5 AArch64 assembly code for division by constant 193.

```
// The following code will calculate w2/193
// It will leave the quotient in x0 and
// the remainder in x1
ldr    x3, =0x54E42524   // x3 = (1/193 << 38)
smull  x4, w2, w3        // x * m
asr    x0, x4, #38
sub    x0, x0, x4, asr #63   // subtract sign of dividend
// calculate remainder in x1
mov    x1, #193          // x1 = divisor
mul    x1, x1, x0        // multiply divisor by product
sub    x1, x2, x1        // subtract that from numerator
```

Example 17 shows how to calculate m and n for division by 193. On the AArch64 processor, division by a constant can be performed very efficiently. Listing 7.5 shows how division by 193 can be implemented using only a few lines of code. In the listing, the numbers are 32 bits in length, so the constant m is much larger than in the example that was multiplied by hand, but otherwise the method is the same.

If we wish to divide by 23 using 32 bits of precision, we compute the multiplier as

$$m = \frac{2^{32+4-1}}{23} + 1 = \frac{2^{35}}{23} + 1 = 1493901669.17 \approx 1493901669 = 590B2165_{16}.$$

That is $01011001000010110010000101100101_2$. Note that there are only 12 non-zero bits, and the pattern 101 1001 appears three times in the 32-bit multiplier. The multiply can be implemented as $2^{24}(2^6x+2^4x+2^3x+2^0x)+2^{13}(2^6x+2^4x+2^3x+2^0x)+2^2(2^6x+2^4x+2^3x+2^0x)$. So the following code sequence can be used on processors that do not have the multiply instruction:

Listing 7.6 AArch64 assembly code for division of a variable by a constant without using a multiply instruction.

```
// The following code will calculate w2/23
// It will leave the quotient in x0 and does
// not calculate a remainder.
sxtw   x2, w2            // Sign extend w2
mov    x0, x2            // Copy into x0
// calculate 2^6x+2^4x+2^3x+2^0x
add    x3, x2, x0, lsl #3
add    x3, x3, x0, lsl #4
add    x3, x3, x0, lsl #6
// now perform three 64-bit shift-add operations
lsl    x3, x3, #2
```

```
12   add   x0, x0, x3
13   lsl   x3, x3, #11    // shift 11 + 2 = 13
14   add   x0, x0, x3
15   lsl   x3, x3, #11    // shift 11 + 13 = 24
16   add   x0, x0, x3
17                .
18   asr   x1, x0, #35
19   sub   x0, x1, x0, asr #63    // subtract sign of dividend
```

7.3.4 Dividing large numbers

Section 7.2.5 showed how large numbers can be multiplied by breaking them into smaller numbers and using a series of multiplication operations. There is no similar method for synthesizing a large division operation with an arbitrary number of digits in the dividend and divisor. However, there is a method for dividing a large dividend by a divisor given that the division operation can operate on numbers with at least the same number of digits as in the divisor.

Suppose we wish to perform division of an arbitrarily large dividend by a one digit divisor using a basic division operation that can divide a two digit dividend by a one digit divisor. The operation can be performed in multiple steps as follows:

1. Divide the most significant digit of the dividend by the divisor. The result is the most significant digit of the quotient.
2. Prepend the remainder from the previous division step to the next digit of the dividend, forming a two-digit number, and divide that by the divisor. This produces the next digit of the result.
3. Repeat from step 2 until all digits of the dividend have been processed.
4. Take the final remainder as the modulus.

The following example shows how to divide 6189 by 7 using only 2-digits at a time:

$$
\begin{array}{cccc}
0 & 8 & 8 & 4 \\
7\overline{)6} & 7\overline{)61} & 7\overline{)58} & 7\overline{)29} \\
 & \underline{56} & \underline{56} & \underline{28} \\
 & 5 & 2 & 1
\end{array}
$$

$$x = 6189 \div 7 = 0884 \text{ remainder } 1$$

This method can be applied in any base and with any number of digits. The only restriction is that the basic division operation must be capable of dividing a $2n$ digit dividend by an n

digit divisor and producing a $2n$ digit quotient and an n digit remainder. for example, the AArch64 processor is capable of dividing a 64-bit dividend by a 64-bit divisor, producing a 64-bit quotient. The remainder can be calculated by multiplying the quotient by the divisor and subtracting the product from the dividend. Using this division operation it is possible to divide an arbitrarily large dividend by a 32-bit divisor.

We have seen that, given a divide operation capable of dividing an n digit dividend by an n digit divisor, it is possible to divide a dividend with any number of digits by a divisor with $\frac{n}{2}$ digits. Unfortunately, there is no similar method to deal with an arbitrarily large divisor, or to divide an arbitrarily large dividend by a divisor with more than $\frac{n}{2}$ digits. In those cases the division must be performed using a general division algorithm as shown previously.

7.4 Big integer ADT

For some programming tasks, it may be helpful to deal with arbitrarily large integers. For example, the factorial function and Ackerman's function grow very quickly and will overflow a 32-bit integer for even small input values. Even a 64 bit integer is insufficient for holding the results of these functions with even relatively small inputs. The GNU C compiler is able to do some big number arithmetic using __uint128_t, but even a 128 bit integer is insufficient for these functions.

In this section, we will examine an abstract data type which provides basic operations for arbitrarily large integer values. Listing 7.7 shows the C header for this ADT, and Listing 7.8 shows the C implementation. Listing 7.9 shows a small program that runs *regression tests* on the big integer ADT. A regression test is a (relatively) small test that is used to verify that changes to the code have not introduced bugs. Regression tests are very valuable tools in software engineering, and are typically used in larger companies by code testing specialists.

Listing 7.7 Header file for a big integer abstract data type.

```
1  #ifndef BIGINT_H
2  #define BIGINT_H
3  #include <stdio.h>
4
5  struct bigint_struct;
6
7  /* Define bigint to be a pointer to a bigint_struct */
8  typedef struct bigint_struct* bigint;
9
10 /* There are three ways to create a bigint */
11 bigint bigint_from_str(char *s);
12 bigint bigint_from_int(int i);
13 bigint bigint_copy(bigint source);
```

```
/* Bigints can be converted to integers.  If it won't fit in an
   integer, the program exits */
int bigint_to_int(bigint b);

/* To print a bigint, you must convert it to a string */
char *bigint_to_str(bigint b);

/* Memory management */
bigint bigint_alloc(int chunks);
void bigint_free(bigint b);

/* There are seven arithmetic operations */
bigint bigint_add(bigint l, bigint r);
bigint bigint_sub(bigint l, bigint r);
bigint bigint_mul(bigint l, bigint r);
bigint bigint_div(bigint l, bigint r);
bigint bigint_negate(bigint b);
bigint bigint_abs(bigint b);
bigint bigint_sqrt(bigint b);

/* Two shift operations (all bigints are signed, so there is no
   such thing as an unsigned shift right) */
bigint bigint_shift_left(bigint l, int shamt);
bigint bigint_shift_right(bigint l, int shamt);

/* There are seven comparison operations.  They return 1 for true,
   or 0 for false. */
int bigint_is_zero(bigint b);
int bigint_le(bigint l, bigint r);
int bigint_lt(bigint l, bigint r);
int bigint_ge(bigint l, bigint r);
int bigint_gt(bigint l, bigint r);
int bigint_eq(bigint l, bigint r);
int bigint_ne(bigint l, bigint r);

/* Low-level comparison. bigint_cmp compares two bigints
   returns -1 if l<r
   returns 0 if l==r
   returns 1 if l>r */
int bigint_cmp(bigint l, bigint r);

/* Functions for binary input/output */
void bigint_write_binary(FILE *f, bigint x);
bigint bigint_read_binary(FILE *f);

#endif
```

Listing 7.8 C source code file for a big integer abstract data type.

```
1   #include <bigint.h>
2   #include <string.h>
3   #include <math.h>
4   #include <stdlib.h>
5   #include <stdio.h>
6   #include <ctype.h>
7   #include <stdint.h>
8
9   /* A big integer is an array of "chunks" of bits.  The following
10  section defines the chunk, and defines a bigchunk to have twice as
11  many bits as a chunk. It also defines a mask that can be used to
12  select bits out of a bigchunk.  When compiling this code, you can
13  define the chunk size by defining EIGHT_BIT, SIXTEEN_BIT, or
14  SIXTYFOUR_BIT.  If none of those are defined, then the chunk size
15  is THIRTYTWO_BIT.
16  */
17  #ifdef EIGHT_BIT
18  typedef uint8_t chunk;
19  typedef int8_t schunk;
20  typedef uint16_t bigchunk;
21  #define CHUNKMASK 0xFFU
22  #else
23  #ifdef SIXTEEN_BIT
24  typedef uint16_t chunk;
25  typedef int16_t schunk;
26  typedef uint32_t bigchunk;
27  #define CHUNKMASK 0xFFFFU
28  #else
29  #ifdef SIXTYFOUR_BIT
30  typedef uint64_t chunk;
31  typedef int64_t schunk;
32  typedef __uint128_t bigchunk;
33  #define CHUNKMASK 0xFFFFFFFFFFFFFFFFULL
34  #else
35  typedef uint32_t chunk;
36  typedef int32_t schunk;
37  typedef uint64_t bigchunk;
38  #define CHUNKMASK 0xFFFFFFFFU
39  #endif
40  #endif
41  #endif
42
43  /* Macros */
44  #define BITSPERCHUNK ((sizeof(chunk)<<3))
45  #define MAX(a,b) ((a>b)?b:a)
46
47  /* A bigint is an array of chunks of bits */
```

```
struct bigint_struct {
  chunk *blks;         /* pointer to array of bit chunks */
  int size;            /* number of chunks in the array  */
};

/* Private function prototypes */
static bigint bigint_adc(bigint l, bigint r, chunk carry);
static bigint bigint_shift_left_chunk(bigint l, int chunks);
static bigint bigint_shift_right_chunk(bigint l, int chunks);
static bigint bigint_trim(bigint b);
static bigint bigint_mul_uint(bigint l, chunk r);
static bigint bigint_extend(bigint b, int nchunks);
static unsigned bigint_smallmod(bigint b, chunk num);

/**********************************************************************/
/* Initialization, conversion, and copy functions            */
/**********************************************************************/

/* convert string to bigint */
bigint bigint_from_str(char *s) {
  bigint d;
  bigint power;
  bigint ten;
  bigint tmp;
  bigint currprod;
  int i, negative = 0;
  d = bigint_from_int(0);
  ten =   bigint_from_int(10);
  power = bigint_from_int(1);
  if (*s == '-') {
    negative = 1;
    s++;
  }
  for (i = strlen(s)-1; i >= 0; i--) {
    if (!isdigit(s[i])) {
      fprintf(stderr,"Cannot convert string to bigint\n");
      exit(1);
    }
    tmp = bigint_from_int(s[i]-'0');
    currprod = bigint_mul(tmp,power);
    bigint_free(tmp);
    tmp = bigint_adc(currprod,d,0);
    bigint_free(d);
    d = tmp;
    bigint_free(currprod);
    if (i > 0) {
      tmp = bigint_mul(power,ten);
      bigint_free(power);
      power = tmp;
```

```
 97        }
 98      }
 99      if (negative) {
100        tmp = bigint_negate(d);
101        bigint_free(d);
102        d = tmp;
103      }
104      bigint_free(ten);
105      bigint_free(power);
106      return d;
107    }
108
109    /* convert integer to bigint */
110    bigint bigint_from_int(int val) {
111      bigint d, tmp;
112      int i;
113      int nchunks = sizeof(int) / sizeof(chunk);
114      if (nchunks < 1)
115        nchunks++;
116      d = bigint_alloc(nchunks);
117      for (i = 0; i < d->size; i++)
118        d->blks[i] = (val >> (i*BITSPERCHUNK)) & CHUNKMASK;
119      tmp = bigint_trim(d);
120      bigint_free(d);
121      return tmp;
122    }
123
124    /* duplicate a bigint */
125    bigint bigint_copy(bigint source) {
126      bigint r;
127      r = bigint_alloc(source->size);
128      memcpy(r->blks, source->blks, r->size * sizeof(chunk));
129      return r;
130    }
131
132    /* convert a bigint into and integer, if possible */
133    int bigint_to_int(bigint b) {
134      int i, negative=0, result=0;
135      bigint tmp1, tmp2;
136      tmp1 = bigint_trim(b); /* make a trimmed copy */
137      if (tmp1->size * sizeof(chunk) > sizeof(int)) {
138        fprintf(stderr, "Cannot convert bigint to int\n%ld bytes\n",
139          (long)tmp1->size * sizeof(chunk));
140        exit(1);
141      }
142      /* check sign and negate if necessary */
143      if (tmp1->blks[tmp1->size-1] & ((bigchunk)1<<(BITSPERCHUNK-1))) {
144        negative = 1;
145        tmp2 = bigint_negate(tmp1);
```

```
146        bigint_free(tmp1);
147        tmp1 = tmp2;
148      }
149      for (i = tmp1->size-1; i >= 0; i--)
150        result |= (tmp1->blks[i]<<(i*BITSPERCHUNK));
151      bigint_free(tmp1);
152      if (negative)
153        result = -result;
154      return result;
155    }
156
157    /* convert a bigint to a string */
158    char *bigint_to_str(bigint b) {
159      int negative = 0;
160      unsigned remainder;
161      char *s, *r;
162      bigint tmp, tmp2;
163      /* rough estimate of the number of characters needed */
164      int chars = log10(pow(2.0, (b->size * BITSPERCHUNK)))+3;
165      int i = chars-1;
166      if ((s = (char*)malloc(1 + chars * sizeof(char))) == NULL) {
167        perror("bigint_str");
168        exit(1);
169      }
170      s[i] = 0;   /* set last character to ASCII null */
171      tmp = bigint_copy(b);
172      if (b->blks[tmp->size-1] & ((bigchunk)1<< (BITSPERCHUNK-1))) {
173        negative = 1;
174        tmp2 = bigint_negate(tmp);
175        bigint_free(tmp);
176        tmp = tmp2;
177      }
178      if (bigint_is_zero(tmp)) {
179        s[--i] = '0';
180      } else {
181        do {
182          remainder = bigint_smallmod(tmp, 10);
183          s[--i] = remainder + '0';
184        } while(!bigint_is_zero(tmp));
185        if (negative)
186          s[--i] = '-';
187      }
188      r = strdup(s + i);
189      bigint_free(tmp);
190      free(s);
191      return r;
192    }
193
194    /* destroy a bigint */
```

```c
195  void bigint_free(bigint b) {
196      if (b != NULL) {
197          if (b->blks != NULL)
198              free(b->blks);
199          free(b);
200      }
201  }
202
203  /****************************************************/
204  /*  Mathematical operations                         */
205  /****************************************************/
206
207  /* this is the internal add function.  It includes a */
208  /* carry.  Several other functions use it.           */
209  static bigint bigint_adc(bigint l, bigint r, chunk carry) {
210      bigint sum, tmpl, tmpr;
211      int i, nchunks;
212      bigchunk tmpsum;
213      /* allocate one extra chunk to make sure overflow */
214      cannot occur */
215      nchunks = MAX(l->size, r->size) + 1;
216      /* make sure both operands are the same size */
217      tmpl = bigint_extend(l, nchunks);
218      tmpr = bigint_extend(r, nchunks);
219      /* allocate space for the result */
220      sum = bigint_alloc(nchunks);
221      /* perform the addition */
222      for (i = 0; i < nchunks; i++) {
223          /* add the current block of bits */
224          tmpsum = (bigchunk)tmpl->blks[i] + (bigchunk)tmpr->blks[i] + (bigchunk)carry;
225          sum->blks[i] = tmpsum & CHUNKMASK;
226          /* calculate the carry bit for the next block */
227          carry = (tmpsum >> BITSPERCHUNK) & CHUNKMASK;
228      }
229      bigint_free(tmpl);
230      bigint_free(tmpr);
231      tmpl = bigint_trim(sum);
232      bigint_free(sum);
233      return tmpl;
234  }
235
236  /* The add function calls adc to perform an add with */
237  /* initial carry of zero                             */
238  bigint bigint_add(bigint l, bigint r) {
239      return bigint_adc(l, r, 0);
240  }
241
242  /* The complement function returns the 1's complement */
243  bigint bigint_complement(bigint b) {
```

```
244    bigint r = bigint_copy(b);
245    for (int i = 0; i < r->size; i++)
246      r->blks[i] ^= CHUNKMASK;
247    return r;
248  }
249
250  /* The sub function gets the 1's complement of r, and adds it to l
251     with an initial carry of 1. The initial carry is equivalent to
252     adding 1 to the 1's complement to create the 2's complement.
253  */
254  bigint bigint_sub(bigint l, bigint r) {
255    bigint tmp1, tmp2;
256    tmp1 = bigint_complement(r);
257    tmp2 = bigint_adc(l, tmp1, 1);
258    bigint_free(tmp1);
259    return tmp2;
260  }
261
262  /* The mul_uint function multiplies a bigint by an unsigned chunk */
263  static bigint bigint_mul_uint(bigint l, chunk r) {
264    int i, negative = 0;
265    bigchunk tmpchunk;
266    bigint sum = bigint_from_int(0);
267    bigint tmp1, tmp2;
268    /* make sure the left operand is not negative */
269    if (l->blks[l->size-1] & ((bigchunk)1<<(BITSPERCHUNK-1))) {
270      negative ^= 1;
271      l = bigint_negate(l);
272    }
273    /* Perform the multiply (See section 7.2.5) */
274    for (i = 0; i < l->size; i++) {
275      tmpchunk = (bigchunk)l->blks[i] * r;
276      tmp1 = bigint_alloc(3);
277      tmp1->blks[0] = tmpchunk & CHUNKMASK;
278      tmp1->blks[1] = (tmpchunk>>BITSPERCHUNK) & CHUNKMASK;
279      tmp1->blks[2] = 0;
280      tmp2 = bigint_shift_left_chunk(tmp1, i);
281      bigint_free(tmp1);
282      tmp1 = bigint_adc(sum, tmp2, 0);
283      bigint_free(sum);
284      bigint_free(tmp2);
285      sum = tmp1;
286    }
287    /* result may need to be negated */
288    if (negative) {
289      tmp1 = sum;
290      sum = bigint_negate(sum);
291      bigint_free(tmp1);
292    bigint_free(l);
```

```
293       }
294       return sum;
295    }
296
297    /* bigint_mul uses the algorithm from Section 7.2.5    */
298    bigint bigint_mul(bigint l, bigint r) {
299       int i, negative = 0;
300       /* the result may require the sum
301          of the number of chunks in l and r */
302       bigint sum = bigint_from_int(0);
303       bigint tmp1, tmp2;
304       /* make sure the right operand is not negative */
305       if (r->blks[r->size-1] & ((bigchunk)1<<(BITSPERCHUNK-1))) {
306          negative = 1;
307          r = bigint_negate(r);      /* make negated copy of r */
308       }
309       for (i = 0; i < r->size; i++) {
310          tmp1 = bigint_mul_uint(l,r->blks[i]);
311          tmp2 = bigint_shift_left_chunk(tmp1,i);
312          bigint_free(tmp1);
313          tmp1 = sum;
314          sum = bigint_adc(sum,tmp2,0);
315          bigint_free(tmp1);
316          bigint_free(tmp2);
317       }
318       if (negative) {
319          tmp1 = sum;                  /* copy original */
320          sum = bigint_negate(sum); /* create complement */
321          bigint_free(tmp1);          /* free original */
322          bigint_free(r);
323       }
324       return sum;
325    }
326
327    /* bigint_div uses the algorithm from Section 7.3.2. */
328    bigint bigint_div(bigint l, bigint r) {
329       bigint lt = bigint_trim(l);
330       bigint rt = bigint_trim(r);
331       bigint tmp,q = bigint_from_int(0);
332       int shift, chunkshift, negative = 0;
333       if (lt->size < rt->size) {
334          bigint_free(lt);
335          bigint_free(rt);
336          return q;
337       }
338       /* make sure the right operand is not negative */
339       if (r->blks[r->size-1] & ((bigchunk)1<<(BITSPERCHUNK-1))) {
340          negative = 1;  /* track sign of result */
341          tmp = rt;
```

```
342      rt = bigint_negate(rt);
343      bigint_free(tmp);
344    }
345    /* make sure the left operand is not negative */
346    if (l->blks[l->size-1] & ((bigchunk)1<<(BITSPERCHUNK-1))) {
347      negative ^= 1;   /* track sign of result */
348      tmp = lt;
349      lt = bigint_negate(lt);
350      bigint_free(tmp);
351    }
352    /* do shift by chunks */
353    chunkshift = lt->size - rt->size;
354    if (chunkshift > 0) {
355      tmp = rt;
356      rt = bigint_shift_left_chunk(rt, chunkshift);
357      bigint_free(tmp);
358    }
359    /* do remaining shift bit-by-bit */
360    shift = 0;
361    while ((shift < (BITSPERCHUNK-1)) && bigint_lt(rt, lt)) {
362      shift++;
363      tmp = rt;
364      rt = bigint_shift_left(rt, 1);
365      bigint_free(tmp);
366    }
367    shift += (chunkshift * BITSPERCHUNK); /* Calculate total shift */
368    /* loop to shift right and subtract */
369    while (shift >= 0) {
370      tmp = q;
371      q = bigint_shift_left(q, 1);
372      bigint_free(tmp);
373      if (bigint_le(rt, lt)) {
374        /* perform subtraction */
375        tmp = lt;
376        lt = bigint_sub(lt,rt);
377        bigint_free(tmp);
378        /* change lsb from zero to one */
379        q->blks[0] |= 1;
380      }
381      tmp = rt;
382      rt = bigint_shift_right(rt,1);
383      bigint_free(tmp);
384      shift--;
385    }
386    /* correct the sign of the result */
387    if (negative) {
388      tmp = bigint_negate(q);
389      bigint_free(q);
390      q = tmp;
```

```
391      }
392      bigint_free(rt);
393      bigint_free(lt);
394      return q;
395    }
396
397    /* The C version of bigint_negate is very short, because it uses
398       existing functions.  However, it is not very efficient. We also
399       have an assembly version of the negate function. The #ifndef
400       allows us to use the assembly version. When USE_ASM is defined,
401       the C version will not be compiled. */
402
403    #ifndef USE_ASM
404    bigint bigint_negate(bigint b) {
405      bigint r = bigint_complement(b);   /* get 1's complement */
406      bigint tmp1 = bigint_from_int(0);  /* create zero */
407      bigint tmp2 = bigint_adc(r, tmp1, 1); /* add with an initial carry */
408      bigint_free(tmp1);
409      bigint_free(r);
410      return tmp2;
411    }
412    #endif
413
414    /* The add function calls adc to perform an add with     */
415    /* initial carry of zero                                 */
416    bigint bigint_abs(bigint b) {
417      if (b->blks[b->size-1] & ((bigchunk)1<<(BITSPERCHUNK-1)))
418        return bigint_negate(b);
419      else
420        return bigint_copy(b);
421    }
422
423    /* The sqrt function returns floor(sqrt(b)) using the digit-by-digit
424       algorithm.  There are better square root algorithms...  */
425    bigint bigint_sqrt(bigint b) {
426      bigint r = bigint_from_int(0), zero = bigint_from_int(0);
427      bigint num = bigint_copy(b), tmp, resplusbit, bit;
428      if (b->blks[b->size-1] & ((bigchunk)1<<(BITSPERCHUNK-1))) {
429        fprintf(stderr,
430          "Cannot compute square root of negative number.\n");
431        exit(1);
432      }
433      /* initialize bit to the largest power of 4 that is <= b */
434      bit = bigint_alloc(b->size);
435      bit->blks[bit->size-1] = ((bigchunk)1<<(BITSPERCHUNK-2));
436      for (int i = 0; i < bit->size-1; i++)
437        bit->blks[i] = 0;
438      while (bit->blks[bit->size-1] > b->blks[b->size-1])
439        bit->blks[bit->size-1] >>= 2;
```

```
440    if (bit->blks[bit->size-1] == 0)
441      bit->blks[bit->size-2] = ((bigchunk)1<<(BITSPERCHUNK-2));
442    /* this could be more efficient. TODO */
443    while (bigint_gt(bit,zero)) {
444      resplusbit = bigint_add(r, bit);
445      if (bigint_ge(num,resplusbit)) {
446        tmp = num;
447        num = bigint_sub(num,resplusbit);
448        bigint_free(tmp);
449        tmp = bigint_shift_right(r,1);
450        bigint_free(r);
451        r = bigint_add(tmp,bit);
452        bigint_free(tmp);
453      } else {
454        tmp = r;
455        r = bigint_shift_right(r,1);
456        bigint_free(tmp);
457      }
458      bigint_free(resplusbit);
459      tmp = bit;
460      bit = bigint_shift_right(bit,2);
461      bigint_free(tmp);
462    }
463    bigint_free(zero);
464    bigint_free(num);
465    bigint_free(bit);
466    return r;
467  }
468
469  /************************************************************************/
470
471  /* shift left by entire chunks */
472  static bigint bigint_shift_left_chunk(bigint l, int chunks) {
473    bigint tmp = bigint_alloc(l->size + chunks);
474    for (int i = -chunks; i < l->size; i++) {
475      if (i < 0)
476        tmp->blks[i+chunks] = 0;
477      else
478        tmp->blks[i+chunks] = l->blks[i];
479    }
480    return tmp;
481  }
482
483  /* shift right by entire chunks */
484  static bigint bigint_shift_right_chunk(bigint l, int chunks) {
485    bigint tmp = bigint_alloc(l->size - chunks);
486    for (int i = 0; i < tmp->size; i++) {
487      if (i<chunks)
488        tmp->blks[i] = 0;   // should do sign extend // TODO comment
```

```
489        else
490          tmp->blks[i]=l->blks[i-chunks];
491      }
492    return tmp;
493  }
494
495  /* Shift left the given about.  This will shift by chunks as much as
496     it can, then finish off with a sub-chunk shift. */
497  bigint bigint_shift_left(bigint l, int shamt) {
498    bigint tmp;
499    l = bigint_extend(l, l->size+1);
500    int extra = shamt % BITSPERCHUNK;
501    shamt = shamt / BITSPERCHUNK;
502    if (shamt) {
503      tmp = l;
504      l = bigint_shift_left_chunk(l, shamt);
505      bigint_free(tmp);
506    }
507    if (extra) {
508      for (int i = l->size - 1; i > 0; i--) {
509        l->blks[i] =
510          (l->blks[i]<<extra) | (l->blks[i-1]>>(BITSPERCHUNK-extra));
511      }
512      l->blks[0] = (l->blks[0] << extra);
513    }
514    tmp = bigint_trim(l);
515    bigint_free(l);
516    return tmp;
517  }
518
519  /* Arithmetic shift right the given about.  This will shift by
520     chunks as much as it can, then finish off with a sub-chunk
521     shift. */
522  bigint bigint_shift_right(bigint l, int shamt) {
523    bigint tmp;
524    schunk tmpc;
525    int extra = shamt % BITSPERCHUNK;
526    shamt = shamt / BITSPERCHUNK;
527    l = bigint_shift_right_chunk(l, shamt);
528    if (extra) {
529      for (int i = 0; i < l->size-1; i++) {
530        l->blks[i] =
531          (l->blks[i]>>extra) | (l->blks[i+1]<<(BITSPERCHUNK-extra));
532      }
533      /* do a signed shift of the top chunk */
534      tmpc = l->blks[l->size - 1];
535      tmpc >>= extra;
536      l->blks[l->size - 1] = tmpc;
537    }
```

```
538      tmp = bigint_trim(l);
539      bigint_free(l);
540      return tmp;
541    }
542
543    /**********************************************************************/
544    /* Test and compare operatioins                                     */
545    /**********************************************************************/
546
547    /* Compare bigint to zero */
548    inline int bigint_is_zero(bigint b) {
549      for (int i = 0; i < b->size; i++)
550        if (b->blks[i])
551          return 0;
552      return 1;
553    }
554
555    inline int bigint_le(bigint l, bigint r) {
556      return (bigint_cmp(l, r) < 1);
557    }
558
559    inline int bigint_lt(bigint l, bigint r) {
560      return (bigint_cmp(l, r) == -1);
561    }
562
563    inline int bigint_ge(bigint l, bigint r) {
564      return (bigint_cmp(l, r) > -1);
565    }
566
567    inline int bigint_gt(bigint l, bigint r) {
568      return (bigint_cmp(l, r) == 1);
569    }
570
571    inline int bigint_eq(bigint l, bigint r) {
572      return (!bigint_cmp(l, r));
573    }
574
575    inline int bigint_ne(bigint l, bigint r) {
576      return abs(bigint_cmp(l, r));
577    }
578
579    /* bigint_cmp is the core of all of the comparisons */
580    int bigint_cmp(bigint l, bigint r) {
581      bigint d = bigint_sub(l,r);
582      int cmp;
583      if ((d->size == 0) || (d->size == 1 && d->blks[0] == 0)) {
584        cmp = 0;      // d == 0
585      } else if (d->blks[d->size-1] & ((bigchunk)1 << (BITSPERCHUNK-1))) {
586        cmp = -1;     // d < 0 (MSB == 1)
```

```
587        } else {
588          cmp = 1;       // d > 0
589        }
590      bigint_free(d);
591      return cmp;
592    }
593
594    /*******************************************************************/
595    /* Functions for binary input/output                              */
596    /*******************************************************************/
597
598    void bigint_write_binary(FILE *f,bigint x) {
599      if (fwrite(&(x->size), sizeof(x->size), 1, f) != 1) {
600        perror("Write failed");
601        exit(4);
602      }
603      if (fwrite(x->blks, sizeof(chunk), x->size, f) != x->size) {
604        perror("Write failed");
605        exit(4);
606      }
607    }
608
609    bigint bigint_read_binary(FILE *f) {
610      bigint r = (bigint) malloc(sizeof(struct bigint_struct));
611      if (r == NULL) {
612        perror("bigint_read_binary");
613        exit(1);
614      }
615      if (fread(&(r->size), sizeof(r->size), 1, f) != 1) {
616        free(r);
617        return NULL;
618      }
619      r->blks = (chunk *) malloc(r->size * sizeof(chunk));
620      if (r->blks == NULL) {
621        perror("bigint_read_binary");
622        exit(2);
623      }
624      if (fread(r->blks, sizeof(chunk), r->size, f) != r->size) {
625        perror("Unable to read from file");
626        exit(4);
627      }
628      return r;
629    }
630
631    /*******************************************************************/
632    /*  Utility functions                                             */
633    /*******************************************************************/
634
635    bigint bigint_alloc(int chunks) {
```

```
bigint r = (bigint) malloc(sizeof(struct bigint_struct));
if (r == NULL) {
  perror("bigint_alloc");
  exit(1);
}
r->size = chunks;
r->blks = (chunk*) malloc(chunks * sizeof(chunk));
if (r->blks == NULL) {
  perror("bigint_alloc");
  exit(1);
}
return r;
}

static bigint bigint_trim(bigint b) {
  bigint d;
  int i = b->size-1;
  if (i > 0) {
    if (b->blks[i] == 0) {
      // we have a leading block that is all 0
      do
        i--; // search for first block that is not all 0
      while ((i > 0) && (b->blks[i] == 0));
      if (b->blks[i] & ((bigchunk)1<<(BITSPERCHUNK-1)))
        i++;  // if msb of current block is 1, then we went too far
    }
    else if (b->blks[i] == CHUNKMASK) {
      // we have a leading block that is all 1
      do
        i--; // search for first block that is not all 1
      while ((i>0) && (b->blks[i]==CHUNKMASK));
      if (!(b->blks[i] & ((bigchunk)1<<(BITSPERCHUNK-1))))
        i++;  // if msb of current block is 0, then we went too far
    }
  }
  i++; // i is now the number of blocks to copy
  if (i < b->size) {
    d = bigint_alloc(i);
    memcpy(d->blks,b->blks,d->size*sizeof(chunk));
  } else {
    d = bigint_copy(b);
  }
  return d;
}

/* smallmod divides a bigint by a small number
   and returns the modulus. b changes as a SIDE-EFFECT.
   This is used by the to_str function. */
static unsigned bigint_smallmod(bigint b, chunk num) {
```

```
685    bigchunk tmp;
686    int i;
687    if (num >= ((bigchunk)1<<(BITSPERCHUNK-1))) {
688      fprintf(stderr,"bigint_smallmod: divisor out of range\n");
689      exit(1);
690    }
691    /* start with most significant chunk and work down, taking
692       two overlapping chunks at a time */
693    tmp = b->blks[b->size-1];
694    for(i = b->size-1; i > 0; i--) {
695      b->blks[i] = tmp/num;
696      tmp = ((tmp % num) << BITSPERCHUNK) | b->blks[i-1];
697    }
698    b->blks[0] = tmp / num;
699    tmp = tmp % num;
700    return tmp;
701  }
702
703  static bigint bigint_extend(bigint b, int nchunks) {
704    bigint tmp;
705    int i, negative;
706    negative = 0;
707    if (b->blks[b->size-1] & ((bigchunk)1<<(BITSPERCHUNK-1)))
708      negative = 1;
709    tmp = bigint_alloc(nchunks);
710    for (i = 0; i < nchunks; i++) {
711      if (i < b->size)
712        tmp->blks[i] = b->blks[i];
713      else if (negative)
714        tmp->blks[i] = CHUNKMASK;
715      else
716        tmp->blks[i] = 0;
717    }
718    return tmp;
719  }
```

Listing 7.9 Program that runs tests on the big integer ADT.

```
1    #include <stdlib.h>
2    #include <stdio.h>
3    #include <time.h>
4    #include <bigint.h>
5
6    #define NTESTS 25000
7
8    char filename[] = "regression_tests.dat";
9
10   /* Error formatting */
11   void matherr(bigint in1, bigint in2, bigint exp, bigint out,
```

```
         char *msg);
void sqrterr(bigint in, bigint exp, bigint out, char *msg);

/* Timing tests */
clock_t test(int size, char *name, bigint l[], bigint r[],
          bigint (*bigint_op)(bigint l, bigint r), bigint op[]);
clock_t test_negate(int size, bigint l[], bigint op[]);
clock_t test_cmp(int size, bigint l[], bigint r[], int op[]);
clock_t test_division(int size, bigint l[], bigint r[], bigint op[]);
clock_t test_sqrt(int size, bigint l[], bigint op[]);

int main(void) {
  static bigint a[NTESTS], b[NTESTS];
  static bigint add[NTESTS], sub[NTESTS], mul[NTESTS];
  static bigint sqrt[NTESTS], div[NTESTS], neg[NTESTS];
  static int comp[NTESTS];
  int N, nmemb;
  clock_t total = 0;

  FILE *inf = fopen(filename, "r");
  if (inf == NULL) {
    perror("Failed to open file");
    exit(2);
  }

  N = 0;
  do {
    a[N] = bigint_read_binary(inf);
    b[N] = bigint_read_binary(inf);
    add[N] = bigint_read_binary(inf);
    sub[N] = bigint_read_binary(inf);
    mul[N] = bigint_read_binary(inf);
    sqrt[N] = bigint_read_binary(inf);
    div[N] = bigint_read_binary(inf);
    neg[N] = bigint_read_binary(inf);

    nmemb = fread(comp+N, sizeof(int), 1, inf);
    if(nmemb != 1) {
      perror("Unable to read from file");
      exit(100);
    }

    N++;
  } while ((N < NTESTS) && (a[N-1] != NULL) && (b[N-1] != NULL) &&
           (add[N-1] != NULL) && (sub[N-1] != NULL) &&
           (mul[N-1] != NULL) && (div[N-1] != NULL));

  fclose(inf);
```

```
61      // Run tests
62      total += test_negate(N, a, neg);
63      total += test_cmp(N, a, b, comp);
64      total += test(N, "Add", a, b, bigint_add, add);
65      total += test(N, "Sub", a, b, bigint_sub, sub);
66      total += test(N, "Mul", a, b, bigint_mul, mul);
67      total += test_division(N, a, b, div);
68      total += test_sqrt(N, a, sqrt);
69
70      // Total time
71      printf("Total time :\t %lf\n",(total)/(double)CLOCKS_PER_SEC);
72
73      // Free
74      for(int i = 0; i < N; i++) {
75        bigint_free(a[i]);
76        bigint_free(b[i]);
77        bigint_free(add[i]);
78        bigint_free(sub[i]);
79        bigint_free(mul[i]);
80        bigint_free(sqrt[i]);
81        bigint_free(div[i]);
82        bigint_free(neg[i]);
83      }
84
85      return 0;
86    }
87
88    void matherr(bigint in1, bigint in2, bigint exp, bigint out, char *msg) {
89      char *input1, *input2, *expected, *actual;
90      input1 = bigint_to_str(in1);
91      input2 = bigint_to_str(in2);
92      expected = bigint_to_str(exp);
93      actual = bigint_to_str(out);
94      printf("Error in %s:\ninputs:\n%s\n%s\n", msg, input1, input2);
95      printf("correct output: %s\n", expected);
96      printf("output: %s\n", actual);
97      exit(4);
98    }
99
100   void sqrterr(bigint in, bigint exp, bigint out, char *msg) {
101     char *input, *expected, *actual;
102     input = bigint_to_str(in);
103     expected = bigint_to_str(exp);
104     actual = bigint_to_str(out);
105     printf("Error in %s:\ninput: %s\n", msg, input);
106     printf("correct output: %s\n", expected);
107     printf("output: %s\n", actual);
108     exit(4);
109   }
```

```
clock_t test(int size, char *name, bigint l[], bigint r[],
            bigint (*bigint_op)(bigint l, bigint r), bigint op[]) {
  bigint d;
  clock_t stop, start = clock();
  for (int i = 0; i < size; i++) {
    d = bigint_op(l[i], r[i]);
    if (bigint_ne(d, op[i]))
      matherr(l[i], r[i], op[i], d, name);
    bigint_free(d);
  }
  stop = clock();
  printf("%s :\t %lf\n", name, (stop - start)/(double)CLOCKS_PER_SEC);
  return (stop - start);
}

clock_t test_negate(int size, bigint l[], bigint op[]) {
  char *name = "Neg";
  bigint d;
  clock_t stop, start = clock();
  for (int i = 0; i < size; i++) {
    d = bigint_negate(l[i]);
    if (bigint_ne(d, op[i]))
      matherr(l[i], l[i], op[i], d, name);
    bigint_free(d);
  }
  stop = clock();
  printf("%s :\t %lf\n", name, (stop - start)/(double)CLOCKS_PER_SEC);
  return (stop - start);
}

clock_t test_cmp(int size, bigint l[], bigint r[], int op[]) {
  char *name = "Cmp";
  int d;
  clock_t stop, start = clock();
  bigint expected, actual;
  for (int i = 0; i < size; i++) {
    d = bigint_cmp(l[i], r[i]);
    if (d != op[i]) {
      actual = bigint_from_int(d);
      expected = bigint_from_int(op[i]);
      matherr(l[i], r[i], expected, actual, name);
    }
  }
  stop = clock();
  printf("%s :\t %lf\n", name, (stop - start)/(double)CLOCKS_PER_SEC);
  return (stop - start);
}
```

```
159   clock_t test_division(int size, bigint l[], bigint r[], bigint op[]) {
160     char *name = "Div";
161     bigint d, labs, rabs;
162     clock_t stop, start = clock();
163     for (int i = 0; i < size; i++) {
164       /* Use largest absolute value for dividend and smaller absoulute
165        * value as divisor */
166       labs = bigint_abs(l[i]);
167       rabs = bigint_abs(r[i]);
168       if (bigint_gt(labs, rabs)) {
169         d = bigint_div(l[i], r[i]);
170         if (bigint_ne(d, op[i]))
171           matherr(l[i], r[i], op[i], d, name);
172       } else {
173         d = bigint_div(r[i], l[i]);
174         if (bigint_ne(d, op[i]))
175           matherr(r[i], l[i], op[i], d, name);
176       }
177       bigint_free(labs);
178       bigint_free(rabs);
179       bigint_free(d);
180     }
181     stop = clock();
182     printf("%s :\t %lf\n", name, (stop - start)/(double)CLOCKS_PER_SEC);
183     return (stop - start);
184   }
185
186   clock_t test_sqrt(int size, bigint l[], bigint op[]) {
187     char *name = "Sqrt";
188     bigint d, abs;
189     clock_t stop, start = clock();
190     for (int i = 0; i < size; i++) {
191       abs = bigint_abs(l[i]);
192       d = bigint_sqrt(abs);
193       if (bigint_ne(d, op[i]))
194         sqrterr(abs, op[i], d, name);
195       bigint_free(abs);
196       bigint_free(d);
197     }
198     stop = clock();
199     printf("%s :\t %lf\n", name, (stop - start)/(double)CLOCKS_PER_SEC);
200     return (stop - start);
201   }
```

The implementation could be made more efficient by writing some of the functions in assembly language. One opportunity for improvement is in the add function, which must calculate the carry from one chunk of bits to the next. In assembly, the programmer has direct access to the carry bit, so carry propagation should be much faster.

When attempting to speed up a C program by converting selected parts of it to assembly language, it is important to first determine where the most significant gains can be made. A profiler, such as `gprof` or `callgrind`, can be used to help identify the sections of code that will have the greatest impact on performance. It is also important to make sure that the result is not just highly optimized C code. If the code cannot benefit from some features offered by assembly, then it may not be worth the effort of re-writing in assembly. The code should be re-written from a pure assembly language viewpoint.

It is also important to avoid premature assembly programming. Make sure that the C algorithms and data structures are efficient before moving to assembly. if a better algorithm can give better performance, then assembly may not be required at all. Once the assembly is written, it is more difficult to make major changes to the data structures and algorithms. Assembly language optimization is the *final* step in optimization, not the first one.

Well-written C code is modularized, with many small functions. This helps readability, promotes code reuse, and may allow the compiler to better optimization. However, each function call has some associated overhead. If optimal performance is the goal, then calling many small functions should be avoided. For instance, if the piece of code to be optimized is in a loop body, then it may be best to write the entire loop in assembly, rather than writing a function and calling it each time through the loop. Writing in assembly is not a guarantee of performance. Spaghetti code is slow. Load/store instructions are slow. Multiplication and division are slow. The secret to good performance is avoiding things that are slow. Good optimization requires rethinking the code to take advantage of assembly language.

The profiler indicated that `bigint_adc` is used more than any other function. If assembly language can make this function run faster, then it should have a profound effect on the program. We will leave that as an exercise for the student, but will give an example of how to optimize a less critical function. The `bigint_negate` function was re-written in assembly, as shown in Listing 7.10. Note that the original C implementation used the `bigint_complement` function to get the 1's complement, and then used the `bigint_adc` function to add one. The C implementation of the `bigint_adc` function requires extra code to calculate the carry from one chunk to the next. However, the assembly version of `bigint_complement` loads each chunk, complements it, and propagates the carry all in one step. Since it only loads and stores the data once, instead of twice, and requires much less work to propagate the carry bit, we would expect the assembly version to run about twice as fast as the C version.

Listing 7.10 AArch64 assembly implementation if the `bigint_negate` function.

```
    .text
    .type   bigint_negate, %function
    .global bigint_negate
```

```
 4  bigint_negate:
 5          stp     x29, x30, [sp, #-32]!  // Save FP & Link Register x30
 6          stp     x19, x20, [sp, #16]    // Save non-volatile regs
 7          mov     x19, x0                // x19 = bigint b
 8          ldr     w20, [x19, #8]         // w20 = b->size
 9          // initialize bigint struct to hold result
10          mov     w0, 0x10000000         // w0 = -MAX_INT
11          cmp     w20, w0
12          cinc    w20, w20, EQ           // increment size if -MAX_INT
13          mov     w0, w20
14          bl      bigint_alloc           // x0 = bigint_alloc(b->size)
15          str     w20, [x0, #8]          // new->size = b->size
16          ldr     x3, [x19]              // x3 = b->blks (src chunks)
17          ldr     x4, [x0]               // x4 = new->blks(dest chunks)
18          // loop from least significant chunk to most significant
19          cmp     w20, wzr               // for(w20=size; w20>0; w20--)
20          ble     endloop
21          negs    xzr, xzr               // set carry flag to 1
22  loop:
23          #ifdef EIGHT_BIT
24          ldrsb   w6, [x3], #1           // load chunk from source
25          mvn     w6, w6                 // complement it (1's comp.)
26          adc     w6, w6, wzr            // add carry flag
27          tst     w6, 0x10
28          cneg    xzr, xzr, NE           // set carry flag on overflow
29          strb    w6, [x4], #1           // store chunk in destination
30          #else
31          #ifdef SIXTEEN_BIT
32          ldrsh   w6, [x3], #2           // load chunk from source
33          mvn     w6, w6                 // complement it (1's comp.)
34          adc     w6, w6, wzr            // add carry flag
35          tst     w6, 0x10000
36          cneg    xzr, xzr, NE           // set carry flag on overflow
37          strh    w6, [x4], #2           // store chunk in destination
38          #else
39          #ifdef THIRTYTWO_BIT
40          ldr     w6, [x3], #4           // load chunk from source
41          mvn     w6, w6                 // complement it (1's comp.)
42          adcs    w6, w6, wzr            // add carry flag, set flags
43          str     w6, [x4], #4           // store chunk in destination
44          #else
45          #ifdef SIXTYFOUR_BIT
46          ldr     x6, [x3], #8           // load chunk from source
47          mvn     x6, x6                 // complement it (1's comp.)
48          adcs    x6, x6, xzr            // add carry flag, set flags
49          str     x6, [x4], #8           // store chunk in destination
50          #else /* default to 32 bits */
51          ldr     w6, [x3], #4           // load chunk from source
52          mvn     w6, w6                 // complement it (1's comp.)
```

Table 7.1: Performance of `bigint_negate` implementations on an nVidia Jetson TX-1.

Chunk	Version	Negate		Program	
		Time	*Speedup*	*Time*	*Speedup*
32-bit	C	0.097846	1.00	27.898544	1.00
	Assembly	0.049903	1.96	27.322730	1.02
64-bit	C	0.095243	1.03	26.719709	1.04
	Assembly	0.048488	2.02	26.492714	1.05

```
53          adcs    w6, w6, wzr         // add carry flag, set flags
54          str     w6, [x4], #4        // store chunk in destination
55          #endif
56          #endif
57          #endif
58          #endif
59          sub     w20, w20, #1
60          cbnz    w20, loop
61   endloop:
62          // return address of new bigint is already in x0
63          ldp     x19, x20, [sp, #16]     // Restore non-volatile regs
64          ldp     x29, x30, [sp], #32     // Restore FP & LR
65          ret
66          .size   bigint_negate,(. - bigint_negate)
```

The `regression` testing program was executed on an nVidia Jetson TX-1 using the C version of `bigint_negate` and then with the assembly version. This was done for chunk sizes of 32 bits and 64 bits. The results are shown in Table 7.1. The total time required for the negate function tests using the 32-bit C version was 0.097846 seconds. The 32-bit assembly version ran in 0.049903 seconds, for a speedup of 1.96, which means that the assembly version was almost twice as fast as the C version. The 64-bit C version ran in 0.095243 seconds, while the 64-bit assembly code ran in 0.048488 seconds, which gives a 1.96 speedup. Note that the assembly version with a 64-bit chunk size is more than twice as fast as the C version using 32-bit chunks.

The `bigint_negate` function also has a small impact on the functions that rely on it, so the overall time to run the regression tests was also reduced slightly. The 64-bit assembly version was 5% faster overall than the 32-bit C version. Because this function was not called often in the test program, the overall program speedup was modest. Implementing other functions, such as `bigint_add` and `bigint_sub` in assembly would result in much larger overall speedups.

7.5 Chapter summary

Complement mathematics provides a method for performing all basic operations using only the complement, add, and shift operations. Addition and subtraction are fast, but multiplication and division are relatively slow. In particular, division should be avoided whenever possible. The exception to this rule is division by a power of the radix, which can be implemented as a shift. Good assembly programmers replace division by a constant c with multiplication by the reciprocal of c. They also replace the multiply instruction with a series of shifts and add or subtract operations when it makes sense to do so. These optimizations can make a big difference in performance.

Writing sections of a program in assembly can result in better performance, but it is not guaranteed. The chance of achieving significant performance improvement is increased if the following rules are used:

1. Only optimize the parts that really matter.
2. Design data structures with assembly in mind.
3. Use efficient algorithms and data structures.
4. Write the assembly code last.
5. Ignore the C version and write good, clean, assembly.
6. Reduce function calls wherever it makes sense.
7. Avoid unnecessary memory accesses.
8. Write good code. The compiler will beat poor assembly every time, but good assembly will beat the compiler every time.

Understanding the basic mathematical operations can enable the assembly programmer to work with integers of any arbitrary size with efficiency that cannot be matched by a C compiler. However, it is best to focus the assembly programming on areas where the greatest gains can be made.

Exercises

7.1. Multiply -90 by 105 using *signed* 8-bit binary multiplication to form a signed 16-bit result. Show all of your work.

7.2. Multiply 166 by 105 using *unsigned* 8-bit binary multiplication to form an unsigned 16-bit result. Show all of your work.

7.3. Write a section of AArch64 assembly code to multiply the value in x1 by 13_{10} using only shift and add operations.

7.4. The following code will multiply the value in x0 by a constant C. What is C?

```
add     x1, x0, x0, lsl #1
add     x0, x1, x0, lsl #2
```

7.5. Show the optimally efficient instruction(s) necessary to multiply a number in register **x0** by the constant 67_{10}.

7.6. Show how to divide 78_{10} by 6_{10} using binary long division.

7.7. Demonstrate the division algorithm using a sequence of tables as shown in Section 7.3.2 to divide 155_{10} by 11_{10}.

7.8. When dividing by a constant value, why is it desirable to have *m* as large as possible?

7.9. Modify your program from Exercise 5.13. in Chapter 5 to produce a 128-bit result, rather than a 64-bit result.

7.10. Modify your program from Exercise 5.13. in Chapter 5 to produce a 128-bit result, rather than a 64-bit result. How would you do this in C?

7.11. Write the `bigint_shift_left_chunk` function shown in Listing 7.8 in AArch64 assembly, and measure the performance improvement.

7.12. Write the `bigint_mul_uint` function in AArch64 assembly, and measure the performance improvement.

7.13. Write the `bigint_mul` function in AArch64 assembly, and measure the performance improvement.

## CHAPTER 8

Non-integral mathematics

Chapter 7 introduced methods for performing computation using integers. Although many problems can be solved using only integers, it is often necessary (or at least more convenient) to perform computation using real numbers or even complex numbers. For our purposes, a non-integral number is any number that is not an integer. Many systems are only capable of performing computation using binary integers, and have no hardware support for non-integral calculations. In this chapter, we will examine methods for performing non-integral calculations using only integer operations.

8.1 Base conversion of fractional numbers

Section 1.3.2 explained how to convert integers in a given base into any other base. We will now extend the methods to convert fractional values. In *radix point notation*, a fractional number can be viewed as consisting of an integer part, a *radix point*, and a fractional part. In base 10, the radix point is also known as the decimal point. In base 2, it is called the binimal point. For base 16, it is the heximal point, and in base 8 it is an octimal point. The term *radix point* is used as a general term for a location that divides a number into integer and fractional parts, without specifying the base.

8.1.1 Arbitrary base to decimal

The procedure for converting fractions from a given base b into base ten is very similar to the procedure used for integers. The only difference is that the digit to the left of the radix point is weighted by b^0 and the exponents become increasingly negative for each digit right of the radix point. The basic procedure is the same for any base b. For example, the value 101.01011_2 can be converted to base ten by expanding it as follows:

$$1 \times 2^2 + 0 \times 2^1 + 1 \times 2^0 + 0 \times 2^{-1} + 1 \times 2^{-2} + 0 \times 2^{-3} + 1 \times 2^{-4}$$
$$= 4 + 0 + 1 + 0 + \frac{1}{4} + 0 + \frac{1}{16}$$
$$= 5.3125_{10}$$

Likewise, the hexadecimal fraction 4F2.9A0 can be converted to base ten by expanding it as follows:

$$4 \times 16^2 + 15 \times 16^1 + 2 \times 16^0 + 9 \times 16^{-1} + 10 \times 16^{-2} + 0 \times 16^{-3}$$

$$= 1024 + 240 + 2 + \frac{9}{16} + \frac{10}{256} + \frac{0}{4096}$$

$$= 1266.6015625_{10}$$

8.1.2 Decimal to arbitrary base

When converting from base ten into another base, the integer and fractional parts are treated separately. The base conversion for the integer part is performed in exactly the same was as in Section 1.3.2, using repeated division by the base b. The fractional part is converted using repeated multiplication. For example, to convert the decimal value 5.6875_{10} to a binary representation:

1. Convert the integer portion, 5_{10} into its binary equivalent, 101_2.
2. Multiply the decimal fraction by two. The integer part of the result is the first binary digit to the right of the radix point.
   Because $x = 0.6875 \times 2 = 1.375$, the first binary digit to the right of the binimal point is 1. So far, we have $5.625_{10} = 101.1_2$.
3. Multiply the fractional part of x by 2 once again.
   Because $x = 0.375 \times 2 = 0.75$, the second binary digit to the right of the binimal point is 0. So far, we have $5.625_{10} = 101.10_2$.
4. Multiply the fractional part of x by 2 once again.
   Because $x = 0.75 \times 2 = 1.50$, the third binary digit to the right of the binimal point is 1. So now we have $5.625 = 101.101$.
5. Multiply the fractional part of x by 2 once again.
   Because $x = 0.5 \times 2 = 1.00$, the fourth binary digit to the right of the binimal point is 1. So now we have $5.625 = 101.1011$.
6. Since the fractional part of x is now zero, we know that all remaining digits will be zero.

The procedure for obtaining the fractional part can be accomplished easily using a table, as shown below:

Operation	Result	
	Integer	*Fraction*
$.6875 \times 2 = 1.375$	1	.375
$.375 \times 2 = 0.75$	0	.75
$.75 \times 2 = 1.5$	1	.5
$.5 \times 2 = 1.0$	1	.0

Putting it all together, $5.6875_{10} = 101.1011_2$. After converting a fraction from base 10 into another base, the result should be verified by converting back into base 10. The results from the previous example can be expanded as follows:

$$1 \times 2^2 + 0 \times 2^1 + 1 \times 2^0 + 1 \times 2^{-1} + 0 \times 2^{-2} + 1 \times 2^{-3} + 1 \times 2^{-4}$$
$$= 4 + 0 + 1 + \frac{1}{2} + 0 + \frac{1}{8} + \frac{1}{16}$$
$$= 5.6875_{10}$$

Converting decimal fractions to base sixteen is accomplished in a very similar manner. To convert 842.234375_{10} into base 16, we first convert the integer portion by repeatedly dividing by 16 to yield 34A. We then repeatedly multiply the fractional part, extracting the integer portion or the result each time as shown in the table below:

Operation	Result	
	Integer	*Fraction*
.234375 × 16 = 3.75	3	.75
.75 × 16 = 12.0	12	.0

In the second line, the integer part is 12, which must be replaced with a hexadecimal digit. The hexadecimal digit for 12_{10} is C, so the fractional part is 3C. Therefore, $842.234375_{10} = 34A.3C_{16}$ The result is verified by converting back into base 10 as follows:

$$3 \times 16^2 + 4 \times 16^1 + 10 \times 16^0 + 3 \times 16^{-1} + 12 \times 16^{-2}$$
$$= 768 + 64 + 10 + \frac{3}{16} + \frac{12}{256}$$
$$= 842.234375_{10}$$

This tabular method works for fractional notation as well as radix-point notation. For example, when converting $15\frac{11}{24}{}_{10}$ to base six, we convert the integer portion by repeatedly dividing by six, which results on 23. Then we convert the fractional portion as follows:

Operation	Result	
	Integer	*Fraction*
$\frac{11}{24} \times 6 = 2\frac{3}{4}$	2	$\frac{3}{4}$
$\frac{3}{4} \times 6 = 4\frac{1}{2}$	4	$\frac{1}{2}$
$\frac{1}{2} \times 6 = 3$	3	0

Therefore, $15\frac{11}{24}{}_{10} = 23.243_6$. To check the result, we can convert it back into base ten as follows:

$$1 \times 6^1 + 3 \times 16^0 + 2 \times 6^{-1} + 4 \times 6^{-2} + 3 \times 6^{-2}$$
$$= 15 + \frac{2}{6} + \frac{4}{36} + \frac{3}{216}$$
$$= 15 + \frac{72}{216} + \frac{24}{216} + \frac{3}{216}$$
$$= 15 + \frac{99}{216}$$
$$= 15\frac{11}{24}$$

8.1.2.1 Bases that are powers-of-two

Converting fractional values between binary, hexadecimal, and octal can be accomplished in the same manner as with integer values. However, care must be taken to align the radix point properly. As with integers, converting from hexadecimal or octal to binary is accomplished by replacing each hex or octal digit with the corresponding binary digits from the appropriate table shown in Fig. 1.3.

For example, to convert $5AC.43B_{16}$ to binary, we just replace "5" with "0101," replace "A" with "1010," replace "C" with "1100," replace "4" with "0100," replace "3" with "0011," replace "B" with "1011," So, using the table, we can immediately see that $5AC.43B_{16} = 010110101100.010000111011_2$. This method works exactly the same for converting from octal to binary, except that it uses the table on the right side of Fig. 1.3.

Converting fractional numbers from binary to hexadecimal or octal is also very easy using the tables. The procedure is to split the binary string into groups of bits, working outwards from the radix point, then replace each group with its hexadecimal or octal equivalent. For example, to convert 01110010.1010111_2 to hexadecimal, just divide the number into groups of four bits, starting at the radix point and working outwards in both directions. It may be necessary to pad with zeroes to make a complete group on the left or right, or both. Our example is grouped as follows: $|0111|0010.1010|1110|_2$. Now each group of four bits is converted to hexadecimal by looking up the corresponding hex digit in the table on the left side of Fig. 1.3. This yields $72.AE_{16}$. For octal, the binary number would be grouped as follows: $|001|110|010.101|011|100|_2$. Now each group of three bits is converted to octal by looking up the corresponding digit in the table on the right side of Fig. 1.3. This yields 162.534_8. Note that the conversion to octal required the addition of leading and trailing zeroes.

8.2 Fractions and bases

One interesting phenomenon that is often encountered when using radix point notation is that fractions which terminate in one base may become non-terminating repeating fractions in another base. For example, the radix point binary representation of the decimal fraction $\frac{1}{10}$ is a non-terminating repeating fraction, as shown in Example 18.

Example 18. A non-terminating, repeating binimal.

$$
\begin{aligned}
.1 \times 2 &= 0.2 \\
.2 \times 2 &= 0.4 \\
.4 \times 2 &= 0.8 \\
.8 \times 2 &= 1.6 \\
.6 \times 2 &= 1.2 \\
.2 \times 2 &= 0.4
\end{aligned}
$$

The resulting fractional part from the last step performed is exactly the same as in the second step. Therefore, the sequence will repeat. If we continue, we will repeat the sequence of steps 2-5 forever. Hence, the final binary representation will be:

$$
\begin{aligned}
0.1_{10} &= .00011001100110011\ldots_2 \\
&= .0\overline{0011}_2
\end{aligned}
$$

Because of this phenomenon, it is impossible to exactly represent 1.10_{10} (and many other fractional quantities) as a binary fraction in a finite number of bits using radix point notation.

8.2.1 Rounding errors

The fact that some base 10 fractions cannot be exactly represented in binary has lead to many subtle software bugs and incorrect results when programmers attempt to work with currency (and other quantities) as real-valued numbers. In this section, we explore the idea that the representation problem can be avoided by working in some base other than base 2. If that is the case, then we can simply build hardware (or software) to work in that base, and will then be able to represent any fractional value precisely using a finite number of digits.

For brevity, we will refer to a binary fractional quantity as a *binimal* and a decimal fractional quantity as a *decimal*. We would like to know whether there are more non-terminating decimals than binimals, more non-terminating binimals than decimals, or neither. Since there are an infinite number of non-terminating decimals and an infinite number of non-terminating binimals, we could be tempted to conclude that they are equal. However, that is an oversimplification. If we ask the question differently, we can discover some important information. A better way to ask the question is as follows:

Question: Is the set of terminating decimals a subset of the set of terminating binimals, or vice versa, or neither.

We start by introducing a lemma which can be used to predict whether or not a terminating fraction in one base will terminate in another base. We introduce the notation $x \mid y$ (read as "x divides y") to indicate that y can be evenly divided by x.

Lemma 1. *If x, $0 < x < 1$, terminates in some base B (a product of primes), then $x = \frac{N_x}{D_x}$, and $D_x = p_1^{k_1} p_2^{k_2} \dots p_n^{k_n}$, where the p_i are the prime factors of B.*

Proof. Let $x = \frac{N_x}{D_x}$, and $D_x = p_1^{k_1} p_2^{k_2} \dots p_n^{k_n}$, where the p_i are the prime factors of B. Then $D_x \mid N_x \times B^{k_{max}}$, where $k_{max} = max(k_1, k_2, \dots k_n)$, so $x = \frac{N_x}{D_x x}$ terminates after k_{max} or fewer divisions.

Let $x = \frac{N_x}{D_x}$ terminate after k divisions. Then $D_x \mid N_x \times B^k$. Since D_x does not evenly divide N_x, D_x must be composed of some combination of the prime factors of B. Thus, D_x can be expressed as $p_1^{k_1} p_2^{k_2} \dots p_n^{k_n}$. $\square$

Theorem 1. *The set of terminating binimals is a subset of the set of terminating decimals.*

Proof. Let b be a terminating binimal. Then, by Lemma 1, $b = \frac{N_b}{D_b}$, such that $D_b = 2^k$, for some $k \geq 0$. Therefore, $D_b = 2^k 5^m$, for some $k, m > 0$, and again by the Lemma, b is also a terminating decimal. $\square$

Theorem 2. *The set of terminating decimals is not a subset of the set of terminating binimals.*

Proof. Let d be a terminating decimal such that $d = \frac{N_d}{D_d}$, where $D_d = 2^k 5^m$. If $m > 0$, then by the Lemma, d is a non-terminating binimal. $\square$

Answer: The set of terminating binimals is a subset of the set of terminating decimals, but the set of terminating decimals is not a subset of the set of terminating binimals.

8.2.2 Implications

Theorem 1 implies that any binary fraction can be expressed exactly as a decimal fraction in radix point notation, but Theorem 2 implies that there are decimal fractions which cannot be expressed exactly in binary radix point notation using a finite number of bits. Every fraction (when expressed in lowest terms) which has a non-zero power of five in its denominator cannot be represented in binary with a finite number of bits. Another implication is that some fractions can not be expressed exactly in either binary or decimal. For example, let $B = 30 = 2 * 3 * 5$. Then any number with denominator $2^{k_1} 3^{k_2} 5^{k_3}$ terminates in base 30. However if $k_2 \neq 0$, then the fraction will terminate in neither base two nor base ten, because three is not a prime factor of ten or two.

Another implication of the theorem is that the more prime factors we have in our base, the more fractions we can express exactly using radix point notation. For instance, the smallest base that has two, three, and five as prime factors is base 30. Using that base, we can exactly express fractions in radix notation that cannot be expressed in base ten or in base two with a finite number of digits. For example, in base 30, the fraction $\frac{11}{15}$ will terminate after one division since $15 = 3^1 5^1$. To see what the number will look like, let's extend the hexadecimal system of using letters to represent digits beyond 9. So we get this chart for base 30:

$$
\begin{array}{lllll}
0_{10} \to 0_{30} & 1_{10} \to 1_{30} & 2_{10} \to 2_{30} & 3_{10} \to 3_{30} & 4_{10} \to 4_{30} \\
5_{10} \to 5_{30} & 6_{10} \to 6_{30} & 7_{10} \to 7_{30} & 8_{10} \to 8_{30} & 9_{10} \to 9_{30} \\
10_{10} \to A_{30} & 11_{10} \to B_{30} & 12_{10} \to C_{30} & 13_{10} \to D_{30} & 14_{10} \to E_{30} \\
15_{10} \to F_{30} & 16_{10} \to G_{30} & 17_{10} \to H_{30} & 18_{10} \to I_{30} & 19_{10} \to J_{30} \\
20_{10} \to K_{30} & 21_{10} \to L_{30} & 22_{10} \to M_{30} & 23_{10} \to N_{30} & 24_{10} \to O_{30} \\
25_{10} \to P_{30} & 26_{10} \to Q_{30} & 27_{10} \to R_{30} & 28_{10} \to S_{30} & 29_{10} \to T_{30}
\end{array}
$$

Since $\frac{11}{15} = \frac{22}{30}$, the fraction can be expressed precisely as $0.M_{30}$. Likewise, the fraction $\frac{13}{45}$ is $0.2\overline{8}_{10}$ but terminates in base 30. Since $45 = 3^3 5^1$, this number will have three or fewer digits following the radix point. To compute the value, we will have to raise it to higher terms. Using 30^2 as the denominator gives us:

$$
\frac{13}{45} = \frac{260}{900}
$$

Now we can convert it to base 30 by repeated division. $\frac{260}{30} = 8$ with remainder 20. Since $20 < 30$, we cannot divide again. Therefore, $\frac{13}{45}$ in base 30 is $0.8K$.

Although base 30 can represent all fractions that can be expressed in base two and base ten, there are still fractions that cannot be represented in base 30. For example, $\frac{1}{7}$ has the prime factor seven in its denominator, and therefore will only terminate in bases were seven is a prime factor of the base. The fraction $\frac{1}{7}$ will terminate in base 7, base 14, base 21, base 42, and an infinite number of other bases, but not in base 30. Since there are an infinite number of primes, no number system is immune from this problem. No matter what base the computer works in, there are fractions that cannot be expressed exactly with a finite number of digits. Therefore, it is incumbent upon programmers and hardware designers to be aware of round-off errors and take appropriate steps to minimize their effects.

For example, there is no reason why the hardware clocks in a computer should work in base ten. They can be manufactured to measure time in base two. Instead of counting seconds in tenths, hundredths or thousandths, they could be calibrated to measure in fourths, eighths, sixteenths, 1024ths, etc. This would eliminate the round-off error problem in keeping track of time. For this reason, the most common crystal oscillators used for driving watches, clocks, and other time-keeping devices are carefully calibrated to run at a frequency of 32768 Hz.

8.3 Fixed point numbers

As shown in the previous section, given a finite number of bits, a computer can only *approximately* represent non-integral numbers. It is often necessary to accept that limitation and perform computations involving approximate values. With due care and diligence, the results will be accurate within some acceptable error tolerance. One way to deal with real-valued numbers is to simply treat the data as *fixed point* numbers. Fixed point numbers are treated as integers, but the programmer must keep track of the radix point during each operation. We will present a systematic approach to designing fixed point calculations.

When using fixed point arithmetic, the programmer needs a convenient way to describe the numbers that are being used. Most languages have standard data types for integers and floating point numbers, but very few have support for fixed point numbers. Notable exceptions include PL/1 and Ada, which provide support for fixed point binary and fixed point decimal numbers. There is also an ISO standard extension to C which gives some support for fixed point numbers. A few other languages also support fixed point. We will focus on fixed point binary, but the techniques presented can also be applied to fixed point numbers in any base. Using this approach, any language which supports integers can be used to perform fixed point computation.

8.3.1 Interpreting fixed point numbers

Fixed point numbers are stored as integers, and integer operations are performed on them. However, the programmer assigns a radix point location to each number, and tracks the radix point through every operation. Each fixed point binary number has three important parameters that describe it:

1. whether the number is signed or unsigned,
2. the position of the radix point in relation to the right side of the sign bit (for signed numbers) or the position of the radix point in relation to the most significant bit (for unsigned numbers), and
3. the number of fractional bits stored.

Unsigned fixed point numbers will be specified as U(i, f), where i is the position of the radix point in relation to the *left* side of the most significant bit, and f is the number of bits stored in the fractional part.

For example, U(10, 6) indicates that there are six bits of precision in the fractional part of the number, and the radix point is ten bits to the right of the most significant bit stored. The layout for this number is shown graphically as:

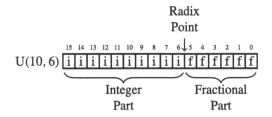

where i is an integer bit and f is a fractional bit. Very small numbers with no integer part may have a negative i. For example, U(−8, 16) specifies an unsigned number with no integer part, eight leading zero bits which are not actually stored, and 16 bits of fractional precision. The layout for this number is shown graphically as:

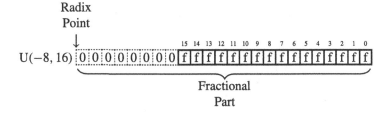

Likewise, signed fixed point numbers will be specified using the following notation: S(i, f), where i is the position of the radix point in relation to the *right* side of the sign bit, and f is the number of fractional bits stored. As with integer two's-complement notation, the sign bit is always the leftmost bit stored. For example, S(9, 6) indicates that there are six bits in the fractional part of the number, and the radix point is nine bits to the right of the sign bit. The layout for this number is shown graphically as:

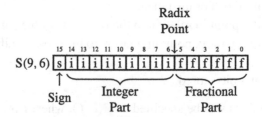

where i is an integer bit and f is a fractional bit. Very small numbers with no integer part may have a negative i. For example, $S(-7, 16)$ specifies a signed number with no integer part, six leading sign bits which are not actually stored, a sign bit that *is* stored and 15 bits of fraction. The layout for this number is shown graphically as:

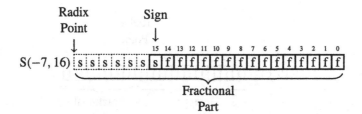

Note that the "hidden" bits in a signed number are assumed to be copies of the sign bit, while the "hidden" bits in an unsigned number are assumed to be zero.

The following figure shows an unsigned fixed point number with seven bits in the integer part and nine bits in the fractional part. It is a $U(7, 9)$ number. Note that the total number of bits is $7 + 9 = 16$

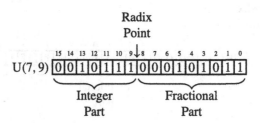

The value of this number in base 10 can be computed by summing the values of each non-zero bit as follows:

$$2^{13-9} + 2^{11-9} + 2^{10-9} + 2^{9-9} + 2^{5-9} + 2^{3-9} + 2^{1-9} + 2^{0-9}$$

$$= 2^4 + 2^2 + 2^1 + 2^0 + 2^{-4} + 2^{-6} + 2^{-8} + 2^{-9}$$
$$= 16 + 4 + 2 + 1 + \frac{1}{16} + \frac{1}{64} + \frac{1}{256} + \frac{1}{512}$$
$$= 23.083984375_{10}$$

Likewise, the following figure shows a signed fixed point number with nine bits in the integer part and six bits in the fractional part. It is as $S(9, 6)$ number. Note that the total number of bits is $9 + 6 + 1 = 16$.

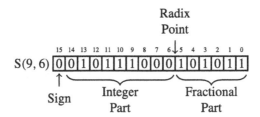

The value of this number in base 10 can be computed by summing the values of each non-zero bit as follows:

$$2^{13-6} + 2^{11-6} + 2^{10-6} + 2^{9-6} + 2^{5-6} + 2^{3-6} + 2^{1-6} + 2^{0-6}$$
$$= 2^7 + 2^5 + 2^4 + 2^3 + 2^{-1} + 2^{-3} + 2^{-5} + 2^{-6}$$
$$= 128 + 32 + 16 + 8 + \frac{1}{2} + \frac{1}{8} + \frac{1}{32} + \frac{1}{64}$$
$$= 184.671875_{10}$$

Note that in the above two examples, the patterns of bits are identical. The value of a number depends upon how it is interpreted. The notation that we have introduced allows us to easily specify exactly how a number is to be interpreted. For signed values, if the first bit is non-zero, then the two's complement should be taken before the number is evaluated. For example the following figure shows an $S(8, 7)$ number that has a negative value.

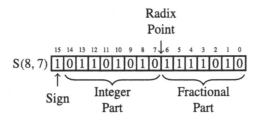

The value of this number in base 10 can be computed by taking the two's complement, summing the values of the non-zero bits, and adding a negative sign to the result. The two's complement of 1011010101111010 is $0100101010000101 + 1 = 0100101010000110$. The value of this number is:

$$
\begin{aligned}
& -\left(2^{14-7} + 2^{11-7} + 2^{9-7} + 2^{7-7} + 2^{2-7} + 2^{1-7}\right) \\
= \ & -\left(2^{7} + 2^{4} + 2^{2} + 2^{0} + 2^{-5} + 2^{-6}\right) \\
= \ & -\left(128 + 16 + 4 + 1 + \frac{1}{32} + \frac{1}{64}\right) \\
= \ & -149.0468750_{10}
\end{aligned}
$$

For a final example we will interpret this bit pattern as an S(−5, 16). In that format, the layout is:

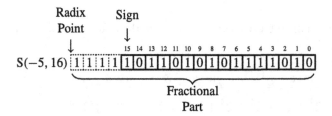

The value of this number in base ten can be computed by taking the two's complement, summing the values of the non-zero bits, and adding a negative sign to the result. The two's complement is:

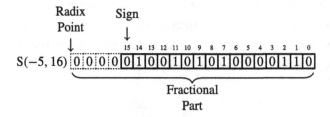

The value of this number interpreted as an S(−5, 16) is:

$$
-\left(2^{-6} + 2^{-9} + 2^{-11} + 2^{-13} + 2^{-18} + 2^{-19}\right) = -0.0181941986083984375
$$

8.3.2 Q notation

Fixed point number formats can also be represented using Q notation, which was developed by Texas Instruments. Q notation is equivalent to the S/U format used in this book, except that the integer portion is not always fully specified. In general, Q formats are specified as Qm, n where m is the number of integer bits, and n is the number of fractional bits. If a fixed word size w is being used then m may be omitted, and is assumed to be $w - n$. For example, a Q10 number has ten fractional bits, and the number of integer bits is not specified, but is assumed to be the number of bits required to complete a word of data. A Q2,4 number has two integer bits and four fractional bits in a six bit word. There are two conflicting conventions for dealing with the sign bit. In one convention, the sign bit is included as part of m, and in the other convention, it is not. When using Q notation, it is important to state which convention is being used. Additionally, a U may be prefixed to indicate an unsigned value. For example UQ8.8 is equivalent to U(8, 8), and Q7,9 is equivalent to S(7, 9).

8.3.3 Properties of fixed point numbers

Once the decision has been made to used fixed point calculations, the programmer must make some decisions about the specific representation of each fixed point variable. The combination of size and radix will affect several properties of the numbers, including:

Precision: the maximum number of non-zero bits representable,

Resolution: the smallest non-zero magnitude representable,

Accuracy: the magnitude of the maximum difference between a true real value and it's approximate representation,

Range: the difference between the largest and smallest number that can be represented, and

Dynamic range: the ratio of the maximum absolute value, and the minimum positive absolute value representable.

Given a number specified using the notation introduced previously, we can determine its properties. For example, an $S(9, 6)$ number has the following properties:

Precision: $P = 16$ bits

Resolution: $R = 2^{-6} = 0.015625$

Accuracy: $A = \frac{R}{2} = 0.0078125$

Range: Minimum value is $1000000000.000000 = -512$
Maximum value is $0111111111.111111 = 511.9921875$
Range is $G = 511.9921875 + 512 = 1023.9921875$

Dynamic range: For a signed fixed-point rational representation, $S(i, f)$, the dynamic range is

$$D = 2 \times \frac{2^i}{2^{-f}} = 2^{i+f+1} = 2^P.$$

Therefore, the dynamic range of an $S(9, 6)$ is $2^{16} = 65536$.

Being aware of these properties, the programmer can select fixed point representations that fit the task that they are trying to solve. This allows the programmer to strive for very efficient code by using the smallest fixed point representation possible, while still guaranteeing that the results of computations will be within some limits for error tolerance.

8.4 Fixed point operations

Fixed point numbers are actually stored as integers, and all of the integer mathematical operations can be used. However, some care must be taken to track the radix point at each stage of the computation. The advantages of fixed point calculations are that the operations are very fast and can be performed on any computer, even if it does not have special hardware support for non-integral numbers.

8.4.1 Fixed point addition and subtraction

Fixed point addition and subtraction work exactly like their integer counterparts. Fig. 8.1 gives some examples of fixed point addition with signed numbers. Note that in each case, the numbers are aligned so that they have the same number of bits in their fractional part. This requirement is the only difference between integer and fixed point addition. In fact, integer arithmetic is just fixed point arithmetic with no bits in the fractional part. The arithmetic that was covered in Chapter 7 was fixed point arithmetic using only $S(i, 0)$ and $U(i, 0)$ numbers. In other words, integers and natural numbers can be considered as fixed point numbers where the number of fractional bits, f, is zero. Now we are simply extending our knowledge to deal with numbers where $f \neq 0$. There are some rules which must be followed to ensure that the results are correct. The rules for subtraction are the same as the rules for addition. Since we are using two's complement math, subtraction is performed using addition.

Suppose we want to add an $S(7, 8)$ number to an $S(7, 4)$ number. The radix points are at different locations, so we cannot simply add them. Instead, we must shift one of the numbers,

2.25	00010.010
+ 1.50 =	+ 00001.100
3.75	00011.110

11.125	01011.001
− 5.625 =	+ 11010.011
5.500	00101.100

−12.375	10011.101
+ 5.250 =	+ 00101.010
− 7.125	11000.111

Figure 8.1: Examples of fixed point signed arithmetic.

changing its format, until the radix points are aligned. The choice of which one to shift depends on what format we desire for the result. If we desire eight bits of fraction in our result, then we would shift the $S(7,4)$ left by four bits, converting it into an $S(7,8)$. With the radix points aligned, we simply use an integer addition operation to add the two numbers. The result will have it's radix point in the same location as the two numbers being added.

8.4.2 Fixed point multiplication

Recall that the result of multiplying an n bit number by an m bit number is an $n + m$ bit number. In the case of fixed point numbers, the size of the fractional part of the result is the sum of the number of fractional bits of each number, and the total size of the result is the sum of the total number of bits in each number. Consider the following example where two $U(5,3)$ numbers are multiplied together:

$$
\begin{array}{r}
00011.110 \\
\times\ 00010.100 \\
\hline
0001.1110 \\
000111.10 \\
\hline
0000001001.011000 \\
\end{array}
$$

The result is a $U(10,6)$ number. The number of bits in the result is the sum of all of the bits of the multiplicand and the multiplier. The number of fractional bits in the result is the sum of the number of fractional bits in the multiplicand and the multiplier. There are three simple rules to predict the resulting format when multiplying any two fixed point numbers.

Listing 8.1 Examples of fixed point multiplication in AArch64 assembly.

```
1   // Multiply two S(10,5) numbers and produce an S(10,5) result.
2   smull x0,w1,w2      // x = a * b -> S(21,10)
3   asr   x0,x0,#5      // shift back to S(10,5)
4
5   // Multiply two U(12,4) numbers and produce a U(22,6) result.
6   umull x3,w4,w5      // x = a * b -> U(24,8)
7   lsr   x3,x3,#2      // shift back to U(22,6)
8
9   // Multiply two S(16,15) numbers and produce an S(16,15) result.
10  smull x0,w2,w3      // x = a * b -> S(33,30)
11  asr   x0,x0,#15     // shift back to S(16,15)
12
13  // Multiply two U(10,22) numbers and produce a U(10,22) result.
14  umull x0,w2,w3      // x = a * b -> U(20,44)
15  lsr   x0,x0,#22     // shift back to U(10,22)
```

Unsigned Multiplication

The result of multiplying two unsigned numbers $U(i_1, f_1)$ and $U(i_2, f_2)$ is a $U(i_1 + i_2, f_1 + f_2)$ number.

Mixed Multiplication

The result of multiplying a signed number $S(i_1, f_1)$ and an unsigned number $U(i_2, f_2)$ is an $S(i_1 + i_2, f_1 + f_2)$ number.

Signed Multiplication

The result of multiplying two signed numbers $S(i_1, f_1)$ and $S(i_2, f_2)$ is an $S(i_1 + i_2 + 1, f_1 + f_2)$ number.

Note that this rule works for integers as well as fixed-point numbers, since integers are really fixed point numbers with $f = 0$. If the programmer desires a particular format for the result, then the multiply is followed by an appropriate shift.

Listing 8.1 gives some examples of fixed point multiplication using the ARM multiply instructions. In each case, the result is shifted to produce the desired format. It is the responsibility of the programmer to know what type of fixed point number is produced after each multiplication and to adjust the result by shifting if necessary.

8.4.3 Fixed point division

The rule for determining the format of the result of division is more complicated than the one for multiplication. We will first consider only unsigned division of a dividend with format $U(i_1, f_1)$ by a divisor with format $U(i_2, f_2)$.

8.4.3.1 Results of fixed point division

Consider the results of dividing two fixed point numbers, using integer operations with limited precision. The value of the least significant bit of the dividend N is 2^{-f_i} and the value of the least significant bit of the divisor D is 2^{-f_2}. In order to perform the division using integer operations, it is necessary to multiply N by 2^{f_i} and multiply D by 2^{f_2} so that both numbers are integers. Therefore the division operation can be written as:

$$Q = \frac{N \times 2^{f_1}}{D \times 2^{f_2}} = \frac{N}{D} \times 2^{f_1 - f_2}.$$

Note that no multiplication is actually performed. Instead, the programmer mentally shifts the radix point of the divisor and dividend, then computes the radix point of the result. For example, given two U(5, 3) numbers, the division operation is accomplished by converting them both to integers, performing the division, then computing the location of the radix point:

$$Q = \frac{N \times 2^3}{D \times 2^3} = \frac{N}{D} \times 2^0.$$

Note that the result is an integer. If the programmer wants to have some fractional bits in the result, then the dividend must be shifted to the left before the division is performed.

If the programmer wants to have f_q fractional bits in the quotient, then the amount that the dividend must be shifted can easily be computed as

$$s = f_q + f_1 - f_2.$$

For example, suppose the programmer wants to divide 01001.011 stored as a U(28, 3) by 00011.110 which is also stored as a U(28, 3), and wishes to have six fractional bits in the result. The programmer would first shift 01001.011 to the left by six bits, then perform the division and compute the position of the radix in the result as shown:

$$01001.011 \div 00011.110 = (0000001001011000000 \div 00011110) \times 2^{-6-3+3}$$

$$
\begin{array}{r}
10100000 \times 2^{-6} = 10.100000 \\
\hline
11110{\overline{)}\,1001011000000} \\
111100000000 \\
\hline
1111000000 \\
1111000000 \\
\hline
0
\end{array}
$$

Since the divisor may be between zero and one, the quotient may actually require *more* integer bits than there are in the dividend. Consider that the largest possible value of the dividend is $N_{max} = 2^{i_1} - 2^{-f_1}$, and the smallest positive value for the divisor is $D_{min} = 2^{-f_2}$. Therefore, the maximum quotient is given by:

$$Q_{max} = \frac{2^{i_1} - 2^{-f_1}}{2^{-f_2}} = 2^{i_1+f_2} - 2^{f_1-f_2}.$$

Taking the limit of the previous equation,

$$\lim_{f_1-f_2\to-\infty} Q_{max} = 2^{i_1+f_2},$$

provides the following bound on how many bits are required in the integer part of the quotient:

$$Q_{max} < 2^{i_1+f_2}.$$

Therefore, in the worst case, the quotient will require $i_1 + f_2$ integer bits. For example, if we divide a U(3, 5), $a = 111.11111 = 7.96875_{10}$, by a U(5, 3), $b = 00000.001 = 0.125_{10}$, we end up with a U(6, 2) $q = 111111.11 = 63.75_{10}$.

The same thought process can be used to determine the results for signed division as well as mixed division between signed and unsigned numbers. The results can be reduced to the following three rules:

Unsigned Division

   The result of dividing an unsigned fixed point number U(i_1, f_1) by an unsigned number U(i_2, f_2) is a U($i_1 + f_2$, $f_1 - f_2$) number.

Mixed Division

   The result of dividing two fixed point numbers where one of them is signed and the other is unsigned is an S($i_1 + f_2$, $f_1 - f_2$) number.

Signed Division

   The result of dividing two signed fixed point numbers is an S($i_1 + f_2 + 1$, $f_1 - f_2$) number.

Consider the results when a U(2, 3), $a = 00000.001 = 0.125_{10}$ is divided by a U(4, 1), $b = 1000.0 = 8.0_{10}$. The quotient is $q = 0.000001$, which requires six bits in the fractional part. However, if we simply perform the division, then according to the rules shown above, the result will be a U(8, −2). This indicates that the result would have to be shifted two bits to the left to restore it to an integer, and the result is only accurate to within ±4.

When $f_2 > f_1$, blindly applying the rules will sometimes result in a negative fractional part. To avoid this, the dividend can be shifted left so that it has at least as many fractional bits as the divisor. This leads to the following rule: If $f_2 > f_1$ then convert the divisor to an S(i_1, x), where $x \geq f_2$, then apply the appropriate rule. For example, dividing an S(5, 2) by a U(3, 12) would result in an S(17, −10). But shifting the S(5, 2) 16 bits to the left will result in an S(5, 18), and dividing that by a U(3, 12) will result in an S(17, 6).

8.4.3.2 *Maintaining precision*

Recall that integer division produces a result and a remainder. In order to maintain precision, it is necessary to perform the integer division operation in such a way that all of the significant bits are in the result and only insignificant bits are left in the remainder. The easiest way to accomplish this is by shifting the dividend to the left before the division is performed.

To find a rule for determining the shift necessary to maintain full precision in the quotient, consider the worst case. The minimum positive value of the dividend is $N_{min} = 2^{-f_1}$ and the largest positive value for the divisor is $D_{min} = 2^{i_2} - 2^{-f_2}$. Therefore, the minimum positive quotient is given by:

$$
\begin{aligned}
Q_{min} &= \frac{2^{-f_1}}{2^{i_2} - 2^{-f_2}} \\
&= \frac{\frac{1}{2^{f_1}}}{\frac{2^{i_2 + f_2}}{2^{f_2}}} \\
&= \frac{2^{f_2}}{2^{f_1 + i_2 + f_2}} \\
&= \frac{1}{2^{f_1 + i_2}} \\
&= 2^{-(i_2 + f_1)}
\end{aligned}
$$

Therefore, in the worst case, the quotient will require $i_2 + f_1$ fractional bits to maintain precision. However, fewer bits can be reserved if full precision is not required.

Recall that the least significant bit of the quotient will be $2^{-(i_2 + f_1)}$. Shifting the dividend left by $i_2 + f_2$ bits will convert it into a $U(i_1, i_2 + f_1 + f_2)$. Using the rule above, when it is divided by a $U(i_2, f_2)$, the result is a $U(i_1 + f_2, i_2 + f_1)$. This is the minimum size which is guaranteed to preserve all bits of precision. The general method for performing fixed point division *while maintaining maximum precision* is as follows:

1. shift the dividend left by $i_2 + f_2$ then
2. perform integer division.

The result will be a $U(i_1 + f_2, i_2 + f_1)$ for unsigned division, or an $S(i_1 + f_2 + 1, i_2 + f_1)$ for signed division. The result for mixed division is left as an exercise for the student.

8.4.4 *Division by a constant*

Section 7.3.3 introduced the idea of converting division by a constant into multiplication by the reciprocal of that constant. In that section it was shown that by pre-multiplying the reciprocal by a power of two (a shift operation), then dividing the final result by the same power of two (a shift operation), division by a constant could be performed using only integer operations with a more efficient multiply replacing the (usually) very slow divide.

This section presents an alternate way to achieve the same results, by treating division by an integer constant as an application of fixed point multiplication. Again, the integer constant divisor is converted into its reciprocal, but this time the process is considered from the viewpoint of fixed-point mathematics. Both methods will achieve exactly the same results, but some people tend to grasp the fixed point approach more easily than the purely integer approach.

When writing code to divide by a constant, the programmer must strive to achieve the largest number of significant bits possible, while using the shortest (and most efficient) representation possible. On modern computers, this usually means using 32-bit integers and integer multiply operations which produce 64-bit results. That would be extremely tedious to show in a textbook, so the principals will be demonstrated here using 8-bit integers and an integer multiply which produces a 16-bit result.

8.4.4.1 Division by constant 23

Suppose we want efficient code to calculate $x \div 23$ using only 8-bit signed integer multiplication. The reciprocal of 23, in binary, is

$$R = \frac{1}{23} = 0.0000101100100001011\ldots_2 .$$

If we store R as an S(1, 11), it would look like this:

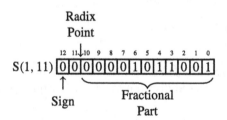

Note that in this format, the reciprocal of 23 has five leading zeros. We can store R in eight bits by shifting it left to remove some of the leading zeros. Each shift to the left changes the format of R. After removing the first leading zero bit, we have:

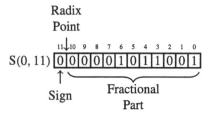

After removing the second leading zero bit, we have:

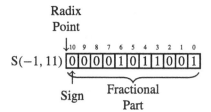

After removing the third leading zero bit, we have:

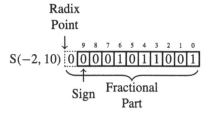

Note that the number in the previous format has a "hidden" bit between the radix point and the sign bit. That bit is not actually stored, but is assumed to be identical to the sign bit. Removing the fourth leading zero produces:

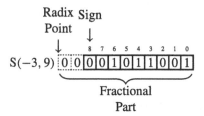

The number in the previous format has two "hidden" bits between the radix point and the sign bit. Those bits are not actually stored, but are assumed to be identical to the sign bit. Removing the fifth leading zero produces:

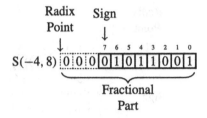

We can only remove five leading zero bits, because removing one more would change the sign bit from 0 to 1, resulting in a completely different number. Note that the final format has three "hidden" bits between the radix point and the sign bit. These bits are all copies of the sign bit. It is an $S(-4, 8)$ number because the sign is four bits to the right of the radix point (resulting in the three "hidden" bits). According to the rules of fixed point multiplication given earlier, an $S(7, 0)$ number x multiplied by an $S(-4, 8)$ number R will yield an $S(4, 8)$ number y. The value y will be $2^3 \times \frac{x}{23}$ because we have three "hidden" bits to the right of the radix point. Therefore,

$$\frac{x}{23} = R \times x \times 2^{-3},$$

indicating that after the multiplication, we must shift the result right by three bits to restore the radix. Since $\frac{1}{23}$ is positive, the number R must be increased by one to avoid round-off error. Therefore, we will use $R + 1 = 01011010 = 90_{10}$ in our multiply operation. To calculate $y = 101_{10} \div 23_{10}$, we can multiply and perform a shift as follows:

$$
\begin{array}{r}
.\,0\,1\,1\,0\,0\,1\,0\,1 \\
\times\ 0\,1\,0\,1\,1\,0\,1\,0 \\
\hline
0.\,1\,1\,0\,0\,1\,0\,1\,0 \\
0\,1\,1.\,0\,0\,1\,0\,1 \\
0\,1\,1\,0.\,0\,1\,0\,1 \\
0\,1\,1\,0\,0\,1.\,1\,1 \\
\hline
0\,0\,1\,0\,0\,1\,0\,0.\,0\,0\,0\,0\,0\,0\,1\,0
\end{array}
$$

Because our task is to implement *integer* division, everything to the right of the radix point can be immediately discarded, keeping only the upper eight bits as the integer portion of the result. The integer portion, 100011_2, shifted right three bits, is $100_2 = 4_{10}$. If the modulus is required, it can be calculated as: $101 - (4 \times 23) = 9$. Some processors, such as the Motorola HC11, have a special multiply instruction which keeps only the upper half of the result. This method would be especially efficient on that processor. Listing 8.2 shows

Listing 8.2 Dividing x by 23 with 8 bits of precision.

```
// Assume that w0 already contains x, where -129 < x < 128
// and x1 is available to hold 1/23 * 2^3
ldr    w1,=0b01011010   // Load 1/23 * 2^3 into r1
umull  x1,w0,w1         // Perform multiply
asr    x1,x1,#11        // shift result right by 8+3 bits
add    x1,x1,x1,lsr #63 // add one if result is negative
```

Listing 8.3 Dividing x by 23 Using Only Shift and Add

```
// Assume that x0 already contains x, where -129 < x < 128
// and x1 is available to hold 1/23 * 2^3
add    x0,x0,x0,lsl #2 // r0 <- x + x*4 = 5x
add    x0,x0,x0,lsl #3 // r0 <- 5x + 5x*8 = 45x
asr    x1,x1,#10        // shift result right
add    x1,x1,x1,lsr #63 // add one if result is negative
```

how the 8-bit division code would be implemented in AArch64 assembly. Listing 8.3 shows an alternate implementation which uses shift and add operations rather than a multiply.

8.4.4.2 Division by constant −50

The procedure is exactly the same for dividing by a negative constant. Suppose we want efficient code to calculate $\frac{x}{-50}$ using 16-bit signed integers. We first convert $\frac{1}{50}$ into binary:

$$\frac{1}{50} = 0.00000\overline{10100011110}$$

The two's complement of $\frac{1}{50}$ is

$$\frac{1}{-50} = 1.11111\overline{01011100001}$$

We can represent $\frac{1}{-50}$ as the following S(1, 21) fixed point number:

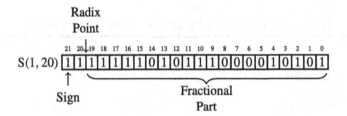

Note that the upper seven bits are all one. We can remove six of those bits and adjust the format as follows. After removing the first leading one, the reciprocal is:

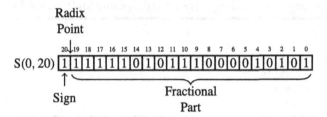

Removing another leading one changes the format to:

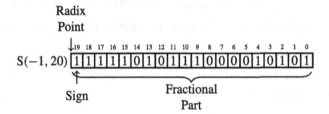

On the next step, the format is:

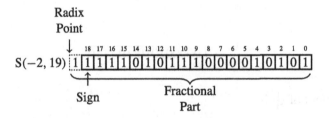

Note that we now have a "hidden" bit between the radix point and the sign bit. The hidden bit is not actually part of the number that we store and use in the computation, but it is assumed to be the same as the sign bit.

After three more leading ones are removed, the format is:

Listing 8.4 Dividing x by -50 with 16 bits of precision.

```
// Assume that r0 already contains x, where -32769 < x < 32768
// and r1 is available to hold 1/-50 * 2^4
ldr    w1,=0xAE15      // Load 1/-50 * 2^4 into r1
mul    x1,w0,w1        // Perform multiply
asr    x1,x1,#20       // shift result right by 16+4 bits
add    x1,x1,x1,lsr #63 // add one if result is negative
```

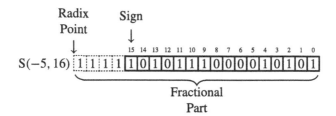

Note that there are four "hidden" bits between the radix point and the sign. Since the reciprocal $\frac{1}{-50}$ is negative, we do not need to round by adding one to the number R. Therefore, we will use $R = 1010111000010101_2 = AE15_{16}$ in our multiply operation.

Since we are using 16-bit integer operations, the dividend, x, will be an $S(15, 0)$. The product of an $S(15, 0)$ and an $S(-5, 16)$ will be an $S(11, 16)$. We will remove the 16 fractional bits by shifting right. The four "hidden" bits indicate that the result must be shifted additional four bits to the right, resulting in a total shift of 20 bits. Listing 8.4 shows how the 16-bit division code would be implemented in ARM assembly.

8.5 Fixed point input and output

When working with fixed point, it is often convenient to define a few functions for conversion between base ten numbers and their fixed point representation. Listing 8.5 provides an example of converting from a base 10 number represented as a string of characters to a fixed point binary number, and an example showing how to print out a fixed point number.

Listing 8.5 Functions to convert from a string to fixed point, and to print out a fixed point number

```
/***************************************************************
 Name: fixedfuncs.c
 Author: Larry Pyeatt
 Date: 2/22/2018
 ***************************************************************/
```

```c
#include <stdlib.h>
#include <stdio.h>
#include <string.h>

#define MAX_DECIMAL_DIGITS 16
/* Multiply an unpacked BCD number by 2. Return 1 if there is a
   carry out of the most significant digit, or 0 otherwise.  The
   resulting number is returned in n1.
*/
int base10double(char n1[MAX_DECIMAL_DIGITS])
{
  int i, tmp, carry=0;
  for(i=0;i<MAX_DECIMAL_DIGITS;i++)
    {
      n1[i] += (n1[i] + carry);
      if(n1[i] > 9)
        {
          n1[i] -= 10;
          carry = 1;
        }
      else
        carry=0;
    }
  return carry;
}

/* Convert a string into a signed fixed-point binary
   representation with up to 32 bits of fractional part.
*/
int strtoSfixed(char *s, int frac_bits)
{
  char *point = s;
  unsigned int value;
  int i,negative=0;
  char digits[MAX_DECIMAL_DIGITS];
  /* get the integer portion*/
  if(*s=='-')
    {
      negative=1;
      s++;
    }
  value = atoi(s);
  /* find the decimal point */
  while((*point != '.')&&(*point != 0))
    point++;
  /* if there is nothing after the decimal point, or there is
     not a decimal point, then shift and return what we already
     have */
  if(( *point == 0 ) || ( *(point+1) == 0 ))
```

```
      {
        if(negative)
          value = -value;
        return value << frac_bits;
      }
    ++point;
    /* convert the remaining part into an unpacked BCD number. */
    for(i=(MAX_DECIMAL_DIGITS-1);i>=0;i--)
      {
        if(*point == 0)
          digits[i] = 0;
        else
          {
            digits[i] = *point - '0';
            ++point;
          }
      }
    /* convert the unpacked BCD number into binary */
    while(frac_bits > 0)
      {
        value <<= 1;
        if(base10double(digits))
          value |= 1;
        frac_bits--;
      }
    /* negate if there was a leading '-' */
    if(negative)
      value = -value;
    return value;
}

// Round a packed BCD fraction
int roundit(int *digits, int current)
{
    while(current > 0)
      {
        if(digits[current] > 9)
          {
            digits[current] %= 10;
            digits[current-1]++;
          }
        current--;
      }
    if((current == 0)&&(digits[current] > 9))
      {
        digits[current] %= 10;
        return 1;
      }
    return 0;
```

```
104   }
105
106   /* Print a signed fixed point number with the given number of
107      bits in the fractional part.  NOTE: frac_bits must be between
108      0 and 31 for this function to work properly.
109   */
110   #define MAX_DIGITS 8
111   void printS( int num, int frac_bits )
112   {
113     unsigned int mask = (1 << frac_bits) - 1;
114     unsigned int fracpart;
115     int intpart;
116     int count = 0;
117     int frac_digits[MAX_DIGITS+1];
118
119     if(num < 0)
120       {
121         printf("-");
122         num = -num;
123       }
124     else
125       printf(" ");
126     /* Remove the integer part and keep the fraction part */
127     fracpart = num & mask;
128     intpart = num>>frac_bits;
129
130     /* Print all of the digits in the fraction part into frac_digits.*/
131     for(count=0;count<MAX_DIGITS+1;count++)
132       {
133         fracpart *= 10;
134         frac_digits[count] = fracpart >> frac_bits;
135         fracpart &= mask ;
136       }
137
138     /* Round the fractional part */
139     frac_digits[MAX_DIGITS] += 5;
140     intpart += roundit(frac_digits,MAX_DIGITS);
141     /* Print the integer part (with the sign, if it is negative) */
142     printf("%d.",intpart);
143     /* Print the fractional digits */
144     for(count=0;count<MAX_DIGITS;count++)
145       printf("%d",frac_digits[count]);
146   }
```

These functions are not meant to be particularly efficient, but are given as examples of the types of functions that would be needed for input and output of fixed point numbers. Another way to approach fixed point I/O would be to provide only functions to convert fixed point values of various sizes to and from C strings, then use scanf and printf to read and write the strings. For some applications it could also be useful to have functions for converting between IEEE floating point (covered later in this chapter) and fixed point formats.

8.6 Computing sine and cosine

It has been said, and is commonly accepted, that "you can't beat the compiler." The meaning of this statement is that using hand-coded assembly language is futile and/or worthless because the compiler is "smarter" than a human. This statement is a myth, as will now be demonstrated.

There are many mathematical functions that are useful in programming. Two of the most useful functions are $\sin x$ and $\cos x$. However, these functions are not always implemented in hardware, particularly for fixed point representations. If these functions are required for fixed point computation, then they must be written in software. These two functions have some nice properties that can be exploited. In particular:

- If we have the $\sin x$ function, then we can calculate $\cos x$ using the relationship

$$\cos x = \sin \frac{\pi}{2 - x}. \tag{8.1}$$

  Therefore, we only need to get the sine function working, and then we can implement cosine with only a little extra effort.
- $\sin x$ is cyclical, so $\ldots \sin -2\pi = \sin 0 = \sin 2\pi \ldots$. This means that we can limit the domain of our function to the range $[-\pi, \pi]$.
- $\sin x$ is symmetric, so that $\sin -x = -\sin x$. This means that we can further restrict the domain to $[0, \pi]$.
- After we restrict the domain to $[0, \pi]$, we notice another symmetry, $\sin x = \sin(\pi - x)$, $\frac{\pi}{2} \le x \le \pi$ and we can further restrict the domain to $[0, \frac{\pi}{2}]$.
- the range of both functions, $\sin x$ and $\cos x$, is in the range $[-1, 1]$.

If we exploit all of these properties, then we can write a single shared function to be used by both sine and cosine. We will name this function `sinq`, and choose the following fixed point formats:

- `sinq` will accept x as an $S(1, 30)$, and
- `sinq` will return an $S(1, 30)$.

These formats were chosen because $S(1, 30)$ is a good format for storing a signed number between zero and $\frac{\pi}{2}$, and also the optimal format for storing a signed number between one and negative one.

Table 8.1: Natural and truncated formats for the powers of x needed by the Taylor series.

Term	Format	32-bit
x	$S(1, 30)$	$S(1, 30)$
x^3	$S(3, 90)$	$S(3, 28)$
x^5	$S(5, 150)$	$S(5, 26)$
x^7	$S(7, 210)$	$S(7, 24)$
x^9	$S(9, 270)$	$S(9, 22)$
x^{11}	$S(11, 330)$	$S(11, 20)$
x^{13}	$S(13, 390)$	$S(13, 18)$
x^{15}	$S(15, 450)$	$S(15, 16)$
x^{17}	$S(17, 510)$	$S(17, 14)$

The sine function will map x into the domain accepted by sinq and then call sinq to do the actual work. If the result should be negative, then the sine function will negate it before returning. The cosine function will use the relationship previously mentioned, and call the sine function.

We have now reduced the problem to one of approximating $\sin x$ within the range $[0, \frac{\pi}{2}]$. An approximation to the function $\sin x$ can be calculated using the Taylor Series:

$$\sin x = \sum_{n=0}^{\infty} (-1)^n \frac{x^{2n+1}}{(2n+1)!} \tag{8.2}$$

The first few terms of the series should be sufficient to achieve a good approximation. The maximum value possible for the seventh term is $\frac{(0.5 \times \pi)^{13}}{13!} \approx 0.0000000005_{10}$, which indicates that our function should be accurate to at least 25 bits using seven terms. If more accuracy is desired, then additional terms can be added. We will use the first nine terms of the Taylor series in our function.

8.6.1 Formats for the powers of x

The numerators in the first nine terms of the Taylor series approximation are: x, x^3, x^5, x^7, x^9, x^{11}, x^{13}, x^{15}, and x^{17}. Given an $S(1, 30)$ format for x, we can predict the format for the numerator of each successive term in the Taylor series. If we simply perform successive multiplies, then we would get the formats shown in Table 8.1. The middle column in the table shows that the format for x^{17} would require 528 bits if all of the fractional bits are retained. Dealing with a number at that level of precision would be slow and impractical. We will, of necessity, need to limit the number of bits used. Since the ARM processor provides a multiply instruction involving two 32-bit numbers, we choose to truncate the numerators to 32 bits. The third column in the table indicates the resulting format for each term if precision is limited to 32 bits.

On further consideration of the Taylor series, we notice that each of the above terms will be divided by a constant. Instead of dividing, we can multiply by the reciprocal of the constant. We will create a similar table holding the formats and constants for the factorial terms. With a bit of luck, the division (implemented as multiplication) in each term will result in a reasonable format for each resulting term.

8.6.2 Formats and constants for the factorial terms

The first term of the Taylor series is $\frac{x}{1!}$, so we can simply skip the division. The second term is $-\frac{x^3}{3!} = x^3 \times -\frac{1}{3!}$ and the third term is $\frac{x^5}{5!} = x^5 \times -\frac{1}{5!}$ We can convert $-\frac{1}{3!}$ to binary as follows:

Multiplication	Result	
	Integer	*Fraction*
$\frac{1}{6} \times 2 = \frac{2}{6}$	0	$\frac{2}{6}$
$\frac{2}{6} \times 2 = \frac{4}{6}$	0	$\frac{4}{6}$
$\frac{4}{6} \times 2 = \frac{8}{6}$	1	$\frac{2}{6}$
$\frac{2}{6} \times 2 = \frac{4}{6}$	0	$\frac{4}{6}$
$\frac{8}{6} \times 2 = \frac{8}{6}$	1	$\frac{2}{6}$

Note that the fraction in the fifth row of the table is the same as the fraction in the third row of the table. Since the pattern repeats, we can conclude that $\frac{1}{3!} = 0.0\overline{01}_2$. Since we need a negative number, we take the two's complement, resulting in $-\frac{1}{3!} = \dots 111.1\overline{10}_2$. The fraction $\frac{1}{3!}$ can be represented as an $S(1, 33)$, as shown below:

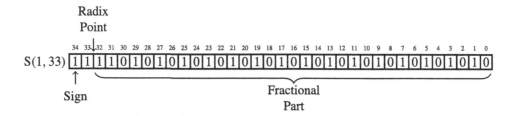

Since the first four bits are one, we can remove three bits and store it as:

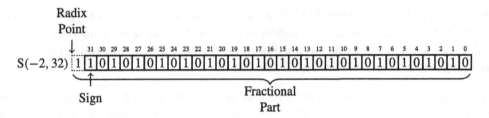

In hexadecimal, this is AAAAAAAA$_{16}$. We will store this value, along with the remaining recip-rocals, in a table. The sine function can quickly retrieve these values and use them as needed to compute the Taylor series approximation. By pre-computing these values and storing them in a look-up table, we will gain a great deal of performance, compared to computing them on-the-fly. This approach is often used to speed up the implementation of numerical functions.

Performing the same operations, we find that $\frac{1}{5!}$ can be converted to binary as follows:

Multiplication	Result	
	Integer	*Fraction*
$\frac{1}{120} \times 2 = \frac{2}{120}$	0	$\frac{2}{120}$
$\frac{2}{120} \times 2 = \frac{4}{120}$	0	$\frac{4}{120}$
$\frac{4}{120} \times 2 = \frac{8}{120}$	0	$\frac{8}{120}$
$\frac{8}{120} \times 2 = \frac{16}{120}$	0	$\frac{16}{120}$
$\frac{16}{120} \times 2 = \frac{32}{120}$	0	$\frac{32}{120}$
$\frac{32}{120} \times 2 = \frac{64}{120}$	0	$\frac{64}{120}$
$\frac{64}{120} \times 2 = \frac{128}{120}$	1	$\frac{8}{120}$

Since the fraction in the seventh row is the same as the fraction in the third row, we know that the table will repeat forever. Therefore, $\frac{1}{5!} = 0.0000\overline{0001}_2$. Since the first six bits to the right of the radix are all zero, we can remove the first five bits. Also adding one to the least significant bit to account for rounding error yields the following S(-6, 32):

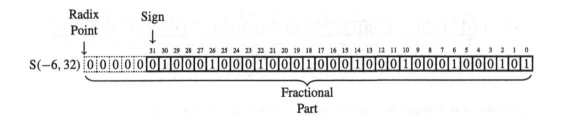

**Table 8.2: Reciprocals for the terms two
through nine of the Taylor series.**

Term	Reciprocal Format	Reciprocal Value (Hex)
$-\frac{1}{3!}$	$S(-2, 32)$	AAAAAAAA
$\frac{1}{5!}$	$S(-6, 32)$	44444445
$-\frac{1}{7!}$	$S(-12, 32)$	97F97F97
$\frac{1}{9!}$	$S(-18, 32)$	5C778E96
$-\frac{1}{11!}$	$S(-25, 32)$	9466EA60
$\frac{1}{13!}$	$S(-32, 32)$	5849184F
$-\frac{1}{15!}$	$S(-40, 32)$	94603063
$\frac{1}{17!}$	$S(-48, 32)$	654B1DC1

In Hexadecimal, the number to be multiplied is 44444445_{16}. Note that since $\frac{1}{5!}$ is a positive number, the reciprocal was incremented by one to avoid round-off errors. We can apply the same procedure to the remaining terms. Table 8.2 shows the reciprocals and their formats for terms two through nine. The first term is just one, so it is not included in the table.

8.6.3 Putting it all together

We want to keep as much precision as is reasonably possible for our intermediate calculations. However, we also want to provide the best performance possible. Using 64 bits of precision for all intermediate calculations will give a good trade-off between performance and precision. Keeping more than 64 bits would require more time to perform the multiplication and addition steps, and would not result in any appreciable difference in our final 32-bit result. The integer portion should never require more than two bits, so we choose an $S(2, 61)$ as our intermediate representation. If we combine Table 8.1 and Table 8.2, we can determine what the format of each complete term will be. This is shown in Table 8.3.

Note that the formats were truncated to fit in a 64 bit result. We can now see that the formats for the first nine terms of the Taylor series are reasonably similar. They all require exactly 64 bits, and the radix points can be shifted so that they are aligned for addition. In order to make the shifting and adding process easier, we will pre-compute the shift amounts and store them in a look-up table. In fact, we will create a look-up table to hold the reciprocal values and the shift amounts together, so that our Taylor series code is greatly simplified and extremely efficient. The procedure will involve calculating the next power of x, multiplying it by the

Table 8.3: Formats for the powers of x and the constant reciprocals for the first nine terms of the Taylor series.

Term	Numerator		Reciprocal			Result
	Value	Format	Value	Format	Hex	Format
1	x	$S(1, 30)$	Extend to 64 bits and shift right			$S(2, 61)$
2	x^3	$S(3, 28)$	$-\frac{1}{3!}$	$S(-2, 32)$	AAAAAAAA	$S(2, 61)$
3	x^5	$S(5, 26)$	$\frac{1}{5!}$	$S(-6, 32)$	44444444	$S(0, 63)$
4	x^7	$S(7, 24)$	$-\frac{1}{7!}$	$S(-12, 32)$	97F97F97	$S(-4, 64)$
5	x^9	$S(9, 22)$	$\frac{1}{9!}$	$S(-18, 32)$	5C778E96	$S(-8, 64)$
6	x^{11}	$S(11, 20)$	$-\frac{1}{11!}$	$S(-25, 32)$	9466EA60	$S(-13, 64)$
7	x^{13}	$S(13, 18)$	$\frac{1}{13!}$	$S(-32, 32)$	5849184F	$S(-18, 64)$
8	x^{15}	$S(15, 16)$	$-\frac{1}{15!}$	$S(-40, 32)$	94603063	$S(-24, 64)$
9	x^{17}	$S(17, 14)$	$\frac{1}{17!}$	$S(-48, 32)$	654B1DC1	$S(-30, 64)$

reciprocal in the table, shifting it by the amount also stored in the table, and adding it to our running total.

The following table shows the shifts that are necessary to convert each term to an $S(2, 61)$ so that it can be added to the running total:

Term Number	Original Format	Shift Amount	Resulting Format
1	$S(1, 30)$	1	$S(2, 61)$
2	$S(2, 61)$	0	$S(2, 61)$
3	$S(0, 63)$	2	$S(2, 61)$
4	$S(-4, 64)$	6	$S(2, 61)$
5	$S(-8, 64)$	10	$S(2, 61)$
6	$S(-13, 64)$	15	$S(2, 61)$
7	$S(-18, 64)$	20	$S(2, 61)$
8	$S(-24, 64)$	26	$S(2, 61)$
9	$S(-30, 64)$	32	$S(2, 61)$

Note that the ninth term will be shifted 32 bits to the right. This indicates that it contributes very little to the final 32-bit sum which is stored in the upper 33 bits of the running total. We now have all of the information that we need in order to implement the function. Listing 8.6 shows how the sine and cosine function can be implemented in C. Listing 8.7 shows how the sine and cosine function can be implemented in ARM assembly using fixed point computation, and Listing 8.8 shows a main program which prints a table of values and their sine and cosines.

Listing 8.6 C implementation of $\sin x$ and $\cos x$ using fixed point calculations.

```
#include <stdio.h>
//*********************************************************************
// Name: sincos.c
// Author: Larry Pyeatt
// Date: 2/22/2018
//*********************************************************************
// This file provides functions for calculating sine and
// cosine using fixed-point arithmetic. It uses the first
// nine terms in the Taylor series.

#define pi 0x3243F6A8     // pi as an S(3,28)
#define pi_2 0x1921FB54   // pi/2 as an S(3,28)
#define pi_x2 0x6487ED51  // 2*pi as an S(3,28)

// Define a structure that holds coefficient and shift.
struct tabentry{
  int coeff;
  int shift;
};

// create a table of coefficients and shift amounts.
#define TABSIZE 8
static struct tabentry sintab[]={
  {0xAAAAAAAA,  0}, // term 2
  {0x44444445,  2}, // term 3
  {0x97F97F97,  6}, // term 4
  {0x5C778E96, 10}, // term 5
  {0x9466EA60, 15}, // term 6
  {0x5849184F, 20}, // term 7
  {0x94603063, 26}, // term 8
  {0x654B1DC1, 32}  // term 9
};

//*********************************************************************
// sinq is used internally by the sine and cosine functions
//    input: x as an S(1,30) s.t. 0 <= x <= pi/2
//    returns sin(x) as an S(3,28)
static int sinq(int x)
{
  long long sum;
  long long tmp;
  long long curpower = x;
  long long xsq;
```

```
44      int i=0;
45      sum = (long long)x << 31;      // initialize 64-bit sum to x
46      xsq = (long long)x*(long long)x; // calculate x^2 as an S(3,60)
47      xsq >>= 31;                    // convert x^2 to an S(2,29)
48      do
49        {
50          // calculate x^(2n-1) as an S(3,28)
51          curpower = ((curpower * xsq) >> 31);
52          // multiply x^(2n-1) by coefficient from the table
53          tmp = curpower * sintab[i].coeff;
54          if(tmp < 0) // if resulting term in negative
55            tmp++;    //   add one to avoid round-off error
56          tmp >>= sintab[i].shift; // convert the term to S(2,61)
57          sum += tmp; // add it to running total
58        }
59      while(++i<TABSIZE);
60      return (sum >> 33);   // convert result to S(3,28) and return
61    }
62
63    //**********************************************************
64    // fixed_sin_C applies symmetries to reduce the range of
65    // the input, then calls sinq and adjusts the result.
66    int fixed_sin_C(int x)
67    {
68      while(x<0)
69        x += pi_x2;
70      while(x>pi_x2)
71        x -= pi_x2;
72      if(x<=pi_2)
73        return sinq(x<<2);
74      if(x<=pi)
75        return sinq((pi-x)<<2);
76      if(x<=(pi+pi_2))
77        return -sinq((x-pi)<<2);
78      return  -sinq((pi_x2 -x)<<2);
79    }
80
81    //**********************************************************
82    // fixed_cos_C applies the sin/cos relation to
83    // the input, then calls fixed_sin_C
84    int fixed_cos_C(int x)
85    {
86      if(x<=0)
87        x += pi_x2;
88      x = pi_2 - x;
89      return fixed_sin_C(x);
90    }
```

Listing 8.7 ARM assembly implementation of $\sin x$ and $\cos x$ using fixed point calculations.

```
//**********************************************************
// Name: sincos.S
// Author: Larry Pyeatt
// Date: 2/22/2018
//**********************************************************
/*
This is a version of the sin/cos functions that uses symmetry
to enhance precision.  The actual sin and cos routines convert
the input to lie in the range 0 to pi/2, then pass it to the
worker routine that computes the result.  The result is then
converted back to correspond with the original input.

We calculate sin(x) using the first seven terms of the Taylor
Series:  sin(x) = x - x^3/3! + x^5/5! - x^7/7! + x^9/9! - ...
and we calculate cos(x) using the relationship:
cos(x) = sin(pi/2-x)

We start by defining a helper function, which we call sinq.
The sinq function calculates sin(x) for 0<=x<=pi/2.  The
input, x, must be an S(1,30) number. The factors of x that
sinq will use are: x, x^3, x^5, x^7, x^9, x^11, and x^13.

Dividing by (2n+1)! is changed to a multiply by a coefficient
as we compute each term, we will add it to the sum, stored as
an S(2,61). Therefore, we want the product of each power of x
and its coefficient to be converted to an S(2,61) for the add.
It turns out that this just requires a small shift.

We build a table to decide how much to shift each product
before adding it to the total.  x^2 will be stored as an
S(2,29), and x is given as an S(1,30).  After multiplying x by
x^2, we will shift left one bit, so the procedure is:
x    will be an  S(1,30) - multiply by x^2 and shift left
x^3 will be an  S(3,28) - multiply by x^2 and shift left
x^5 will be an  S(5,26) - multiply by x^2 and shift left
x^7 will be an  S(7,24) - multiply by x^2 and shift left
x^9 will be an  S(9,22) - multiply by x^2 and shift left
x^11 will be an S(11,20)- multiply by x^2 and shift left
x^13 will be an S(13,18)- multiply by x^2 and shift left
x^15 will be an S(15,16)- multiply by x^2 and shift left
x^17 will be an S(17,14)
*/
/*
The following table shows the constant coefficients
needed for calculating each term.
```

```
47          -1/3! =  AAAAAAAA as an S(-2,32)
48           1/5! =  44444445 as an S(-6,32)
49          -1/7! =  97F97F97 as an S(-12,32)
50           1/9! =  5C778E96 as an S(-18,32)
51          -1/11! = 9466EA60 as an S(-25,32)
52           1/13! = 5849184F as an S(-32,32)
53          -1/15! = 94603063 as an S(-40,32)
54           1/17! = 654B1DC1 as an S(-48,32)
55
56       Combining the two tables of power and coefficient formats, we
57       can now determine how much shift we need after each step in
58       order to do all sums in S(2,61) format:
59
60       power  powerfmt     coef  coeffmt      resultfmt   right shift
61       x      S(1,30)    *   1 (skip the multiply)        1 -> S(2,61)
62       x^3    S(3,28)    *  -1/3! S(-2,32)   = S(2,61)    0 -> S(2,61)
63       x^5    S(5,26)    *   1/5! S(-6,32)   = S(0,63)    2 -> S(2,61)
64       x^7    S(7,24)    *  -1/7! S(-12,32)  = S(-4,64)   6 -> S(2,61)
65       x^9    S(9,22)    *   1/9! S(-18,32)  = S(-8,64)   10-> S(2,61)
66       x^11   S(11,20)   *  -1/11! S(-25,32) = S(-13,64)  15-> S(2,61)
67       x^13   S(13,18)   *   1/13! S(-32,32) = S(-18,64)  20-> S(2,61)
68       x^15   S(15,16)   *  -1/15! S(-40,32) = S(-24,64)  26-> S(2,61)
69       x^17   S(17,14)   *   1/17! S(-48,32) = S(-30,64)  32-> S(2,61)
70       Note: the last row has a shift of 32, which indicates that
71             it can contribute no more than 1/2 bit of precision
72             to the 32-bit result.
73       */
74
75       // We will define a few constants that may be useful
76       .data
77       .global pi
78  pi:      .word   0x3243F6A8    // pi as an S(3,28)
79           .equ    PI_LO,0xF6A8
80           .equ    PI_HI,0x3243
81  pi_2:    .word   0x1921FB54    // pi/2 as an S(3,28)
82           .equ    PI_2_LO,0xFB54
83           .equ    PI_2_HI,0x1921
84  pi_x2:   .word   0x6487ED51    // 2*pi as an S(3,28)
85           .equ    PI_X2_LO,0xED51
86           .equ    PI_X2_HI,0x6487
87
88       // This is the table of coefficients and shifts for sinq to use
89       .align  3               // align to double word boundary
90  sintab: .word 0xAAAAAAAA,  0   // -1/3!  as an S(-2,32)
91          .word 0x44444445,  2   //  1/5!  as an S(-6,32)
92          .word 0x97F97F97,  6   // -1/7!  as an S(-12,32)
93          .word 0x5C778E96, 10   //  1/9!  as an S(-18,32)
94          .word 0x9466EA60, 15   // -1/11! as an S(-25,32)
95          .word 0x5849184F, 20   //  1/13! as an S(-32,32)
```

```
 96        .word 0x94603063,  26   // -1/15! as an S(-40,32)
 97        .word 0x654B1DC1,  32   // 1/17! as an S(-48,32)
 98        .equ tablen,(.-sintab)  // set tablen to size of table
 99        // The '.' refers to the current address counter value.
100        // Subtracting the address of sintab from the current
101        // address gives the size of the table.
102
103        .text
104 //-------------------------------------------------------------
105 // sinq(x)
106 // input: x -> S(1,30) s.t. 0 <= x <= pi/2
107 // returns sin(x) -> S(3,28)
108 // x0 : Sum of terms
109 // x1 : pointer to table
110 // x2 : next coefficient
111 // x3 : next shift
112 // x4 : x^2
113 // x5 : x^(2n-1)
114 // x6 : next term
115 // x7 : pointer to end of table
116 sinq:    smull   x4,w0,w0         // w4 will hold x^2
117        // x^2 is now an S(3,60) in x4 (0<= x^2 <= 2.467)
118        mov     x5,x0            // x5 will keep x^(2n-1). Start with x
119        // x5 now contains x as an S(1,30)
120        // The first term in the Taylor series is simply x, so
121        // convert x to an S(2,61) by shifting it left
122        lsl     x0,x0,#31        // x0 holds the sum
123        ldr     x1,=(sintab+8)   // get pointer to beginning of table
124                                 // but skip first entry
125        mov     w2,#0xAAAAAAAA   // Since first coefficient is a pattern
126        mov     x3,#0            // we can load it more quickly this way
127        asr     x4,x4,#31        // convert x^2 to an S(2,29)
128        add     x7,x1,#(tablen-8) // get pointer to end of table
129        b       firstmul         // skip the first load
130
131        // We know that we will always execute the loop 6 times,
132        // so we use a post-test loop.
133 sloop:   ldpsw   x2,x3,[x1],#8    // Load two values from the table
134        // x2 now has 1/(2n+1)! sign extended to 64 bits
135        // x3 contains the correcting shift  sign extended to 64 bits
136        //  The multiply will take time, so start it now
137 firstmul:
138        smull   x5,w4,w5         // x5 <- x^(2n+1) as an S(4,59)
139        cmp     x1, x7           // perfomance: do loop test early
140        asr     x5,x5,#31        // convert x^(2n-1) to S(3,28)
141        smull   x6,w5,w2         // multiply by value from the table
142        add     x6,x6,x6,lsr #63 // if the result is negative, then add one
143        asr     x6,x6,x3         // apply shift to make an S(2,61)
144        add     x0,x0,x6         // add to running total
```

```
145         blt      sloop              // Repeat for every table entry
146         // convert from S(2,61) to S(3,28) and return
147         asr      x0,x0,#33
148         ret                         // return the result
149
150
151 //------------------------------------------------------------
152 // cos(x)  NOTE: The cos(x) function does not return.
153 //              It is an alternate entry point to sin(x).
154 // input: x -> S(3,28)
155 // returns cos(x) -> S(3,28)
156         .global fixed_cos
157 fixed_cos:
158         cmp      x0,#0              // Add 2*pi to x if needed, to make
159         bge      cosgood            // sure x does not become too negative.
160         ldr      x1,=pi_x2          // load pointer to 2*pi
161         ldrsw    x1,[x1]            // load 2*pi
162         add      x0,x0,x1
163 cosgood:ldr      x1,=pi_2           // load pointer to pi/2
164         ldrsw    x1,[x1]            // load pi/2
165         sub      x0,x1,x0           // cos(x) = sin(pi/2-x)
166         // now we just fall through into the sin function
167
168 //------------------------------------------------------------
169 // sin(x)
170 // input: x -> S(3,28)
171 // returns sin(x) -> S(3,28)
172         .global fixed_sin
173 fixed_sin:
174         stp      x29,x30,[sp, #-16]! // push FP & LR
175
176         // load w1, w2 and w3 with constants
177         mov      w1,#PI_2_LO
178         movk     w1,#PI_2_HI,lsl #16
179         mov      w2,#PI_LO
180         movk     w2,#PI_HI,lsl #16
181         mov      w3,#PI_X2_LO
182         movk     w3,#PI_X2_HI,lsl #16
183
184         // step 1: make sure x>=0.0 and x<=2pi
185 negl:   cmp      x0,#0              // while(x < 0)
186         bge      nonneg
187         add      x0,x0,x3           //   x = x + 2 * pi
188         b        negl               // end while
189 nonneg: cmp      x0,x3              // while(x > pi/2)
190         ble      inrange
191         sub      x0,x0,x3           //   x = x - 2 * pi
192         b        nonneg             // end while
193         // step 2: find the quadrant and call sinq appropriately
```

```
inrange:cmp      x0,x1
        bgt      chkq2
        // it is in the first quadrant... just shift and call sinq
        lsl      x0,x0,#2
        bl       sinq
        b        sin_done
chkq2:  cmp      x0,x2
        bgt      chkq3
        // it is in the second quadrant... mirror, shift, and call sinq
        sub      x0,x2,x0
        lsl      x0,x0,#2
        bl       sinq
        b        sin_done
chkq3:  add      x1,x1,x2        // we won't need pi/2 again
        cmp      x0,x1           // so use x1 to calculate 3pi/2
        bgt      chkq4
        // it is in the third quadrant... rotate, shift, call sinq,
        // then complement the result
        sub      x0,x0,x2
        lsl      x0,x0,#2
        bl       sinq
        neg      x0,x0
        b        sin_done
        // it is in the fourth quadrant... rotate, mirror, shift,
        // call sinq, then complement the result
chkq4:  sub      x0,x0,x2
        sub      x0,x2,x0
        lsl      x0,x0,#2
        bl       sinq
        neg      x0,x0
sin_done:
        // return the result
        ldp      x29,x30,[sp],#16 // pop FP & LR
        ret
//------------------------------------------------------------
```

Listing 8.8 Example showing how the $\sin x$ and $\cos x$ functions can be used to print a table.

```
//*****************************************************************
// Name: sincosmain.S
// Author: Larry Pyeatt
// Date: 2/22/2014
// *****************************************************************

// This is a short program to print a table of sine and
// cosine values using the fixed point sin/cos functions.
// Compile with:
```

```
        // gcc -o sincos sincos.S sincosmain.S fixedfuncs.c



        .data
fmta:   .asciz  "%14.6f "
head:   .asciz  "      x               sin(x)        cos(x)\n"
line:   .asciz  "    -------------------------------------\n"
newline:.asciz  "\n"
tab:    .asciz  "\t"



        //------------------------------------------------------------

        .text
        .global main
main:   stp     x27,x28,[sp,#-16]!
        stp     x29,x30,[sp,#-16]!
        ldr     x0,=head
        bl      printf
        ldr     x0,=line
        bl      printf

        mov     x27,#0
mloop:
        // load count to r0 and convert it to a number x
        // between 0.0 and pi/2
        mov     x0,x27
        // multiply it by pi
        ldr     x1,=pi
        ldrsw   x1,[x1]
        smull   x0,w0,w1
        lsr     x0,x0,#4
        mov     x28,x0           // save it in x28 for later
        // print x
        mov     x1,#28
        ldr     x2,=fmta
        bl      printS
        ldr     x0,=tab
        bl      printf
        // calculate and print sin(x)
        mov     x0,x28           // retrieve x
        bl      fixed_sin
        mov     x1,#28
        ldr     x2,=fmta
        bl      printS
        ldr     x0,=tab
        bl      printf
        // // calculate and print cos(x)
        mov     x0,x28           // retrieve x
```

```
59        bl       fixed_cos
60        mov      x1,#28
61        ldr      x2,=fmta
62        bl       printS
63
64        ldr      x0,=newline
65        bl       printf
66
67        add      x27,x27,#1
68        cmp      x27,#33
69        blt      mloop
70
71        ldp      x29,x30,[sp],#16
72        ldp      x27,x28,[sp],#16
73        ret
74
75        //-------------------------------------------------------------
```

Table 8.4: Performance of sine function with various implementations.

Optimization	Implementation	CPU seconds
None	32-bit Fixed Point Assembly	5.15
	32-bit Fixed Point C	18.60
	Single Precision floating point C	10.56
	Double Precision floating point C	10.07
–02	32-bit Fixed Point Assembly	4.73
	32-bit Fixed Point C	5.38
	Single Precision floating point C	9.36
	Double Precision floating point C	9.23

8.6.4 Performance comparison

In some situations it can be very advantageous to use fixed point math. Many processors intended for use in embedded systems do not have a hardware floating point unit available. Table 8.4 shows the CPU time required for running a program to compute the sine function on 100,000,000 random values, using various implementations of the sine function. In each case, the program main() function was written in C. The only difference in the four implementations was the data type (which could be fixed point, IEEE single precision, or IEEE double precision), and the sine function that was used. The times shown in the table include only the amount of CPU time actually used in the sine function, and do not include the time required for program startup, storage allocation, random number generation, printing results, or program exit. The four implementations are as follows:

32-bit Fixed Point Assembly   The sine function is computed using the code shown in Listing 8.7.

32-bit Fixed Point C   The sine function is computed using the code shown in Listing 8.6.

Single Precision C   Sine is computed using the floating point sine function which is provided by the C compiler. The C code is written to use IEEE single precision floating point numbers.

Double Precision C   Same as the previous method, but using IEEE double precision instead of single precision.

Each of the four implementations were compiled both with and without compiler optimizations, resulting in a total of eight test cases. All cases were run on an NVIDIA Jetson TX2.

From Table 8.4, it is clear that the (carefully written) fixed point implementation written in Assembly beats the code generated by the compiler in every case. The closest that the compiler can get is the C version of the fixed-point algorithm when the compiler is run with full optimization. Even in that case, the fixed point assembly implementation is almost 14% faster. The fixed point assembly code is 98% faster than the single precision floating point implementation, and has 33% more precision (32 bits versus 24 bits). Note that even with floating point hardware support, the fixed point assembly implementation is almost twice as fast as the optimized floating point implementation provided by the compiler. For processors without hardware floating point support, fixed point arithmetic can be twenty or more times faster than floating point.

Similar results could be obtained on any processor architecture, and any reasonably complex mathematical problem. When developing software for small systems, the developer must weigh the costs and benefits of alternative implementations. For battery powered systems, it is important to realize that choices of hardware and software can affect power consumption even more strongly than computing performance. First, the power used by a system which includes a hardware floating point processor will be consistently higher than that of a system without one. Second, the reduction in processing time required for the job is closely related to the reduction in power required. Therefore, for battery operated systems, a fixed point implementation could greatly extend battery life. The following statements summarize the results from the experiment in this section.

1. With some effort, a competent assembly programmer can beat the assembler, in some cases by a very large margin.
2. If computational performance is critical, then a well-designed fixed point implementation will usually outperform even a hardware-accelerated floating point implementation.
3. If there is no hardware support for floating point, then floating point performance is extremely poor, and fixed point will always provide the best performance.

Listing 8.9 Inefficient representation of a binimal.

```
typedef struct{
   int sign;
   int exponent;
   int mantissa;
} poorfloat;
```

4. If battery life is a consideration, then a fixed point implementation can have an enormous advantage.

Note also from the table that the assembly language version of the fixed-point sine function beats the identical C version by a significant margin. It does take considerable effort to write assembly code that provides higher performance than a good optimizing compiler, but in some situations, the extra effort is warranted. Section 9.8 will demonstrate that a good assembly language programmer who is familiar with the floating point hardware can beat the compiler by an even wider performance margin.

8.7 Floating point numbers

Sometimes we need more range than we can easily get from fixed precision. One approach to solving this problem is to create an aggregate data type that can represent a fractional number by having fields for an exponent, a sign bit, and an integer mantissa. For example, in C, we could represent a fractional number using the data structure shown in Listing 8.9. That data structure, along with some subroutines for addition, subtraction, multiplication and division, would provide the capability to perform arithmetic without explicitly tracking the radix point. The subroutines for the basic arithmetic operations could do that, thereby freeing the programmer to work at a higher level.

The structure shown in Listing 8.9 is a rather inefficient way to represent a fractional number, and may create different data structures on different machines. It would be better to find some way to pack all three fields into 32 bits. The C language includes the notion of bit fields. This allows the programmer to specify exactly how many bits are to be used for each field within a `struct`, Listing 8.10 shows a C data structure that consumes 32 bits on all machines and architectures. It provides the same fields as the structure in Listing 8.9, but specifies exactly how many bits each field consumes. The compiler will compress this data structure into 32 bits, regardless of the natural word size of the machine.

The method of representing fractional numbers as a sign, exponent, and mantissa is very powerful, and IEEE has set standards for various floating point formats. These formats can be

Listing 8.10 Efficient representation of a binimal.

```
typedef struct{
  int sign:1;
  int exponent:8;
  int mantissa: 23;
}IEEEsingle;
```

described using bit fields in C, as described above. Many processors have hardware that is specifically designed to perform arithmetic using the standard IEEE formatted data. The following sections highlight most of the IEEE defined numerical definitions.

The IEEE standard specifies the bitwise representation for numbers, and specifies parameters for how arithmetic is to be performed. The IEEE standard for numbers includes the possibility of having numbers that cannot be easily represented. For example, any quantity that is greater than the most positive representable value is represented as positive infinity, and any quantity that is less than the most negative representable value is represented as negative infinity. There are special bit patterns to encode these quantities. The programmer or hardware designer is responsible for ensuring that their implementation conforms to the IEEE standards. The following sections describe some of the IEEE standard data formats.

8.7.1 IEEE 754 half-precision

The half-precision format gives a 16-bit encoding for fractional numbers with a small range and low precision. There are situations where this format is adequate. If the computation is being performed on a very small machine, then using this format may result in significantly better performance than could be attained using one of the larger IEEE formats. However, in most situations, the programmer can achieve better performance and/or precision by using a fixed-point representation. The format is as follows:

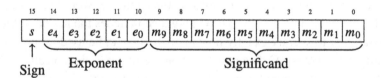

- The Significand (a.k.a. "Mantissa" or "Fractional Part") is stored using a sign-magnitude coding, with bit 15 being the sign bit.
- The exponent is an excess-15 number, i.e. the number stored is 15 greater than the actual exponent.

Table 8.5: Format for IEEE 754 Half-Precision.

Exponent	Significand $= 0$	Significand $\neq 0$	Equation
00000	± 0	subnormal	$-1^{sign} \times 2^{-14} \times 0.significand$
00001 ... 11110	normalized value		$-1^{sign} \times 2^{exp-15} \times 1.significand$
11111	$\pm\infty$	NaN	

- There are 10 bits of significand, but there are 11 bits of significand precision. There is a "hidden" bit, m_{10}, between m_9 and e_0. When a number is stored in this format, it is shifted until its leftmost non-zero bit is in the hidden bit position, and the hidden bit is not actually stored. The exception to this rule is when the number is zero or very close to zero. The radix point is assumed to be between the hidden bit and the first bit stored. The radix point is then shifted by the exponent.

Table 8.5 shows how to interpret IEEE 754 Half-Precision numbers. The exponents 00000 and 11111 have special meaning. The value 00000 is used to represent zero and numbers very close to zero, and the exponent value 11111 is used to represent infinity and NaN. NaN, which is the abbreviation for *not a number*, is a value representing an undefined or unrepresentable value. One way to get NaN as a result is to divide infinity by infinity. Another is to divide zero by zero. The NaN value can help indicate that there is a bug in the program, or to indicate that a calculation must be performed using a different method.

Subnormal means that the value is too close to zero to be completely normalized. The minimum strictly positive (subnormal) value is $2^{-24} \approx 5.96 \times 10^{-8}$. The minimum positive normal value is $2^{-14} \approx 6.10 \times 10^{-5}$. The maximum *exactly* representable value is $(2 - 2^{-10}) \times 2^{15} = 65504$.

8.7.1.1 Examples

The following bit value:

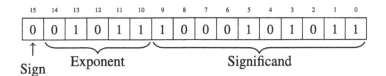

represents

$$+1.1000101011 \times 2^{01011-01111} = 1.1000101011 \times 2^{-4} = .00011000101011$$
$$\approx 0.09637.$$

The following bit value:

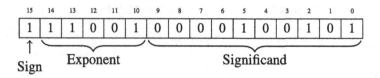

represents

$$-1.0000100101 \times 2^{11001-01111} = -1.0000100101 \times 2^{10} = -10000100101.0$$
$$= -1061_{10}.$$

8.7.2 IEEE 754 single-precision

The single precision format provides a 23-bit mantissa, and an 8-bit exponent. This is enough to represent a reasonably large range, with reasonable precision. This type can be stored in 32 bits, so it is relatively compact. At the time that the IEEE standards were defined, most machines used a 32-bit word, and were optimized for moving and processing data in 32-bit quantities. For many applications this format represents a good trade-off between performance and precision.

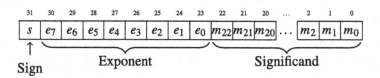

8.7.3 IEEE 754 double-precision

The double-precision format was designed to provide enough range and precision for most scientific computing requirements. It provides a 10-bit exponent and a 53-bit mantissa. When the IEEE 754 standard was introduced, this format was not supported by most hardware. That has changed. Most modern floating point hardware is optimized for the IEEE 754 double-precision standard, and most modern processors are designed to move 64-bit or larger quantities. On modern floating-point hardware, this is the most efficient representation. However processing large arrays of double-precision data requires twice as much memory, and twice as much memory bandwidth, as single-precision.

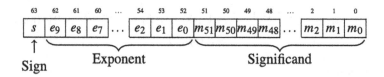

8.7.4 IEEE 754 quad-precision

The IEEE 754 Quad-Precision format was designed to provide enough range and precision for very demanding applications. It provides a 14-bit exponent and a 116-bit mantissa. This format is still not supported by most hardware. The IBM POWER9 CPU fully supports quad precision in hardware. Some other processors, such as SPARC V8 and V9, and PA-RISC, offer partial support. However for mid-range processors such as the Intel x86 family and the ARM, this format is still definitely out of their league. It may be supported by some compilers, but the operations are implemented in software, and can take ten times as long (or more) as a hardware implementation.

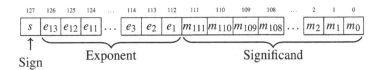

8.8 Floating point operations

Many processors do not have hardware support for floating point. On those processors, all floating point must be accomplished through software. Processors that do support floating point in hardware must have quite sophisticated circuitry to manage the basic operations on data in the IEEE 754 standard formats. Regardless of whether the operations are carried out in software or hardware, the basic arithmetic operations require multiple steps.

8.8.1 Floating point addition and subtraction

The steps required for addition and subtraction of floating point numbers is the same, regardless of the specific format. The steps for adding or subtracting to floating point numbers a and b are as follows:

1. Extract the exponents E_a and E_b.
2. Extract the significands M_a and M_b, and convert them into 2's complement numbers, using the signs S_a and S_b.
3. Shift the significand with the smaller exponent right by $|E_a - E_b|$.
4. Perform addition (or subtraction) on the significands to get the significand of the result, M_r. Remember that the result may require one more significant bit to avoid overflow.
5. If M_r is negative, then take the 2's complement and set S_r to 1. Otherwise set S_r to 0.
6. Shift M_r until the leftmost 1 is in the "hidden" bit position, and add the shift amount to the smaller of the two exponents to form the new exponent E_r.

7. Combine the sign S_r, the exponent E_r, and significand M_r to form the result.

The complete algorithm must also provide for correct handling of infinity and NaN.

8.8.2 Floating point multiplication and division

Multiplication and division of floating point numbers also requires several steps. The steps for multiplication and division of two floating point numbers a and b are as follows:

1. Calculate the sign of the result S_r.
2. Extract the exponents E_a and E_b.
3. Extract the significands M_a and M_b.
4. Multiply (or divide) the significands to form M_r.
5. Add (or subtract) the exponents (in excess-N) to get E_r.
6. Shift M_r until the leftmost 1 is in the "hidden" bit position, and add the shift amount to E_r.
7. Combine the sign S, the exponent E_r, and significand M_r to form the result.

The complete algorithm must also provide for correct handling of infinity and NaN.

8.9 Ethics case study: patriot missile failure

Fixed point arithmetic is very efficient on modern computers. However it is incumbent upon the programmer to track the radix point at all stages of the computation, and to ensure that a sufficient number of bits are provided on both sides of the radix point. The programmer must ensure that all computations are carried out with the desired level of precision, resolution, accuracy, range, and dynamic range. Failure to do so can have dire consequences.

On February 25, 1991, during the Gulf War, an American Patriot Missile battery in Dharan, Saudi Arabia, failed to intercept an incoming Iraqi SCUD missile. The SCUD struck an American army barracks, killing 28 soldiers and injuring around 98 other people. The cause was an inaccurate calculation of the time elapsed since the system was last booted.

The hardware clock on the system counted the time in tenths of a second since the last reboot. Current time, in seconds, was calculated by multiplying that number by $\frac{1}{10}$. For this calculation, $\frac{1}{10}$ was represented as a U(1,23) fixed point number. Since $\frac{1}{10}$ cannot be represented precisely in a fixed number of bits, there was round-off error in the calculations. The small imprecision, when multiplied by a large number, resulted in significant error. The longer the system ran after boot, the larger the error became.

The system determined whether or not it should fire by predicting where the incoming missile would be at a specific time in the future. The time and predicted location were then fed to a second system which was responsible for locking onto the target and firing the Patriot missile. The system would only fire when the potential target was at the proper location at the specified time. If the radar did not detect the incoming missile at the correct time and location, then the system would not fire.

At the time of the failure, the Patriot battery had been up for around 100 hours. We can estimate the error in the timing calculations by considering how the binary number was stored. The binary representation of $\frac{1}{10}$ is $0.0\overline{0011}$. Note that it is a non-terminating, repeating binimal. The 24 bit register in the Patriot could only hold the following set of bits:

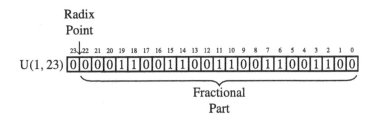

This resulted in an error of $0.00000000000000000000000\overline{1100}_2$. The error can be computed in base 10 as:

$$
\begin{aligned}
e &= 2^{-24} + 2^{-25} + 2^{-28} + 2^{-29} + 2^{-32} + 2^{-33} + \dots \quad &(8.3) \\
&= \sum_{i=0}^{\infty} 2^{-(4i+24)} + 2^{-(4i+25)} \quad &(8.4) \\
&\approx 9.5 \times 10^{-8}. \quad &(8.5)
\end{aligned}
$$

To find out how much error was in the total time calculation, we multiply e by the number of tenths of a second in 100 hours. This gives $9.5 \times 10^{-8} \times 100 \times 60 \times 60 \times 10 = 0.34$ seconds. A SCUD missile travels at about 1,676 meters per second. Therefore it travels about 570 meters in 0.34 seconds. Because of this, the targeting and firing system was expecting to find the Scud at a location that was over half a kilometer from where it really was. This was far enough that the incoming Scud was outside the "range gate" that the Patriot tracked. It did not detect the Scud at its predicted location, so it could not lock on and fire the Patriot.

This is an example of how a seemingly insignificant error can lead to a major failure. In this case, it led to loss of life and serious injury. Ironically, one factor that contributed to the problem was that part of the code had been modified to provide more accurate timing calculations, while another part had not. This meant that the inaccuracies did not cancel each other. Had both sections of code been re-written, or neither section changed, then the issue probably would not have surfaced.

The Patriot system was originally designed in 1974 to be mobile and to defend against aircraft that move much more slowly than ballistic missiles. It was expected that the system would be moved often, and therefore the computer would be rebooted frequently. Also, the slow-moving aircraft would be much easier to track, and the error in predicting where it is expected to be would not be significant. The system was modified in 1986 to be capable of shooting down Soviet ballistic missiles. A SCUD missile travels at about twice the speed of the Soviet missiles that the system was re-designed for.

The system was deployed to Iraq in 1990, and successfully shot down a SCUD missile in January of 1991. In mid-February of 1991, Israeli troops discovered that if the system became inaccurate if it was allowed to run for long periods of time. They claimed that the system would become unreliable after twenty hours of operation. U.S. military did not think the discovery was significant, but on February 16th, a software update was released. Unfortunately, the update could not immediately reach all units because of wartime difficulties in transportation. The Army released a memo on February 21st, stating that the system was not to be run for "very long times," but did not specify how long a "very long time" would be. The software update reached Dhahran one day after the Patriot Missile system failed to intercept a SCUD missile, resulting in the death of 28 Americans and many more injuries.

Part of the reason this error was not found earlier was that the program was written in assembly language, and had been patched several times in its 15-year life. The code was difficult to understand and maintain, and did not conform to good programming practices. The people who worked to modify the code to handle the SCUD missiles were not as familiar with the code as they would have been if it were written more recently, and time was a critical factor. Prolonged testing could have caused a disaster by keeping the system out of the hands of soldiers in a time of war. The people at Raytheon Labs had some tough decisions to make. It cannot be said that Raytheon was guilty of negligence or malpractice. The problem with the system was not necessarily the developers, but that the system was modified often and in inconsistent ways, without complete understanding. However, it should be noted that if the hardware clock had counted time in eighths or sixteenths of a second, there would have been no round off error at all. Using floating point for the calculations would not have solved the problem, because the value $\frac{1}{10}$ cannot be represented precisely using either fixed point or floating point.

8.10 Chapter summary

Sometimes it is desirable to perform calculations involving non-integral numbers. The two common ways to represent non-integral numbers in a computer are fixed point and floating point. A fixed point representation allows the programmer to perform calculations with non-integral numbers using only integer operations. With fixed point, the programmer must track the radix point throughout the computation. Floating point representations allow the radix point to be tracked automatically, but require much more complex software and/or hardware. Fixed point will usually provide better performance than floating point, but requires more programming skill.

Fractional numbers in radix notation may not terminate in all bases. Numbers which terminate in base two will also terminate in base ten, but the converse is not true. Programmers should avoid counting using fractions which do not terminate in base two, because it leads to the accumulation of round-off errors.

Exercises

8.1. Perform the following base conversions:
   a. Convert 10110.001_2 to base ten.
   b. Convert 11000.0101_2 to base ten.
   c. Convert 10.125_{10} to binary.
8.2. Complete the following table (assume all values represent positive fixed-point numbers):

Base 10	Base 2	Base 16	Base 13
49.125			
	101011.011		
		AF.3	
			12

8.3. You are working on a problem involving real numbers between -2 and 2, on a computer that has 16-bit integer registers and has no hardware floating point support. You decide to use 16-bit fixed point arithmetic.
   a. What fixed point format should you use?
   b. Draw a diagram showing the sign, if any, radix point, integer part, and fractional part.
   c. What is the precision, resolution, accuracy, and range of your format?
8.4. What is the resulting type of each of the following fixed point operations?
   a. $S(24, 7) \times S(27, 15)$
   b. $S(3, 4) \div U(4, 20)$
8.5. Convert 26.640625_{10} to a binary $U(18,14)$ representation. Show the AArch64 assembly code necessary to load that value into register **r4**.
8.6. For each of the following fractions, indicate whether or not it will terminate in bases 2, 5, 7, and 10.

  a. $\frac{13}{64}$

  b. $\frac{37}{60}$

  c. $\frac{25}{74}$

  d. $\frac{39}{1250}$

  e. $\frac{17}{343}$

8.7. What is the exact value of the binary number 0011011100011010 when interpreted as an IEEE half-precision number? Give your answer in base ten.

8.8. The "Software Engineering Code of Ethics And Professional Practice" states that a responsible software engineer should "Approve software only if they have well-founded belief that it is safe, meets specifications, passes appropriate tests..." (sub-principle 1.03) and "Ensure adequate testing, debugging, and review of software...on which they work." (sub-principle 3.10).

   The software engineering code of ethics also states that a responsible software engineer should "Treat all forms of software maintenance with the same professionalism as new development."

  a. Explain how the Software Engineering Code of Ethics And Professional Practice were violated by the Patriot Missile system developers.

  b. How should the engineers and managers at Raytheon have responded when they were asked to modify the Patriot Missile System to work outside of its original design parameters?

  c. What other ethical and non-ethical considerations may have contributed to the disaster?

## CHAPTER 9

Floating point

For ARMv7 and previous architectures, floating point is provided by an optional Vector Floating Point (VFP) *coprocessor*. Many ARMv7 processors also support the NEON extensions. All AArch64 processors support a version of NEON referred to as Advanced SIMD. For brevity, this book will often use the term NEON to refer to Advanced SIMD. Advanced SIMD supports not only floating point, but also integer and fixed point operations. The remainder of this chapter will cover the floating point operations that are inherited from, or similar to, the instructions provided by the older VFP coprocessor.

9.1 Floating point overview

Historically, ARM has implemented floating point operations by adding a coprocessor to the CPU. The coprocessor extends the instruction set that is supported by the system. There are five major revisions of the ARM floating point instruction set and coprocessor:

VFPv1: Vector Floating Point coprocessor version 1 is obsolete.

VFPv2: An optional extension to the ARMv5 and ARMv6 processors. VFPv2 has 16 64-bit FPU registers.

VFPv3: An optional extension to the ARMv7 processors. It is backwards compatible with VFPv2, except that it cannot trap floating-point exceptions. VFPv3-D32 has 32 64-bit FPU registers. Some processors have VFPv3-D16, which supports only 16 64-bit FPU registers. VFPv3 adds several new instructions to the VFP instruction set.

VFPv4: Implemented on some Cortex ARMv7 processors. VFPv4 has 32 64-bit FPU registers. It adds both half-precision extensions and multiply-accumulate instructions to the features of VFPv3. Some processors have VFPv4-D16, which supports only 16 64-bit FPU registers.

NEON: Defines Single Instruction Multiple Data (SIMD) extensions to the instruction set. NEON supports the floating point instruction set along with integer operations. The NEON instruction set allows for a single instruction to perform operations on multiple pieces of data.

AArch64 integrates floating point and SIMD more tightly into the processor core, while still maintaining instructions that are similar to the original FP/NEON instruction set.

ARM 64-Bit Assembly Language
https://doi.org/10.1016/B978-0-12-819221-4.00016-X

293

63	32	31	0		127	0	
X0		W0					V0
X1		W1					V1
X2		W2					V2
X3		W3					V3
X4		W4					V4
X5		W5					V5
X6		W6					V6
X7		W7					V7
X8		W8					V8
X9		W9					V9
X10		W10					V10
X11		W11					V11
X12		W12					V12
X13		W13					V13
X14		W14					V14
X15		W15					V15
X16		W16					V16
X17		W17					V17
X18		W18					V18
X19		W19					V19
X20		W20					V20
X21		W21					V21
X22		W22					V22
X23		W23					V23
X24		W24					V24
X25		W25					V25
X26		W26					V26
X27		W27					V27
X28		W28					V28
X29		W29					V29
X30		W30					V30
							V31
XZR		WZR					
SP							FPCR
							FPSR
PC							
PSTATE							

Figure 9.1: AArch64 Integer (left) and FP/NEON (right) User Program Registers.

Fig. 9.1 shows the 32 AArch64 integer registers, and the additional registers provided by the AArch64 Advanced SIMD instruction set. As shown in Fig. 9.2, AArch64 FP/NEON instructions provide the ability to view the register set as thirty-two quadruple-word (128 bit) registers, named q0 through q31, or thirty-two double word (64 bit) registers, named d0 through

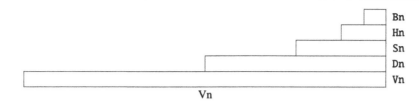

Figure 9.2: With the FP/NEON instructions, parts of register Vn can be accessed by using different views. Each view has a unique name.

d31, or as thirty-two single-word (32 bit) registers, named s0 through s31. Each register can also be referred to as a halfword register (h0 through h31) or a byte register (b0 through b31). Additionally, portions of each register can be referred to by using a subscript notation, such as v1.f64[1] which refers to the upper 64 bits of register v1.

The processor supports IEEE single precision and double precision numbers. Each of the first sixteen double precision registers can be used to store either one 64-bit number or two 32-bit numbers. For example, double precision register d0 may also be referred to as single precision registers s0 and s1. Registers d16 through d31 cannot be used as single precision registers using the basic FP instructions.

The VFP instruction set added about 28 new instructions to the ARM instruction set. The exact number of VFP instructions depended on the specific version of the VFP coprocessor. AArch64 supports instructions equivalent to the VFPv4 instruction set. There 23 basic instructions in the AArch64 FP/NEON instruction set, but some have several variations. Instructions are provided to:

- transfer floating point values between FP registers,
- transfer floating-point values between the FP coprocessor registers and main memory,
- transfer 32-bit values between the FP coprocessor registers and the ARM integer registers,
- perform addition, subtraction, multiplication, and division, involving two source registers and a destination register,
- compute the square root of a value,
- perform combined multiply-accumulate operations,
- perform conversions between various integer, fixed point, and floating point representations, and
- compare floating-point values.

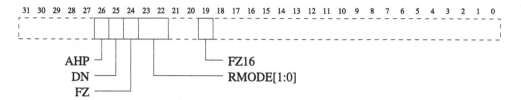

Figure 9.3: Bits in the FPCR.

9.2 Register usage rules

As with the integer registers, there are rules for using the FP/NEON registers. These rules are a convention, and following the convention ensures interoperability between code written by different programmers and compilers.

- Registers v0 through v7 are *volatile*. They are used for passing floating point arguments into a subroutine and/or storing local variables. Register v0 is also used for returning the result of a function (if the result is a floating point number.

- Registers v8 through v15 are *non-volatile*, meaning that the contents of these registers must be preserved across subroutine calls. If the subroutine uses these registers, their contents must first be saved somewhere (usually on the stack) and restored before the subroutine returns.

- Registers v16 through v31 are also considered *volatile*, and can be used for local variables and temporary variables.

In addition to the FPCR, all FP/NEON implementations contain at least two additional system registers. The Floating-point System ID register (FPSID) is a read-only register whose value indicates which FP/NEON implementation is being provided. The contents of the FPSID can be transferred to an ARM integer register, then examined to determine which FP/NEON version is available. There is also a Floating-point Exception register (FPEXC). Two bits of the FPEXC register provide system-level status and control. The remaining bits of this register are defined by the sub-architecture. These additional system registers should not be accessed by user applications.

9.3 Floating point control and status registers

The Floating Point Control Register (FPCR) allows the programmer to control how floating point operations are performed. The layout of the NEON FPCR is shown in Fig. 9.3. The meaning of each field is as follows:

AHP Alternate Half Precision control bit.

    0:    Use IEEE half precision format

    1:    Use alternative half precision format.

DN Default NaN enable:

    0:    Disable Default NaN mode. NaN operands propagate through to the output of a floating-point operation.

    1:    Enable Default NaN mode. Any operation involving one or more NaNs returns the default NaN.

FZ Flush-to-Zero enable:

    0:    Disable Flush-to-Zero mode.

    1:    Enable Flush-to-Zero mode.

Flush-to-Zero mode replaces subnormal numbers with 0. This does not comply with IEEE 754 standard, but may increase performance.

RMODE Rounding mode:

    00   Round to Nearest (RN).

    01   Round towards Plus infinity (RP).

    10   Round towards Minus infinity (RM).

    11   Round towards Zero (RZ).

FZ16 Flush-to-zero mode control bit for half precision data processing instructions.

    0:    Disable Flush-to-Zero mode for half precision.

    1:    Enable Flush-to-Zero mode for half precision.

The FPCR can be read into an AArch64 system register, modified, and written back to change settings. The code sequence for changing settings is as follows:

```
MRS    Xt, FPCR // Read FPCR into Xt
// set/clear bits
MSR    FPCR, Xt // Write Xt to FPCR
```

The Floating Point Status Register (FPSR) allows the programmer to view information about the status of the floating point unit, the FPSR is shown in Fig. 9.4. The meaning of each field is as follows:

N The Negative flag is set by vcmp if the processor is running in AArch32 mode. AArch64 comparisons set the PSTATE.N flag directly.

Z The Zero flag is set by vcmp if the processor is running in AArch32 mode. AArch64 comparisons set the PSTATE.N flag directly.

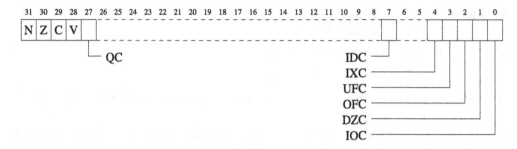

Figure 9.4: Bits in the FPSR.

C The Carry flag is set by vcmp if the processor is running in AArch32 mode. AArch64 comparisons set the PSTATE.N flag directly.

V The oVerflow flag is set by vcmp if the processor is running in AArch32 mode. AArch64 comparisons set the PSTATE.N flag directly.

QC The saturation cumulative flag is set to one by saturating instructions if saturation has occurred. This will be explained with more detail in Chapter 10.

IDC The Input Subnormal Cumulative flag is set to one when an IDE condition has occurred.

IXC The IneXact Cumulative flag is set to one when an IXE condition has occurred.

UFC The UnderFlow Cumulative flag is set to one when a UFE condition has occurred.

OFC The OverFlow Cumulative flag is set to one when an OFE condition has occurred.

DZC The Division by Zero Cumulative flag is set to one when an DZE condition has occurred.

IOC The Invalid Operation Cumulative flag is set to one when an OFE condition has occurred.

The FPSR can be read into an AArch64 system register and examined. The instruction for reading the FPSR is as follows:

```
MRS     Xt, FPSR // Read FPCR into Xt
```

The FPSR bits can also be set by moving data from an integer register to the FPSR. This allows the programmer to clear the exception bits. The instruction is as follows:

```
MSR     FPSR, Xt // Write Xt to FPSR
```

9.4 Load/store instructions

The FP instruction set provides several instructions for moving data between memory and the FP registers. There are instructions for loading and storing single and double precision registers, and for moving multiple registers to or from memory. All of the load and store instructions require a memory address to be in one of the ARM integer registers. The available memory addressing modes are identical to the integer load and store instructions.

9.4.1 Load/store single register

The single register load/store instructions allow moving data between memory and an FP/-NEON register. These instructions can load/store a byte, half-word, word (single precision), double-word (double precision), or a quad-word. The following instructions are used to load or store a single FP/NEON register:

ldr   Load FP/NEON Register, and
str   Store FP/NEON Register.

9.4.1.1 Syntax

```
<op>  Bt, <addr>
<op>  Ht, <addr>
<op>  St, <addr>
<op>  Dt, <addr>
<op>  Qt, <addr>
```

- <op> is either ldr or str.
- <addr> is one of the address specifiers described in Chapter 3, Section 3.3.3 on Page 61. The valid shift amount for the Register Offset mode depends on the size of the register being loaded or stored.

9.4.1.2 Operations

Name	Effect	Description
ldr	Yt ← Mem[addr]	Load Yt, where Yt is any FP register.
str	Mem[addr] ← Yt	Store Yt, where Yt is any FP register.

9.4.1.3 Examples

```
1    ldr    s5,[x0]        // load s5 from address in x0
2    str    d4,[x3],#8     // store d4 using address in
3                          // x3, then add 8 to x3
4    str    s0,[x1,#4]!    // Add 4 to x1, then load s0 from
5                          // the resulting address
```

9.4.2 Load/store single register with unscaled offset

These versions of the single register load/store instructions allow moving data between memory and an FP/NEON register. These instructions can load/store a byte, half-word, word (single precision), double-word (double precision), or a quad-word. These instructions use an unscaled 9-bit offset. The following instructions are used to load or store a single FP/NEON register:

ldur   Load FP/NEON Register, and
stur   Store FP/NEON Register.

9.4.2.1 Syntax

```
    <op>  Bt, <addr>
    <op>  Ht, <addr>
    <op>  St, <addr>
    <op>  Dt, <addr>
    <op>  Qt, <addr>
```

- <op> is either ldur or stur.
- <addr> is register-offset. The offset is not scaled by the size of the data.

9.4.2.2 Operations

Name	Effect	Description
ldur	Yt ← Mem[addr]	Load Yt, where Yt is any FP register.
stur	Mem[addr] ← Yt	Store Yt, where Yt is any FP register.

9.4.2.3 Examples

```
1    ldur   s5,[x0]        // load s5 from address in x0
2    stur   d4,[x3],#8     // store d4 using address in
3                          // x3, then add 8 to x3
```

```
stur   s0,[x1,#4]!   // Add 4 to x1, then load s0 from
                      // the resulting address
```

9.4.3 Load/store pair

These instructions are used to load or store two registers at a time. This can be useful for moving registers onto the stack or for copying data. These two instructions are particularly useful for transferring data in a load-store architecture because each instruction can move twice as much information as the `ldr` and `str` instructions.

ldp   Load Pair, and
stp   Store Pair.

9.4.3.1 Syntax

```
<op>  St, St2, <addr>
<op>  Dt, Dt2, <addr>
<op>  Qt, Qt2, <addr>
```

-    `<op>` is either `ldp` or `stp`.
-    `<addr>` is 7 bits Pre-indexed, Post-indexed, or Signed immediate.
-    Signed immediate `Xt` range: [-0x200, 0x1f8]. `Wt` range: [-0x100, 0xfc].

9.4.3.2 Operations

Name	Effect	Description
ldp	Yt ← Mem[addr] Yt2 ← Mem[addr+size(Yt)]	Load register pair from memory
stp	Mem[addr] ← Yt Mem[addr+size(Yt)] ← Yt2	Store register pair in memory

9.4.4 Load/store non-temporal pair

These are load/store pair instructions which also provide a hint to the memory system that the data being accessed is "non-temporal." Non-temporal data is data which is unlikely to be referenced again in the near future (such as streaming media), and need not be retained in data caches.

ldnp   Load Non-Temporal Pair, and
stnp   Store Non-Temporal Pair.

9.4.4.1 Syntax

```
<op>  St, St2, <addr>
<op>  Dt, Dt2, <addr>
<op>  Qt, Qt2, <addr>
```

- <op> is either ldp or stp.
- <addr> is 7 bits Pre-indexed, Post-indexed, or Signed immediate.
- Signed immediate Xt range: [-0x200, 0x1f8]. Wt range: [-0x100, 0xfc].

9.4.4.2 Operations

Name	Effect	Description
ldnp	Yt ← Mem[addr] Yt2 ← Mem[addr+size(Yt)]	Load register pair from memory
stnp	Mem[addr] ← Yt Mem[addr+size(Yt)] ← Yt2	Store register pair in memory

9.5 Data movement instructions

With the addition of all of the FP registers, there many more possibilities for how data can be moved. There are many more registers, and FP registers may be 32 or 64 bit. This results in several combinations for moving data among all of the registers. The FP instruction set includes instructions for moving data between two FP registers, between FP and integer registers, and between the various system registers.

9.5.1 Moving between data registers

The most basic move instruction involving FP registers simply moves data between two floating point registers, or moves data between an FP register and an Integer register. The instruction is:

fmov   Move Between Data Registers.

9.5.1.1 Syntax

```
fmov   Sd, Sn
fmov   Wd, Sn
fmov   Sd, Wn
```

```
fmov    Dd, Dn
fmov    Xd, Dn
fmov    Dd, Xn
fmov    Xd, Vn.D[1]
fmov    Vn.D[1], Xn
```

- The two registers specified must be the same size.
- Vn.D[1] refers to the top 64 bits of register Vn.

9.5.1.2 Operations

Name	Effect	Description
fmov	Fd ← Fn	Move Fn to Fd

9.5.1.3 Examples

```
fmov.f64 d3,d4          // d3 <- d4
fmov.f32 s5,s12         // s5 <- s12
```

9.5.2 Floating point move immediate

The FP/NEON instruction set provides an instruction for moving an immediate value into a register, but there are some restrictions on what the immediate value can be. The instruction is:

fmov   Floating Point Move Immediate.

9.5.2.1 Syntax

```
fmov    Sd, #fpimm
fmov    Dd, #fpimm
```

- The floating point constant, fpimm, may be specified as a decimal number such as 1.0.
- The floating point value must be expressable as $\pm n \div 16 \times 2^r$, where n and r are integers such that $16 \le n \le 31$ and $-3 \le r \le 4$.
- The floating point number will be stored as a normalized binary floating point encoding with 1 sign bit, 4 bits of fraction and a 3-bit exponent (see Chapter 8, Section 8.7).
- Note that this encoding does not include the value 0.0, however this value may be loaded using the FMOV Sd,WZR instruction.

9.5.2.2 Operations

Name	Effect	Description
fmov	Fd ← fpimm	Move Immediate Data to Fd

9.5.2.3 Examples

```
1   fmov    d3,#5.0         // d3 <- 5.0
2   fmov    s5,#-7.0        // s5 <- -7.0
```

9.6 Data conversion instructions

The ARM FP provides several instructions for converting between various floating point and integer formats. Some FP versions also have instructions for converting between fixed point and floating point formats.

9.6.1 Convert between double, single, and half precision

These instructions are used for converting between IEEE double precision, single precision, and half precision floating point:

fcvt   Convert Between Double, Single, and Half Precision.

9.6.1.1 Syntax

```
    fcvt    Sd, Hn
    fcvt    Dd, Hn
    fcvt    Hd, Sn
    fcvt    Dd, Sn
    fcvt    Hd, Dn
    fcvt    Sd, Dn
```

- Although the data processing instructions do not operate on half precision numbers, they may be used to store data in half the space required for single precision, or one quarter the space required for double precision.

9.6.1.2 Operations

Name	Effect	Description
fcvt	Sd ← *single*(Hn)	Convert half precision number in Hn to single precision and store in Sd.

continued on next page

Name	Effect	Description
fcvt	Dd ← *double*(Hn)	Convert half precision number in Hn to double precision and store in Dd.
fcvt	Hd ← *half*(Sn)	Convert single precision number in Sn to half precision and store in Hd.
fcvt	Dd ← *double*(Sn)	Convert single precision number in Sn to double precision and store in Dd.
fcvt	Hd ← *half*(Dn)	Convert double precision number in Dn to half precision and store in Hd.
fcvt	Sd ← *single*(Dn)	Convert double precision number in Dn to single precision and store in Sd.

9.6.1.3 Examples

```
fcvt  s0,d0    // Convert from half precision to single precision
fcvt  d1,s1    // Convert from single to half precision
```

9.6.2 Convert between floating point and integer

These instructions are used to convert integers to single or double precision floating point, or for converting single or double precision to integer:

fcvtas   Convert Float to Signed Integer, Round Away From Zero.
fcvtau   Convert Float to Unsigned Integer, Round Away From Zero.
fcvtms   Convert Float to Signed Integer, Round Towards $-\infty$.
fcvtmu   Convert Float to Unsigned Integer, Round Towards $-\infty$.
fcvtns   Convert Float to Signed Integer, Round To Even.
fcvtnu   Convert Float to Unsigned Integer, Round To Even.
fcvtps   Convert Float to Signed Integer, Round Towards $+\infty$.
fcvtpu   Convert Float to Unsigned Integer, Round Towards $+\infty$.
fcvtzs   Convert Float to Signed Integer, Round Towards Zero.
fcvtzu   Convert Float to Unsigned Integer, Round Towards Zero.
scvtf   Convert Signed Integer to Float Using FPCR Rounding Mode.
ucvtf   Convert Unsigned Integer to Float Using FPCR rounding mode.

9.6.2.1 Syntax

```
fcvt<mode><u|s>    Rd, Fn
<u|s>cvtf          Fd, Rn
```

- The mode can be one of a, m, n, p, or z, to specify the rounding mode.
- Rd and Rn can be any AArch64 integer register.
- Fd and Fn can be any single or double precision register.

9.6.2.2 Operation

Opcode	Effect	Description
fcvt<mode>s	Rd ← int(Fn)	Convert float to signed integer.
fcvt<mode>u	Rd ← $uint$(Fn)	Convert float to unsigned integer.
scvtf	Fd ← $float$(Rn)	Convert signed integer to float.
ucvtf	Fd ← $ufloat$(Rn)	Convert unsigned integer to float.

9.6.2.3 Examples

```
fcvtas   d5, x7   // Convert signed 64-bit integer to double
fcvtpu   d0, w4   // Convert unsigned 32-bit integer to double
fcvtas   s0, x7   // Convert signed 64-bit integer to single
ucvtf    x3, d0   // Convert double to unsigned integer
```

9.6.3 Convert between fixed point and floating point

VFPv3 and higher coprocessors have additional instructions used for converting between fixed point and single precision floating point. The AArch64 instructions are:

fcvtzs Convert Float to Signed Fixed Point.
fcvtzu Convert Float to Unsigned Fixed Point.
scvtf   Convert Signed Fixed Point to Float.
ucvtf   Convert Unsigned Fixed Point to Float.

9.6.3.1 Syntax

```
fcvtz<s|u>   Rd, Fn, #fbits
<s|u>cvtf    Fd, Rm, #fbits
```

- The fcvtz instructions always round towards zero.
- The cvtf instructions always round using the FPCR rounding mode.
- Rd and Rn can be any AArch64 integer register.

- Fd and Fn can be any single or double precision FP/NEON register.
- The #fbits operand specifies the number of fraction bits in the fixed point number, and must be less than or equal to the size of the integer register specified.

9.6.3.2 Operations

Name	Effect	Description
fcvtzs	Rd ← $floatToFix$(Fn, fbits)	Convert floating point to signed fixed point.
fcvtzu	Rd ← $ufloatToFix$(Fn, fbits)	Convert floating point to unsigned fixed point.
scvtf	Fd ← $fixToFloat$(Fn, fbits)	Convert signed fixed point to float.
scvtf	Fd ← $ufixToFloat$(Fn, fbits)	Convert unsigned fixed point to float.

9.6.3.3 Examples

```
fcvtzu     s0,w4,#4 // Convert from U(12,4) to single
scvtf      w3,s1,#8 // Convert from single to S(23,8)
```

9.7 Data processing instructions

Instructions are provided to perform the four basic arithmetic functions, plus absolute value, negation, and square root. There are also special forms of the multiply instructions that perform multiply-accumulate.

9.7.1 Round to integral

These operations will round a floating point number to an integer. There are several options to control exactly how the rounding is performed:

frinta   Round away from zero
frinti   Round using the mode set in the FPCR.
frintm   Round towards $-\infty$
frintn   Round with halfway cases rounding to even
frintp   Round towards $+\infty$
frintx   Round using mode set in the FPCR and raise an exception if the result does not have the same value as the input.
frintz   Round towards zero

9.7.1.1 Syntax

```
frint<op>  Fd, Fm
```

- <op> is one of a, i, m, n, p, or x.
- Fd and Fm must specify either two single precision registers or two double precision registers.

9.7.1.2 Operations

Name	Effect	Description
frint*x*	Fd ← *roundx*(Fm)	Round

9.7.1.3 Examples

```
frinta    d3, d5    // Round d5 away from zero and store in d3
frintm    s15, s15  // Round s5 towards negative infinity
```

9.7.2 Absolute value, negate, square root

The unary operations require one source register and a destination register. The source and destination can be the same register. There are four unary operations:

fabs Absolute Value,
fneg Negate, and
fsqrt Square Root

9.7.2.1 Syntax

```
f<op>  Fd, Fm
```

- <op> is one of abs, neg, or sqrt.
- Fd and Fm must specify either two single precision registers or two double precision registers.

9.7.2.2 Operations

Name	Effect	Description
fabs	Fd ← |Fn|	Absolute Value
fneg	Fd ← −Fn	Negate
fsqrt	Fd ← $\sqrt{\text{Fn}}$	Square Root

9.7.2.3 Examples

```
fabs    d3, d5    // Store absolute value of d1 in d3
fneg    s15, s15  // Negate s15
```

9.7.3 Add, subtract, multiply, and divide

The basic mathematical operations require two source registers and one destination register. There are five basic mathematical operations:

fadd   Add,
fsub   Subtract,
fmul   Multiply,
fnmul Multiply and Negate, and
fdiv   Divide.

9.7.3.1 Syntax

```
f<op>   Fd, Fn, Fm
```

- <op> is one of add, sub, mul, nmul, or div.
- Fd, Fn, and Fm must specify either three single precision registers or three double precision registers.

9.7.3.2 Operations

Name	Effect	Description
fadd	Fd ← Fn + Fm	Add
fsub	Fd ← Fn − Fm	Subtract
fmul	Fd ← Fn × Fm	Multiply
fnmul	Fd ← −(Fn × Fm)	Multiply and Negate
fdiv	Fd ← Fn ÷ Fm	Divide

9.7.3.3 Examples

```
fadd    d0, d1, d2    // d0 <- d1 + d2
fadd    s0, s1, s2    // s0 <- s1 + s2
fnmul   s10, s10, s14 // s10 <- -(s10 * s14)
fdiv    d0, d7, d8    // d0 <- d7 / d8
```

9.7.4 Multiply and accumulate

These instructions are used to multiply and accumulate the results. A fused multiply accumulate operation does not perform rounding between the multiply and add operations. The two operations are *fused* into one. Using the fused multiply accumulate can speed up and improve the accuracy of many computations that involve the accumulation of products. NEON provides the following fused multiply accumulate instructions.:

fmadd   Fused Multiply Accumulate
fmsub   Fused Multiply Subtract
fnmadd   Fused Multiply Accumulate and Negate
fnmsub   Fused Multiply Subtract and Negate

9.7.4.1 Syntax

```
f<op> Fd, Fn, Fm, Fa
```

- Fd, Fn, Fm, and Fa must specify either four single precision registers or four double precision registers.
- <op> is one of madd, msub, nmadd, or nmsub.

9.7.4.2 Operations

Name	Effect	Description
fmadd	$Fd \leftarrow Fa + Fn \times Fm$	Multiply the values in the operand registers, and add the value to the destination register.
fmsub	$Fd \leftarrow Fa - Fn \times Fm$	Multiply the values in the operand registers, and subtract the value from the destination register.
fnmadd	$Fd \leftarrow -(Fa + Fn \times Fm)$	Multiply the values in the operand registers, and add the value to the destination register, then negate the final result and store it in the destination register
fnmsub	$Fd \leftarrow -(Fa - Fn \times Fm)$	Multiply the values in the operand registers, and subtract the value from the destination register, then negate the final result and store it in the destination register

9.7.4.3 Examples

```
fmla    s1,s6,s8,s1   // Multiply s8 by s6 and add to s1
fmla    d0,d4,d5,d0   // Multiply d4 by d5 and add to d0
```

9.7.5 Min and max

The `min(x,y)` and `max(x,y)` operations compare two registers and set a third register to the minimum or maximum of the two values. The instructions are:

fmax   Max,
fmaxnm   Max Number,
fmin   Min, and
fminnm   Min Number.

9.7.5.1 Syntax

```
f<op>   Fd, Fn, Fm
```

*   <op> is one of `fmax`, `fmaxnm`, `min`, or `minnum`.
*   Fd, Fn, and Fm must specify either three single precision registers or three double precision registers.
*   If one of the two values is a NaN, then `fmaxnm` and `fminnm` return the other value.
*   If either of the two values is a NaN, then `fmax` and `fmin` return NaN.
*   If both values are NaN, then all of these instructions produce a NaN.

9.7.5.2 Operations

Name	Effect	Description
fmax	Fd ← max(Fn, Fm)	Maximum
fmaxnm	Fd ← $maxNum$(Fn, Fm)	Maximum Number
fmin	Fd ← min(Fn, Fm)	Minimum
fminnm	Fd ← $minNum$(Fn, Fm)	Minimum Number

9.7.5.3 Examples

```
fmin    d0, d1, d2    // d0 <- min(d1,d2)
fmaxnm  s0, s1, s2    // s0 <- max(s1,s2)
```

Table 9.1: Condition code meanings for ARM and FP/NEON.

<cond>	ARM data processing instruction	FP `fcmp` instruction
AL	Always	Always
EQ	Equal	Equal
NE	Not Equal	Not equal, or unordered
GE	Signed greater than or equal	Greater than or equal
LT	Signed less than	Less than, or unordered
GT	Signed greater than	Greater than
LE	Signed less than or equal	Less than or equal, or unordered
HI	Unsigned higher	Greater than, or unordered
LS	Unsigned lower or same	Less than or equal
HS	Carry set/unsigned higher or same	Greater than or equal, or unordered
CS	Same as HS	Same as HS
LO	Carry clear/ unsigned lower	less than
CC	Same as LO	Same as LO
MI	Negative	Less than
PL	Positive or zero	Greater than or equal, or unordered
VS	Overflow	Unordered (at least one NaN operand)
VC	No overflow	Not unordered

9.7.6 Compare

The compare instruction subtracts the value in `Fm` from the value in `Fn` and sets the flags in the `PSTATE` register based on the result. The condition code meanings after an `fcmp` instruction are shown in Table 9.1. The comparison instructions are:

`fcmp`   Compare,
`fcmpe`   Compare with Exception,
`fccmp`   Conditional Compare, and
`fccmpe`   Conditional Compare with Exception.

9.7.6.1 Syntax

```
fcmp{e}   Fn, <Fm | #0.0>
fccmp{e}  Fn, <Fm | #0.0>, #uimm4, cond
```

- For the `fcmpe` and `fccmpe` instructions, an exception is raised if any kid of NaN is encountered. Otherwise an exception is raised only for signaling NaNs.

- cond is one of the two character condition codes from Table 3.2.
- #uimm4 is a value that is used to set the NZCV flags if cond is not true.

9.7.6.2 Operations

Name	Effect	Description
fcmp{e}	PSTATE ← $flags$(Fn − Fm)	Compare two registers
fcmp{e}	PSTATE ← $flags$(Fn − 0)	Compare to zero

fccmp{e}	**if** cond is True **then**
	   PSTATE ← $flags$(Fn − Fm)
	else
	   PSTATE ← uimm4
	end if
		Conditional compare two registers

fccmp{e}	**if** cond is True **then**
	   PSTATE ← $flags$(Fn − 0)
	else
	   PSTATE ← uimm4
	end if
		Conditional compare to zero

9.7.6.3 Examples

```
fcmp    s0, s1   // Subtract s1 from s0 and set NZCV flags
```

9.7.7 Conditional select

The conditional select instruction selects a value from one of two registers, based on the given condition, and stores it in the result register. The instruction is:

fcsel   Conditional Select.

9.7.7.1 Syntax

```
fcsel    Fd, Fn, Fm, cond
```

- Fd, Fn, and Fm must specify either three single precision registers or three double precision registers.
- cond is one of the two character condition codes from Table 3.2.

9.7.7.2 Operations

Name	Effect	Description

fcsel	**if** cond is True **then**
	    Fd ← Fn
	else
	    Fd ← Fm
	end if
		Set Fd to Fn or Fm.

9.7.7.3 Examples

```
fcsel    s0, s1, s2, ge // if ge, then s0<-s1 else s0<-s2
```

9.8 Floating point sine function

A fixed point implementation of the sine function was discussed in Section 8.6, and shown to be superior to the floating point sine function provided by GCC. Now that we have covered the FP instructions, we can write an assembly version using floating point that also performs better than the routines provided by GCC.

Listing 9.1 Simple scalar implementation of the $\sin x$ function using IEEE single precision.

```
//*******************************************************
// Name: sincos_a_f.S
// Author: Larry Pyeatt
// Date: 2/22/2018
//*******************************************************
// This is a version of the sin functions that uses single
// precision floating point with the FP/NEON instruction set.
// ----------------------------------------------------------
        .data
        // The following is a table of constants used in the
        // Taylor series approximation for sine
        .align  5               // Align to cache
ctab:   .word   0xBE2AAAAA      // -1.666666e-01
        .word   0x3C088889      //  8.333334e-03
        .word   0xB9500D00      // -1.984126e-04
        .word   0x3638EF1D      //  2.755732e-06
        .word   0xB2D7322A      // -2.505210e-08
        .equ    TERMS,((. - ctab)/4)
// ----------------------------------------------------------
        .text
        .align  2
        // float sin_a_f(x)
        // sin_a_f implements the sine function using IEEE single
```

```
24          // precision floating point. It computes sine by summing
25          // the first six terms of the Taylor series.
26          .global sin_a_f
27 sin_a_f://  register s0 contains x
28          // initialize variables
29          fmul    s1,s0,s0        // s1 <- x^2
30          fmov    s3,s0           // s3 <= x
31          ldr     x0,=ctab        // load pointer to coefficients
32          mov     x3,#TERMS       // load loop counter
33          // loop over table
34 loop:    fmul    s3,s1,s3        // s3 <- x^(2n+1)
35          ldr     s4,[x0],#4      // load coefficient and increment pointer
36          subs    x3,x3,#1        // decrement and test loop count
37          fmadd   s0,s3,s4,s0     // s0 += next term
38          bne     loop            // loop five times
39          ret
```

Listing 9.1 shows a single precision floating point implementation of the sine function, using the ARM FP/NEON instruction set. It works in a similar way to the previous fixed point code. There is a table of constants, each of which is the reciprocal of one of the factorial divisors in the Taylor series for sine. The subroutine calculates the powers of x one-by-one, and multiplies each power by the next constant in the table, summing the results as it goes. Note that the single precision floating point version uses fewer terms of the Taylor series than the fixed point version. This is because there are fewer bits of precision in the IEEE single precision format than in the fixed point format used in the previous chapter.

Listing 9.2 Simple scalar implementation of the $\sin x$ function using IEEE double precision.

```
1  //****************************************************************
2  // Name: sincos_a_f.S
3  // Author: Larry Pyeatt
4  // Date: 2/22/2018
5  //****************************************************************
6  // This is a version of the sin functions that uses double
7  // precision floating point with the FP/NEON instruction set.
8  // ----------------------------------------------------------
9          .data
10         // The following is a table of constants used in the
11         // Taylor series approximation for sine
12         .align 6        // Align for efficient caching
13 ctab:   .word 0x55555555, 0xBFC55555    // -1.666666666666667e-01 (-1/3!)
14         .word 0x11111111, 0x3F811111    //  8.333333333333333e-03 (1/5!)
15         .word 0x1A01A01A, 0xBF2A01A0    // -1.984126984126984e-04 (-1/7!)
16         .word 0xA556C734, 0x3EC71DE3    //  2.755731922398589e-06 (1/9!)
17         .word 0x67F544E4, 0xBE5AE645    // -2.505210838544172e-08
```

```
18          .word 0x13A86D09, 0x3DE61246    //  1.605904383682161e-10
19          .word 0xE733B81F, 0xBD6AE7F3    // -7.647163731819816e-13
20          .word 0x7030AD4A, 0x3CE952C7    //  2.811457254345521e-15
21          .word 0x46814157, 0xBC62F49B    // -8.220635246624329e-18
22          .equ  TERMS,((. - ctab)/8)
23  // -----------------------------------------------------------
24          .text
25          .align 2
26  // double sin_a_d(double x)
27  // sin_a_f_d implements the sine function using IEEE
28  // double precision floating point.  It computes sine
29  // by summing the first ten terms of the Taylor series.
30          .global sin_a_d
31  sin_a_d:// d0 contains x
32          ldr   x0,=ctab          // load pointer to coefficient table
33          // initialize variables
34          fmul  d1,d0,d0          // d1 <- x^2
35          fmov  d3,d0             // d3 <- x
36          mov   x3,#TERMS         // load loop counter
37          // loop over table
38  loop:   fmul  d3,d1,d3          // d4 <- x^(2n+1)
39          ldr   d4,[x0],#8        // load coefficient and increment pointer
40          subs  x3,x3,#1          // decrement and test loop counter
41          fmadd d0,d3,d4,d0       // d0 += next term
42          bne   loop              // loop nine times
43          ret
```

Listing 9.2 shows a double precision floating point implementation of the sine function, using the ARM FP/NEON instruction set. Again, there is a table of constants, each of which is the reciprocal of one of the factorial divisors in the Taylor series for sine. The subroutine calculates the powers of *x* one-by-one, and multiplies each power by the next constant in the table, summing the results as it goes. Note that the table of constants is longer than the fixed point version of the code, because there are more bits of precision in a double precision floating point number than there are in the fixed point representation that was used previously.

9.8.1 Performance comparison

Table 9.2 shows the performance of different implementations of the sine function, with and without compiler optimization. The Single Precision C and Double Precision C implementations are the standard implementations provided by GCC. The comparison tests were performed by timing each of the four cases on the same set of 100,000,000 random numbers, with and without compiler optimization.

Table 9.2: **Performance of sine function with various implementations.**

Optimization	Implementation	CPU seconds
None	Single Precision Scalar Assembly	2.01
	Single Precision C	6.75
	Double Precision Scalar Assembly	2.95
	Double Precision C	6.49
-Ofast	Single Precision Scalar Assembly	1.66
	Single Precision C	4.05
	Double Precision Scalar Assembly	2.45
	Double Precision C	5.83

When compiler optimization is not used, the single precision assembly implementation achieves a speedup of about 3.36 compared to the GCC implementation, and the double precision assembly implementation achieves a speedup of about 2.2 compared to the GCC implementation. When the best possible compiler optimization is used (-Ofast), the single precision assembly implementation achieves a speedup of about 2.44 compared to the GCC implementation. The double precision assembly implementation achieves a speedup of about 2.38 compared to the GCC implementation.

In every case, the assembly versions were significantly faster than the functions provided by GCC. It is clear that writing some functions in assembly can result in large performance gains. One interesting thing to note is that without optimization, the single precision C code is actually *slower* than the double precision C code. This is because, when optimization is not enabled, the C compiler converts single precision numbers to double precision numbers before calling the sine function. When optimization is enabled, the C compiler uses a single precision version of its sine function for single precision numbers.

9.9 Alphabetized list of FP/NEON instructions

Name	Page	Operation
fabs	308	Absolute Value
fadd	309	Add
fccmp	312	Conditional Compare
fccmpe	312	Conditional Compare with Exception
fcmp	312	Compare
fcmpe	312	Compare with Exception
fcsel	313	Conditional Select

continued on next page

Name	Page	Operation
fcvt	304	Convert Between Double, Single, and Half Precision
fcvtas	305	Convert Float to Signed Integer, Round Away From Zero
fcvtau	305	Convert Float to Unsigned Integer, Round Away From Zero
fcvtms	305	Convert Float to Signed Integer, Round Towards $-\infty$
fcvtmu	305	Convert Float to Unsigned Integer, Round Towards $-\infty$
fcvtns	305	Convert Float to Signed Integer, Round To Even
fcvtnu	305	Convert Float to Unsigned Integer, Round To Even
fcvtps	305	Convert Float to Signed Integer, Round Towards $+\infty$
fcvtpu	305	Convert Float to Unsigned Integer, Round Towards $+\infty$
fcvtzs	306	Convert Float to Signed Fixed Point
fcvtzs	305	Convert Float to Signed Integer, Round Towards Zero
fcvtzu	306	Convert Float to Unsigned Fixed Point
fcvtzu	305	Convert Float to Unsigned Integer, Round Towards Zero
fdiv	309	Divide
fmax	311	Max
fmaxnm	311	Max Number
fmin	311	Min
fminnm	311	Min Number
fmov	303	Floating Point Move Immediate
fmov	302	Move Between Data Registers
fmul	309	Multiply
fneg	308	Negate
fnmul	309	Multiply and Negate
frinta	307	Round away from zero
frinti	307	Round using the mode set in the FPCR.
frintm	307	Round towards $-\infty$
frintn	307	Round with halfway cases rounding to even
frintp	307	Round towards $+\infty$
frintx	307	Round using mode set in the FPCR and raise an exception if the result does not have the same value as the input.
frintz	307	Round towards zero
fsqrt	308	Square Root

continued on next page

Name	Page	Operation
fsub	309	Subtract
ldnp	301	Load Non-Temporal Pair
ldp	301	Load Pair
scvtf	306	Convert Signed Fixed Point to Float
scvtf	305	Convert Signed Integer to Float Using FPCR Rounding Mode
stnp	301	Store Non-Temporal Pair
stp	301	Store Pair
ucvtf	306	Convert Unsigned Fixed Point to Float
ucvtf	305	Convert Unsigned Integer to Float Using FPCR rounding mode

9.10 Chapter summary

The AArch64 FP/NEON coprocessor adds a great deal of power to the ARM architecture. The FP/NEON register set can hold over twice the amount of data that can be held in the AArch64 integer registers. The additional instructions allow the programmer to deal directly with the most common IEEE 754 formats for floating point numbers. In the next chapter, we will see that the ability to treat groups of registers as vectors adds a significant performance improvement. The GCC compiler does not make good use of these advanced features, which gives the assembly programmer a big advantage when high-performance code is needed.

Exercises

9.1. How many registers does the FP/NEON coprocessor add to the AArch64 architecture?
9.2. What is the purpose of the AHP, DN, and FZ, RMODE, and FZ16 bits in the FPCR?
9.3. How are floating point parameters passed to subroutines? How is a *pointer* to a floating point value (or array of values) passed to a subroutine?
9.4. Write the following C code in AArch64 assembly:

```
for (x = 0.0; x != 10.0; x += 0.1)
  {
    .
    .
    .
  }
```

9.5. In the previous exercise, the C code contains a subtle bug.
   • What is the bug?

- Show two ways to fix the code in AArch64 assembly. Hint: One way is to change the amount of the increment, which will change the number of times that the loop executes.

9.6. The fixed point sine function from the previous chapter was not compared directly to the hand-coded VFP implementation. Based on the information in Table 9.2 and Table 8.4, would you expect the fixed point sine function from the previous chapter to beat the hand-coded assembly VFP sine function in this chapter? Why or why not?

9.7. 3-D objects are often stored as an array of points, where each point is a vector (array) consisting of four values, x, y, z, and the constant 1.0. Rotation, translation, scaling, and other operations are accomplished by multiplying each point by a 4 × 4 transformation matrix. The following C code shows the data types and the transform operation:

```c
typedef float point[4];     // Point is an array of floats
typedef float matrix[4][4]; // Matrix is a 2-D array of floats
.
.
.
void xform(matrix *m, point* p)
{
  int i,j;
  point result;
  for(i=0;i<4;i++)
    {
       result[i] = 0.0;
       for(j=0;j<4;j++)
         result[i] += *m[j][i] * *p[j];
    }
  for(i=0;i<4;i++)
    *p[i] = result[i];
}
```

Write the equivalent AArch64 assembly code.

9.8. Optimize the AArch64 assembly code you wrote in the previous exercise. Use vector mode if possible.

9.9. Since the fourth element of the point is always 1.0, there is no need to actually store it. This will reduce memory requirements by about 25%, and require one fewer multiply. The C code would look something like this:

```c
typedef float[3] point;     // Point is an array of floats
typedef float[4][4] matrix; // Matrix is a 2-D array of floats
.
.
.
void xform(matrix *m, point* p)
```

```
 7    {
 8      int i,j;
 9      point result;
10      for(i=0;i<3;i++)
11        result[i] = m[3][i];
12      for(i=0;i<3;i++)
13      {
14        for(j=0;j<3;j++)
15          result[i] += m[j][i] * p[j];
16      }
17      *p = result;
18    }
```

Write optimal AArch64 FP/NEON code to implement this function.

9.10. The function in the previous problem would typically be called multiple times to process an array of points, as in the following function:

```
1  void xformall(matrix *m, point* p, int num_points)
2  {
3    int i;
4    for(i=0;i<num_points;i++)
5      xform(m,p+i);
6  }
```

This could be somewhat inefficient. Re-write this function in assembly, so that the transformation of each point is done without resorting to a function call. Make your code as efficient as possible.

## CHAPTER 10

Advanced SIMD instructions

In addition to the FP/NEON instructions described in the previous chapter, AArch64 also supports Advanced SIMD instructions, which allow the programmer to treat the FP/NEON registers as vectors (arrays) of data. Advanced SIMD uses the same set of registers described in Chapter 9, but adds new *views* to provide the ability to access the registers in more ways. Advanced SIMD adds about 125 instructions and pseudo-instructions to support not only floating point, but also integer and fixed point.

A single Advanced SIMD instruction can operate on up to 128 bits, which may represent multiple integer, fixed point, or floating point numbers. For example, if two of the 128-bit registers each contain eight 16-bit integers, then a single Advanced SIMD instruction can add all eight integers from one register to the corresponding integers in the other register, resulting in eight simultaneous additions. For certain applications, this vector architecture can result in extremely fast and efficient implementations. Advanced SIMD is particularly useful at handling streaming video and audio, but also can give very good performance on floating point intensive tasks.

The 32 FP/NEON/Advanced SIMD registers, originally introduced in Chapter 9, can be accessed using various *views*. Some SIMD instructions use the byte, half-word, word, and double-word views from Chapter 9, but most of them use the Advanced SIMD views. Fig. 10.1 shows the different ways of viewing an Advanced SIMD register. Each register can be viewed as containing a vector of 2, 4, 8, or 16 elements, all of the same size and type. Individual elements of each vector can also be accessed by some instructions. The scalar register names and views introduced in Chapter 9 are also used by some instructions. A scalar can be 8 bits, 16 bits, 32 bits, or 64 bits. The instruction syntax is extended to refer to elements of a vector register by using an index, x. For example Vm.4s[x] is element x in register Vm, where Vm is treated as a vector of four single-word (32-bit) elements.

10.1 Instruction syntax

The syntax of Advanced SIMD instructions can be described using an extension of the notation used throughout this book. Each instruction operates on certain types of register(s), and there are many registers. Advanced SIMD instruction syntax may use any of the following register definitions:

ARM 64-Bit Assembly Language
https://doi.org/10.1016/B978-0-12-819221-4.00017-1

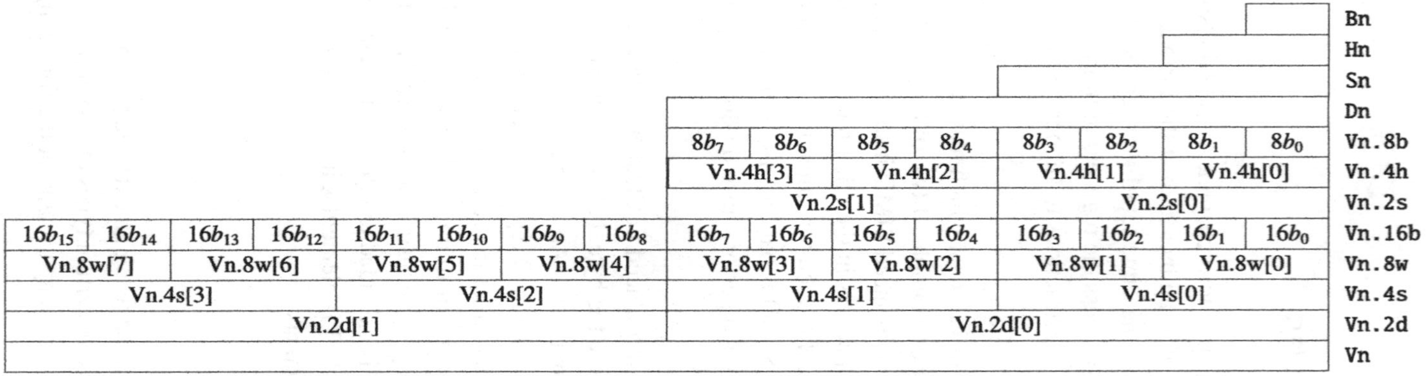

Bn
Hn
Sn
Dn
Vn.8b
Vn.4h
Vn.2s
Vn.16b
Vn.8w
Vn.4s
Vn.2d
Vn

$8b_7$	$8b_6$	$8b_5$	$8b_4$	$8b_3$	$8b_2$	$8b_1$	$8b_0$
Vn.4h[3]		Vn.4h[2]		Vn.4h[1]		Vn.4h[0]	
Vn.2s[1]				Vn.2s[0]			

$16b_{15}$	$16b_{14}$	$16b_{13}$	$16b_{12}$	$16b_{11}$	$16b_{10}$	$16b_9$	$16b_8$	$16b_7$	$16b_6$	$16b_5$	$16b_4$	$16b_3$	$16b_2$	$16b_1$	$16b_0$
Vn.8w[7]		Vn.8w[6]		Vn.8w[5]		Vn.8w[4]		Vn.8w[3]		Vn.8w[2]		Vn.8w[1]		Vn.8w[0]	
Vn.4s[3]				Vn.4s[2]				Vn.4s[1]				Vn.4s[0]			
Vn.2d[1]								Vn.2d[0]							

$8b_7 = Vn.8b[7]$
$8b_6 = Vn.8b[6]$
$8b_5 = Vn.8b[5]$
$8b_4 = Vn.8b[4]$
$8b_3 = Vn.8b[3]$
$8b_2 = Vn.8b[2]$
$8b_1 = Vn.8b[1]$
$8b_0 = Vn.8b[0]$

$16b_{15} = Vn.16b[15]$
$16b_{14} = Vn.16b[14]$
$16b_{13} = Vn.16b[13]$
$16b_{12} = Vn.16b[12]$
$16b_{11} = Vn.16b[11]$
$16b_{10} = Vn.16b[10]$
$16b_9 = Vn.16b[9]$
$16b_8 = Vn.16b[8]$
$16b_7 = Vn.16b[7]$
$16b_6 = Vn.16b[6]$
$16b_5 = Vn.16b[5]$
$16b_4 = Vn.16b[4]$
$16b_3 = Vn.16b[3]$
$16b_2 = Vn.16b[2]$
$16b_1 = Vn.16b[1]$
$16b_0 = Vn.16b[0]$

Figure 10.1: All of the possible views for register Vn. Valid names depend on which instruction is being used.

Xy	Refers to a 64-bit AArch64 integer register.
Wy	Refers to 32-bit AArch64 integer register.
By	Refers to the lower 8-bits, or byte, of an Advanced SIMD register.
Hy	Refers to the lower 16-bits, or half-word, of an Advanced SIMD register.
Sy	Refers to the lower 32-bits, or single-word, of an Advanced SIMD register.
Dy	Refers to the lower 64-bits, or double-word, of an Advanced SIMD register.
Fy	Is used to indicate either a single-word or double-word FP/NEON register. F must be either s for a single word register, d for a double word register.
Vy	A 128-bit Advanced SIMD register. y can be any valid register number.

Vy.T   A 128-bit Advanced SIMD register, treated as a vector of elements of type T, where T may be one of:

     b   A vector of 16 bytes.

     h   A vector of 8 half-words.

     s   A vector of 4 words.

     d   A vector of 2 double-words.

     Some instructions can only allow a subset of these types.

Vy.nT   A 128-bit Advanced SIMD register, or the lower 64 bits of an Advanced SIMD register, treated as a vector of elements of type T, where T may be one of:

     16b   A 128-bit Advanced SIMD register, treated as a vector of sixteen bytes.

     8b   The lower 64 bits of an Advanced SIMD register, treated as a vector of eight bytes.

     8h   A 128-bit Advanced SIMD register, treated as a vector of eight half-words.

     4h   The lower 64 bits of an Advanced SIMD register, treated as a vector of four half-words.

     4s   A 128-bit Advanced SIMD register, treated as a vector of four words.

     2s   The lower 64 bits of an Advanced SIMD register, treated as a vector of two words.

     2d   A 128-bit Advanced SIMD register, treated as a vector of two double-words.

     Some instructions can only allow a subset of these types.

Vy.nT[x]   Element x of an Advanced SIMD register, treated as a vector of type nT.

Each instruction has its own set of restrictions on legal values for the registers and types used. For example, one possible form of the mov instruction is:

```
mov Vd.T[x],Wn
```

which indicates that a 32-bit AArch32 register is used as the source operand, and any element of any Advanced SIMD register may be used as the destination. However, the instruction further requires that T must be either 2s or 4s, in order to match the size of Wn.

Instructions may have several forms. In those cases, the following syntax is used to specify possible forms:

{opt} Braces around a string indicate that the string is optional. For example, several operations have an optional r which indicates that the result is rounded instead of truncated.

(s|u) Parentheses indicate a choice between two or more possible characters or strings, separated by the pipe "|" character. For example, (s|u)shr would describe two forms for the shr instruction: sshr and ushr.

<Tn> A string inside the < and > symbols indicates a choice or special syntax that is too complex to be easily described using the parenthesis and pipe (a|b) syntax, and is described in the following text. It is also used to define a syntactical token when simply using a character would lead to confusion.

The following function definitions are used in describing the effects of many of the instructions:

$\lfloor x \rfloor$ The *floor* function maps a real number, x, to the next smallest integer.

$\lceil x \rfloor$ The *saturate* function limits the value of x to the highest or lowest value that can be stored in the destination register. Saturation is a method used to prevent overflow.

$\|x\|$ The *round* function maps a real number, x, to the nearest integer.

$\succ x \prec$ The *narrow* function reduces a $2n$ bit number to an n bit number, by taking the n least significant bits.

$\prec x \succ$ The *extend* function converts an n bit number to a $2n$ bit number, performing zero extension if the number is unsigned, or sign extension if the number is signed.

10.2 Load and store instructions

These instructions can be used to perform interleaving of data when structured data is loaded or stored. The data should be properly aligned for best performance. These instructions are very useful for common multimedia data types.

For example, image data is typically stored in arrays of pixels, where each pixel is a small data structure such as the pixel struct shown in Listing 5.38. Since each pixel is three bytes, and a d register is 8 bytes, loading a single pixel into one register would be inefficient. It would be much better to load multiple pixels at once, but an even number of pixels will not fit in a register. It will take three doubleword or quadword registers to hold an even number of pixels without wasting space, as shown in Fig. 10.2. This is the way data would be loaded using an Advanced SIMD ldr instruction. Many image processing operations work best if each color "channel" is processed separately. The SIMD load and store instructions can be used to

$green_2$	red_2	$blue_1$	$green_1$	red_1	$blue_0$	$green_0$	red_0	d0
red_5	$blue_4$	$green_4$	red_4	$blue_3$	$green_3$	red_3	$blue_2$	d1
$blue_7$	$green_7$	red_7	$blue_6$	$green_6$	red_6	$blue_5$	$green_5$	d2

Figure 10.2: Pixel data interleaved in thee doubleword registers.

red_7	red_6	red_5	red_4	red_3	red_2	red_1	red_0	d0
$green_7$	$green_6$	$green_5$	$green_4$	$green_3$	$green_2$	$green_1$	$green_0$	d1
$blue_7$	$blue_6$	$blue_5$	$blue_4$	$blue_3$	$blue_2$	$blue_1$	$blue_0$	d2

Figure 10.3: Pixel data de-interleaved in thee doubleword registers.

split the image data into color channels, where each channel is stored in a different register, as shown in Fig. 10.3.

Other examples of interleaved data include stereo audio, which is two interleaved channels, and surround sound, which may have up to nine interleaved channels. In all of these cases, most processing operations are simplified when the data is separated into non-interleaved channels.

10.2.1 Load or store single structure using one lane

These instructions are used to load and store structured data across multiple registers:

ld<n>   Load Structured Data, and
st<n>   Store Structured Data.

They can be used for interleaving or deinterleaving the data as it is loaded or stored, as shown in Fig. 10.3.

10.2.1.1 Syntax

```
<op><n> <list>[index],[Xn]
<op><n> <list>[index],[Xn],Xm
<op><n> <list>[index],[Xn],#imm
```

- <op> must be either ld or st.
- <n> must be one of 1, 2, 3, or 4.

- <list> specifies the list of registers. There are four list formats:
  1. {Vt.T}
  2. {Vt.T, V(t+1).T} or {Vt.T-V(t+1).T}
  3. {Vt.T, V(t+1).T, V(t+2).T} or {Vt.T-V(t+2).T}
  4. {Vt.T, V(t+1).T, V(t+2).T, V(t+3).T} or {Vt.T-V(t+3).T}

  The registers must be consecutive. Register 0 is consecutive to register 31.
- T must be b, h, s, or d.
- The immediate index specifies which element of each register is to be used, and must be appropriate for the data size specified by T. The same element will be used for all registers.
- Xn is the AARCH64 register containing the base address.
- Xm is the AARCH64 register containing an offset.
- If a register or immediate offset is given, then the base register, Xn, will be post-incremented.
- The post-increment immediate offset, if present, must be 8, 16, 24, 32, 48, or 64, depending on the number of elements transferred and the size specified by T.

10.2.1.2 *Operations*

Name	Effect	Description

ld<n>	$tmp \leftarrow Xn$
	$incr \leftarrow byteSize(T)$
	for $V \in regs($<list>$)$ **do**
	  $V[index] \leftarrow Mem[tmp]$
	  $tmp \leftarrow tmp + incr$
	end for
	if #imm is present **then**
	  $Xn \leftarrow Xn + imm$
	else
	  if Xm is specified **then**
	    $Xn \leftarrow Xn + Xm$
	  end if
	end if
		Load one or more data items into a single lane of one or more registers.

continued on next page

Name	Effect	Description

st\<n\>	$tmp \leftarrow Xn$
	$incr \leftarrow byteSize(T)$
	for $V \in regs(\texttt{<list>})$ **do**
	$\quad Mem[tmp] \leftarrow V[index]$
	$\quad tmp \leftarrow tmp + incr$
	end for
	if #imm is present **then**
	$\quad Xn \leftarrow Xn + imm$
	else
	$\quad$ **if** Rm is specified **then**
	$\qquad Xn \leftarrow Xn + Xm$
	$\quad$ **end if**
	end if
		Store one or more data items from a single lane of one or more registers.

10.2.1.3 Examples

```
1   ld3    {v0.b ,v1.b ,v2.b}[0],[x0],#3 // load first pixel
2   ld3    {v0.b ,v1.b ,v2.b}[1],[x0],#3
3   ld3    {v0.b ,v1.b ,v2.b}[2],[x0],#3
4   ld3    {v0.b ,v1.b ,v2.b}[3],[x0],#3
5   ld3    {v0.b ,v1.b ,v2.b}[4],[x0],#3
6   ld3    {v0.b ,v1.b ,v2.b}[5],[x0],#3
7   ld3    {v0.b ,v1.b ,v2.b}[6],[x0],#3
8   ld3    {v0.b ,v1.b ,v2.b}[7],[x0],#3 //load eigth pixel
```

10.2.2 Load or store multiple structures

These instructions are used to load and store multiple data structures across multiple registers with interleaving or deinterleaving:

ld\<n\>   Load Multiple Structured Data, and
st\<n\>   Store Multiple Structured Data.

10.2.2.1 Syntax

```
<op><n> <list>,[Xn]
<op><n> <list>,[Xn],Xm
<op><n> <list>,[Xn],#imm
```

- <op> must be either ld or st.
- <n> must be one of 1, 2, 3, or 4.
- <list> specifies the list of registers. There are four list formats:
  1. {Vt.T}
  2. {Vt.T, V(t+1).T} or {Vt.T-V(t+1).T}
  3. {Vt.T, V(t+1).T, V(t+2).T} or {Vt.T-V(t+2).T}
  4. {Vt.T, V(t+1).T, V(t+2).T, V(t+3).T} or {Vt.T-V(t+3).T}

  The registers must be consecutive. Register 0 is consecutive to register 31.
- T must be 16b, 8b, 8h, 4h, 4s, 2s, or 2d. If <n> is 1, then T can be 1d.
- Xn is the AARCH64 register containing the base address.
- Xm is the AARCH64 register containing an offset.
- If a register or immediate offset is given, then the base register, Xn, will be post-incremented.
- The post-increment immediate offset, if present, must be 8, 16, 24, 32, 48, or 64, depending on the number of elements transferred and the size specified by T.

10.2.2.2 Operations

Name	Effect	Description

ld<n>	$tmp \leftarrow$ Xn
	 $incr \leftarrow byteSize(T)$
	 for $0 \leq x < nLanes(T)$ **do**
	   for $V \in regs(\text{<list>})$ **do**
	     $V[x] \leftarrow Mem[tmp]$
	     $tmp \leftarrow tmp + incr$
	   end for
	 end for
	 if #imm is present **then**
	   Xn $\leftarrow$ Xn + imm
	 else
	   if Rm is specified **then**
	     Xn $\leftarrow$ Xn + Xm
	   end if
	 end if
		Load multiple structures into all lanes of one or more registers.

continued on next page

Name	Effect	Description

st\<n\>	$tmp \leftarrow Xn$
	$incr \leftarrow byteSize(T)$
	for $0 \leq x < nLanes(T)$ **do**
	  for $V \in$\<list\> **do**
	    $Mem[tmp] \leftarrow V[x]$
	    $tmp \leftarrow tmp + incr$
	  end for
	end for
	if #imm is present **then**
	  $Xn \leftarrow Xn + imm$
	else
	  if Rm is specified **then**
	    $Xn \leftarrow Xn + Xm$
	  end if
	end if
		Save multiple structures from all lanes of one or more registers.

10.2.2.3 Examples

```
@ Load eight rgb pixel structures into
@ d0(red),d1(green),and d2(blue)
ld3    {v0.8b,v1.8b,v2.8b},[x0],#24 // load eight pixels

@ Load sixteen rgb pixel structures into
@ v0(red),v1(green),and v2(blue)
ld3    {v0.16b-v2.16b},[x0],#48    // load sixteen pixels
```

10.2.3 Load copies of a structure to all lanes

This instruction is used to load multiple copies of structured data across multiple registers:

ld\<n\>r   Load Copies of Structured Data.

The data is copied to all lanes. This instruction is useful for initializing vectors for use in later instructions.

10.2.3.1 Syntax

```
ld<n>r <list>,[Xn]
ld<n>r <list>,[Xn],Xm
ld<n>r <list>,[Xn],#imm
```

- <n> must be one of 1, 2, 3, or 4.
- <list> specifies the list of registers. There are four list formats:
  1. `{Vt.T}`
  2. `{Vt.T, V(t+1).T}` or `{Vt.T-V(t+1).T}`
  3. `{Vt.T, V(t+1).T, V(t+2).T}` or `{Vt.T-V(t+2).T}`
  4. `{Vt.T, V(t+1).T, V(t+2).T, V(t+3).T}` or `{Vt.T-V(t+3).T}`

  The registers must be consecutive. Register 0 is consecutive to register 31.
- T must be 16b, 8b, 8h, 4h, 4s, 2s, or 2d. If <n> is 1, then T can be 1d.
- Xn is the AARCH64 register containing the base address.
- Xm is the AARCH64 register containing an offset.
- If a register or immediate offset is given, then the base register, Xn, will be post-incremented.
- The post-increment immediate offset, if present, must be 1, 2, 3, 4, 6, 8, 12, 16, 24, or 32, depending on the number of elements transferred and the size specified by T.

10.2.3.2 Operations

Name	Effect	Description

ld<n>r	$tmp \leftarrow Xn$
	 $incr \leftarrow byteSize(T)$
	 for $V \in regs(\text{<list>})$ **do**
	   for $0 \leq x < nLanes(T)$ **do**
	     $V[x] \leftarrow Mem[tmp]$
	   end for
	   $tmp \leftarrow tmp + incr$
	 end for
	 if #imm is present **then**
	   $Xn \leftarrow Xn + imm$
	 else
	   if Rm is specified **then**
	     $Xn \leftarrow Xn + Xm$
	   end if
	 end if
		Load one structure into all lanes of one or more registers.

10.2.3.3 Examples

```
// Load eight copies of an rgb struct into
// v0(red),v1(green),and v2(blue)
ld3r    {v0.8b-v2.8b},[x0]  // load 8 copies
```

10.3 Data movement instructions

With the additional register views added by Advanced SIMD, there are many more ways to specify data movement. Instructions are provided to move data using the Advanced SIMD views, the FP/NEON views, and the AArch64 integer register views. This results in a large number of possible move instructions.

10.3.1 Duplicate scalar

The duplicate instruction copies a scalar into every element of the destination vector. The scalar can be in an Advanced SIMD register or an AARCH64 integer register. The instruction is:

dup   Duplicate Scalar.

10.3.1.1 Syntax

```
dup     Vd.2d, Xn
dup     Vd.Td1, Wn
dup     Vd.Td2, Vn.Ts[index]
dup     Fd, Vn.Td3[index]
```

- Td1 may be 8b, 16b, 4h, 8h, 2s, or 4s. The lowest n bits of Wn will be used, where n is the number of bits specified by Td1.
- Td2 may be 8b, 16b, 4h, 8h, 2s, 4s, or 2d.
- Ts can be one of b, h, s, or d, and must match Td2.
- Fd may be any of the FP/NEON register names used in Chapter 9.
- Td3 may be b, h, s, or d, and must match Fd.
- The immediate index must be valid for the type of vector element specified.
- MOV Fd Vn.Td[index] is an alias for DUP Fd,Vn.Td[index].
- The MOV Fd,Fn instruction, which was introduced in Chapter 9, is an alias for DUP Fd,Vn.Td[0].

10.3.1.2 Operations

Name	Effect	Description
dup Vd.2D,Xn	Vd[] ← Xn	Copy Xn to both 64-bit elements of Vd.

continued on next page

Name	Effect	Description
dup Vd.Td1,Wn	Vd[] ← $bits$(Wn)	Copy $bitSize$(Td1) least significant bits of Wn to all elements of Vd.
dup Vd.Td2,Vn.Ts[x]	Vd[] ← Vn[x]	Copy element x of Vn to all elements of Vd.
dup Fd,Vn.Td3[x]	Fd[] ← Vn[x]	Copy element x of Vn to all elements of Fd.

10.3.1.3 Examples

```
    dup     v0.8b,w1     // copy 8 bits from w1 to
                         // 8 8-bit elements of v0
    dup     v3.8h,V2.h[1] // copy second 16 bit half word
                         // from v2 to 8 16-bit elements of v3
```

10.3.2 Move vector element

These instructions copy one element into a vector:

mov   Copy element into vector,

umov   Copy unsigned integer element from vector to AARCH64 register, and

smov   Copy signed integer element from vector to AARCH64 register.

10.3.2.1 Syntax

```
    mov Vd.T[index], Wn
    mov Vd.2d[index], Xn
    mov Vd.T2[index], Vn.T2[index2]
    smov Wd, Vn.T3[index]
    umov Wd, Vn.Ts[index]
    smov Xd, Vn.T4[index]
    umov Xd, Vn.d[index]
```

- T may be 8b, 16b, 4h, 8h, 2s, or 4s.
- The lowest n bits of Wn will be used, where n is the number of bits specified by T.
- The type T2 may be b, h, s, or d.

- The type T3 may be b or h.
- Ts may be 8b, 16b, 4h, 8h, 2s, or 4s.
- T4 may be b, h, or s.
- Both immediates, `index` and `index2`, must be valid for the type of vector element specified.
- `INS Vd.T[index],Wn` is an alias for `MOV Vd.T[index],Wn`.
- `INS Vd.D[index],Xn` is an alias for `MOV Vd.2d[index],Xn`.
- `INS Vd.T[index],Vn.T[index2]` in an alias for `MOV Vd.T[index], Vn.T[index2]`.

10.3.2.2 Operations

Name	Effect	Description
`mov Vd.T[x],Wn`	$Vd.T[x] \leftarrow Wn$	Copy least-significant bits from Wn to specified element of Vd.
`mov Vd.2d[x],Xn`	$Vd.2d[x] \leftarrow Xn$	Copy from Xn to specified element of Vd.
`mov Vd.T2[x],Vn.T2[x2]`	$Vd.T[x] \leftarrow Vn.T[x2]$	Copy element x2 from Vn to element x of Vd.
`smov Wd, Vn.T3[x]`	$Wd \leftarrow signExtend(Vn[x])$	Move sign-extended element from Vn to Wd.
`umov Wd, Vn.Ts[x]`	$Wd \leftarrow zeroExtend(Vn[x])$	Move zero-extended element from Vn to Wd.
`smov Xd, Vn.T4[x]`	$Xd \leftarrow signExtend(Vn[x])$	Move sign-extended element from Vn to Xd.
`umov Xd, Vn.d[x]`	$Xd \leftarrow zeroExtend(Vn[x])$	Move zero-extended element from Vn to Xd.

10.3.2.3 Examples

```
    mov    v3.2s[1],v4.2s[0]
    umov   x3,v4.d[1]
```

10.3.3 Move immediate

These instructions are used to load immediate data into the vector registers:

movi	Vector Move Immediate,
mvni	Vector Move NOT Immediate, and
fmov	Vector Floating Point Move Immediate.

10.3.3.1 Syntax

```
movi  Vn.T, #uimm8{, sop #shift}
mvni  Vn.T, #uimm8{, sop #shift}
movi  Vn.2D, #uimm64
movi  Dn, #uimm64
fmov  Vn.Td, #fpimm
```

- `sop` may be `lsl` or `msl`, where `msl` is a left shift which fills the low order bits with ones instead of zeros.
- If `sop` is *not* present, then the shift is assumed to be an `lsl` with `shift` amount of zero. The valid combinations of T and `shift` are given by the following table:

sop	T	shift	Description
lsl	4h or 8h	0 or 8	Replicate LSL(`uimm8`,`shift`) into each 16-bit element.
lsl	2s or 4s	0, 8, 16, or 24	Replicate LSL(`uimm8`,`shift`) into each 32-bit element.
msl	2s or 4s	8 or 16	Replicate MSL(`uimm8`,`shift`) into each 32-bit element.

- For `movi`, if `sop` is *not* present, then T may be 8b or 16b, in addition to the values shown in the previous table.
- `uimm64` may be either `0` or `0xFFFFFFFFFFFFFFFF`.
- Td may be 2s, 4s, or 2d.
- `fpimm` may be specified either in decimal notation or in hexadecimal using its IEEE754 encoding. The value must be the expressable as $\pm n \div 16 \times 2^r$, where n and r are integers such that $16 \leq n \leq 31$ and $-3 \leq r \leq 4$. It is encoded as a normalized binary floating point number with sign, 4 bits of fraction, and a 3-bit exponent.

10.3.3.2 Operations

Name	Effect	Description

movi	**if** uimm64 is present **then**
	$\quad c \leftarrow$ uimm64
	else
	$\quad c \leftarrow shift(\text{sop}, \text{uimm8}, \text{shift})$
	end if
	$n \leftarrow$ # of elements
	for $0 \leq i < n$ **do**
	$\quad$ Vn$[i] \leftarrow c$
	end for
		Replicate immediate data, possibly shifted, into each element.

mvni	$c \leftarrow \neg shift(\text{sop}, \text{uimm8}, \text{shift})$
	$n \leftarrow$ # of elements
	for $0 \leq i < n$ **do**
	$\quad$ Vn$[i] \leftarrow c$
	end for
		Replicate complement of immediate data, possibly shifted, into each element.

fmov	$n \leftarrow$ # of elements
	for $0 \leq i < n$ **do**
	$\quad$ Vn$[i] \leftarrow$ fpimm
	end for
		Replicate immediate data into each element.

10.3.3.3 Examples

```
    movi  v0.16b, #0xAA
    mvni  v0.8h, #0xAA, lsl #8
```

10.3.4 Transpose matrix

Advanced SIMD provides two versions of the transpose instruction that can be used together for transposing 2×2 matrices. Fig. 10.4 shows two examples of this instruction. The instruction is:

trn   Transpose Matrix.

10.3.4.1 Syntax

```
    trn(1|2)   Vd.T, Vn.T, Vm.T
```

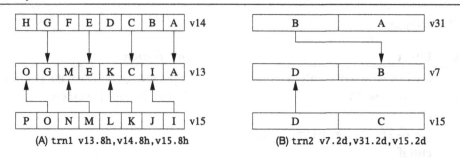

(A) trn1 v13.8h,v14.8h,v15.8h   (B) trn2 v7.2d,v31.2d,v15.2d

Figure 10.4: Examples of the trn instruction.

- T must be 8b, 16b, 4h, 8h, 2s, 4s, or 2d.
- Larger matrices can be transposed using a divide-and-conquer approach.

10.3.4.2 Operation

Name	Effect	Description

trn1	$n \leftarrow$ # of elements
	for $0 \leq i < n$ **do**
	   if $even(i)$ **then**
	      $Vd[i] \leftarrow Vn[i]$
	   else
	      $Vd[i] \leftarrow Vm[i-1]$
	   end if
	end for
		Copy even elements of Vn and Vm into Vd, interleaved.

trn2	$n \leftarrow$ # of elements
	for $0 \leq i < n$ **do**
	   if $even(i)$ **then**
	      $Vd[i] \leftarrow Vn[i+1]$
	   else
	      $Vd[i] \leftarrow Vm[i]$
	   end if
	end for
		Copy odd elements of Vn and Vm into Vd, interleaved.

10.3.4.3 Examples

```
// Transpose a 2x2 matrix of 32-bit elements in v3:v4, store
// transposed matrix in v0:v1
trn1      v0.2s,v3.2s,v4.2s
trn2      v1.2s,v3.2s,v4.2s
// Transpose eight 2x2 matrices of 16-bit elements
```

Figure 10.5: Transpose of a 3 x 3 matrix.

```
trn1        v8.8h,v10.8h,v11.8h
trn2        v9.8h,v10.8h,v11.8h
```

Fig. 10.5 shows how the `trn` instruction can be used to transpose a 3×3 matrix.

10.3.5 Vector permute

These instructions are used to interleave or deinterleave the data from two vectors, or to extract bits from a vector:

zip   Zip Vectors,
uzp   Unzip Vectors, and
ext   Byte Extract.

Fig. 10.6 gives an example of the `zip` instruction. The `uzp` instruction performs the inverse operation.

10.3.5.1 Syntax

```
zip(1|2)    Vd.T, Vn.T, Vm.T
uzp(1|2)    Vd.T, Vn.T, Vm.T
ext         Vd.Ta, Vn.Ta, Vm.Ta, #index
```

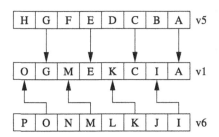

Figure 10.6: Example of `zip1 v1.8h,v5.8h,v6.8h`. The `zip2` instruction does the same thing, but uses the odd elements of the source registers, rather than the even elements.

- T is 8b, 16b, 4h, 8h, 2s, 4s, or 2d.
- For zip and uzp:
  - If 1 is present, use lower half of source registers.
  - If 2 is present, use upper half of source registers.
- Ta is either 8b (use only 64 bits of each register) or 16b (use all 128 bits of each register).
- index is an immediate value in the range 0 to $nelem(T) - 1$.

10.3.5.2 Operations

Name	Effect	Description

zip	$o \leftarrow 0$
	$n \leftarrow$ # of elements
	if 2 is present **then**
	   $o \leftarrow n \div 2$
	end if
	for $0 \le i < n$ **do**
	  if $even(i)$ **then**
	    $Vd[i] \leftarrow Fn[i \div 2 + o]$
	  else
	    $Vd[i] \leftarrow Fm[\lfloor i \div 2 \rfloor + o]$
	  end if
	end for
		Interleave data from two vectors.

uzp	$o \leftarrow 0$
	if 2 is present **then**
	   $o \leftarrow 1$
	end if
	$n \leftarrow$ # of elements
	for $0 \le i < n$ **do**
	  if $i < n \div 2$ **then**
	    $Vd[i] \leftarrow Fn[i \times 2 + o]$
	  else
	    $Vd[i] \leftarrow Fm[i \times 2 + o]$
	  end if
	end for
		Deinterleave data from two vectors.

continued on next page

Name	Effect	Description

ext	$n \leftarrow$ # of elements
	for $0 \leq i < n$ **do**
	  if $i < n - index$ **then**
	    $Vd[i] \leftarrow Vn[i + index]$
	  else
	    $Vd[i] \leftarrow Vm[i - (n - index)]$
	  end if
	end for
		Copy bytes from Vn followed by Vm, starting at Vn[index], until Vd is full.

10.3.5.3 Examples

```
zip1   v0.16b,v3.16b,v4.16b  // interleave even elements
zip2   v1.16b,v3.16b,v4.16b  // interleave odd elements
```

10.3.6 Table lookup

The table lookup instructions use indices held in one vector to lookup values from a table held in one or more other vectors. The resulting values are stored in the destination vector. The table lookup instructions are:

tbl   Table Lookup, and
tbx   Table Lookup with Extend.

10.3.6.1 Syntax

```
<op>   Vd.T, <list>, Vm.T
```

- <op> is one of tbl or tbx.
- T may be 8b or 16b.
- <list> specifies the list of registers. There are five list formats:
  1. {Vn.16b},
  2. {Vn.16b, V(n+1).16b},
  3. {Vn.16b, V(n+1).16b, V(n+2).16b}, or
  4. {Vn.16b, V(n+1).16b, V(n+2).16b, V(n+3).16b}.
- A dash "-" can be used to specify a range of registers, as shown in the examples below.
- Vm is the register holding the indices.
- The table can contain up to 64 bytes.

10.3.6.2 Operations

Name	Effect	Description

tbl	$Minr \leftarrow$ first register
	 $Maxr \leftarrow$ last register
	 for $0 \leq i < 16$ **do**
	   $r \leftarrow Minr + (\text{Vm}[i] \div 16)$
	   if $r > Maxr$ **then**
	     $\text{Vd}[i] \leftarrow 0$
	   else
	     $e \leftarrow \text{Vm}[i] \mod 16$
	     $\text{Vd}[i] \leftarrow \text{Vr}[e]$
	   end if
	 end for
		Use indices Vm to look up values in a table and store them in Vd. If the index is out of range, zero is stored in the corresponding destination.

tbx	$Minr \leftarrow$ first register
	 $Maxr \leftarrow$ last register
	 for $0 \leq i < 16$ **do**
	   $r \leftarrow Minr + (\text{Vm}[i] \div 16)$
	   if $r \leq Maxr$ **then**
	     $e \leftarrow \text{Vm}[i] \mod 8$
	     $\text{Vd}[i] \leftarrow \text{Vr}[e]$
	   end if
	 end for
		Use indices Vm to look up values in a table and store them in Vd. If the index is out of range, the corresponding destination is unchanged.

10.3.6.3 Examples

```
1   tbl V1.8b,{v2.16b},v3.8b
2   tbl V1.8b,{v2.16b-v3.16b},v3.8b
3   tbl V1.8b,{v2.16b-v4.16b},v3.8b
4   tbl V1.8b,{v2.16b-v5.16b},v3.8b
```

10.4 Data conversion

When high precision is not required, The IEEE half-precision format can be used to store floating point numbers in memory. This can reduce memory requirements by up to 50%. This can also result in a significant performance improvement, since only half as much data needs to be moved between the CPU and main memory. However, on most processors half-precision data must be converted to single or double precision before it is used in calculations. Advanced SIMD provides instructions to support conversion to and from IEEE half precision.

There are also functions to perform integer or fixed-point to floating-point conversions, and convert between IEEE single and double precision.

10.4.1 Convert between integer or fixed point and floating point

These instructions can be used to perform data conversion between floating point and fixed point (or integer) on each element in a vector:

fcvt   Vector convert floating point to integer or fixed point, and
cvtf   Vector convert integer or fixed point to floating point.

The elements in the result vector must be the same size as the elements in the source vector. An out of range integer or fixed-point result will be saturated to the destination size.

Fixed point (or integer) arithmetic operations are up to twice as fast as floating point operations. In some cases it is much more efficient to make this conversion, perform the calculations, then convert the results back to floating point.

10.4.1.1 Syntax

```
    fcvt<x>(s|u)    Vd.T, Vn.T
    fcvtz(s|u)      Vd.T, Vn.T, #<fbits>
    (s|u)cvtf       Vd.T, Vn.T{, #<fbits>}
```

- <x> is a single character which specifies the rounding mode:
  N: round to nearest with ties to even,
  A: round to nearest with ties away from zero,
  P: round towards $+\infty$,
  M: round towards $-\infty$, or
  Z: round towards zero.
- T may be either 2s, 4s, or 2d.
- <type> must be either s32 or u32.
- <fbits> specifies the number of fraction bits for a fixed point number, and must be between 1 and the number of bits specified by T. If it is omitted, then it is assumed to be zero.

10.4.1.2 Operations

Name	Effect	Description
fcvt<x>s	Vd[] ← $fixed$(Vm[], $fbits$)	Convert single precision to 32-bit signed fixed point or integer.
fcvt<x>u	Vd[] ← $ufixed$(Vm[], $fbits$)	Convert single precision to 32-bit unsigned fixed point or integer.
scvtf	Vd[] ← $float$(Vm[])	Convert signed 32-bit fixed point or integer to single precision
ucvtf	Vd[] ← $single$(Vm[])	Convert unsigned 32-bit fixed point or integer to single precision

10.4.1.3 Examples

```
ucvtf   v0.2s,v0.2s,#4   // Convert from U(28,4) to IEEE single
fcvtzs  v1.2d,v1.2d      // Convert two IEEE doubles to signed integer
```

10.4.2 Convert between half, single, and double precision

The following instructions can be used to convert between floating point formats:

fcvtl   Vector convert from half to single precision,
fcvtn   Vector convert from single to half precision, and
fcvtxn   Vector convert from double to single precision.

These instructions operate on vectors. There are additional conversion instructions available, but they only operate on scalar values.

10.4.2.1 Syntax

```
fcvtl     Vd.Td, Vn.Ts
fcvtl2    Vd.Td2, Vn.Ts2
fcvtn     Vd.Td3, Vn.Ts3
fcvtn2    Vd.Td4, Vn.Ts4
fcvtxn    Vd.2s, Vn.2d
fcvtxn2   Vd.4s, Vn.2d
```

• If 2 is present, then the upper 64 bits of the register containing the smaller elements will be used. Otherwise, the lower 64 bits are used.

- Td/Ts may be 4s/4h or 2d/2s.
- Td2/Ts2 may be 4s/8h or 2d/4s.
- Td3/Ts3 may be 4h/4s or 2s/2d.
- Td4/Ts4 may be 8h/4s or 4s/2d.

10.4.2.2 Operations

Name	Effect	Description
fcvtl	Vd[] ← $single$(Vn[])	Convert half precision to single precision.
fcvtn	Vd[] ← $half$(Vn[])	Convert single precision to half precision.
fcvtxn	Vd[] ← $single$(Vd[])	Convert double precision to single precision.

10.4.2.3 Examples

```
fcvtl  v0.2d,v1.2s    // convert halfs to singles
fcvtxn v3.2s,v4.2d    // convert doubles to singles
```

10.4.3 Round floating point to integer

The following instruction can be used to round a vector of floating point values to integers:

frint   Round Floating Point to Integer.

10.4.3.1 Syntax

```
frint<x>    Vd.T, Vn.T
```

- T is 2s, 4s, or 2d.
- <x> selects the rounding mode. It may be one of the following:
  N: round to nearest with ties to even,
  A: round to nearest with ties away from zero,
  P: round towards $+\infty$,
  M: round towards $-\infty$,
  Z: round towards zero,
  I: round using FPCR rounding mode, and
  X: round using FPCR rounding mode with exactness test.

10.4.3.2 Operations

Name	Effect	Description
frint*x*	Vd[] ← *roundx*(Vn[], *x*)	Round to integer using specified rounding mode.

10.4.3.3 Examples

```
frintn v0.2d,v1.2d      // Round to nearest with ties to even
```

10.5 Bitwise logical operations

Advanced SIMD provides instructions to perform bitwise logical operations on the vector register set. These operations add a great deal of power to the AARCH64 processor.

10.5.1 Vector logical operations

The bitwise logical operations are:

and   Vector bitwise AND,
orr   Vector bitwise OR,
orn   Vector bitwise NOR,
eor   Vector bitwise Exclusive-OR,
bic   Vector bit clear,
bif   Vector insert if false,
bit   Vector insert if true, and
bsl   Vector bitwise select.

10.5.1.1 Syntax

```
v<op>  Vd.T, Vn.T, Vm.T
```

- <op> must be one of **and**, **orr**, **orn**, **eor**, **bic**, **bif**, **bit**, or **bsl**.
- T is 8b or 16b (though an assembler should accept any other equivalent format).

10.5.1.2 Operations

Name	Effect	Description
and	Vd ← Vn ∧ Vm	Bitwise AND

continued on next page

Name	Effect	Description
orr	Vd ← Vn ∨ Vm	Bitwise OR
orn	Vd ← ¬(Vn ∨ Vm)	Complement of bitwise OR
eor	Vd ← Vn ⊕ Vm	Exclusive OR
bic	Vd ← Vn ∧ ¬Vm	Bit Clear
bif	Vd ← Vd ∨ ((¬Vm) ∧ Vn)	Insert if false
bit	Vd ← Vd ∨ (Vm ∧ Vn)	Insert if false
bsl	Vd ← (Vd ∧ Vm) ∨ (¬Vd ∧ Vm)	Set each bit in Vd to the corresponding bit from Vn when the original bit in Vd is 1, otherwise set it to the corresponding bit in Vm.

10.5.1.3 Examples

```
      and   v0.8b,v1.8b,v2.8b     // d0 <= d1 & d2
      bic   v3.16b,v3.16b,v5.16b  // v3 <= v3 & NOT v5
      orr   v0.16b,v1.16b,v2.16b  // v0 <= v1 | v2
      orr   v0.8b,v1.8b,v2.8b     // d0 <= d1 | d2
```

10.5.2 Bitwise logical operations with immediate data

Advanced SIMD provides vector versions of the logical OR and bit clear instructions:

orr   Vector bitwise OR immediate, and
bic   Vector Bit clear immediate.

10.5.2.1 Syntax

```
    <op>    Vn.T, #uimm8{, lsl #shift}
```

- <op> must be either orr, or bic.
- T may be either 2s, 4s, 4h, or 8h.
- If T is 2s or 4s, then shift may be 0, 8, 16, or 24.
- If T is 4h or 8h, then shift may be 0 or 8.
- If shift is not specified, then it is assumed to be zero.
- uimm8 is an 8-bit unsigned immediate value, which is shifted left by shift bits to create the desired pattern for the orr, or bic operation on each vector element.

10.5.2.2 Operations

Name	Effect	Description
vorr	$Vd[] \leftarrow Vd[] \vee (uimm8 \ll shift)$	Logical OR
vbic	$Vd[] \leftarrow Vd[] \wedge (uimm8 \ll shift)$	Bit Clear

10.5.2.3 Examples

```
bic  v0.2s, #0xFF         // clear least significant bytes
                          // of v0.s[0] and v0.s[1]
orr  v3.8h, #0x80,lsl #8  // set sign bit of every halfword
                          // in v3
```

10.6 Basic arithmetic instructions

Advanced SIMD provides many instructions for addition, subtraction, and multiplication, but does not provide an integer divide instruction. When division cannot be avoided, it is performed by multiplying the reciprocal, as was described in Chapter 7 and Chapter 8. When dividing by a constant, the reciprocal can be calculated in advance. For dividing by variables, Advanced SIMD provides instructions for quickly calculating the reciprocals of the elements in a vector. In most cases, this is faster than using a divide instruction. For floating point numbers, the FP/NEON divide instructions can be used.

10.6.1 Vector add and subtract

The add and fadd instructions add corresponding elements in two vectors and store the results in the corresponding elements of the destination register. The sub and fsub instructions subtract elements in one vector from corresponding elements in another vector and store the results in the corresponding elements of the destination register. Other versions of the add and subtract instructions allow mismatched operand and destination sizes, and the saturating versions prevent overflow by limiting the range of the results. The following instructions perform vector addition and subtraction:

add   Vector integer add,
fadd   Vector floating point add,
qadd   Vector saturating add,
addl   Vector add long,

addw	Vector add wide,
sub	Vector integer subtract,
fsub	Vector floating point subtract,
qsub	Vector saturating subtract,
subl	Vector subtract long, and
subw	Vector subtract wide.

10.6.1.1 Syntax

```
<op>      Vd.T, Vn.T, Vm.T
<op>      Dd, Dn, Dm
f<op>     Vd.T, Vn.T, Vm.T
sq<op>    Vd.T, Vn.T, Vm.T
uq<op>    Vd.T, Vn.T, Vm.T
<sop>l{2} Vd.Td, Vn.Ts, Vm.Ts
<sop>w{2} Vd.Td, Vn.Td, Vm.Ts
```

- <op> can be add or sub.
- If double word registers are specified (Dd, Dn, Dm) then the operation is a simple add or subtract of scalar 64-bit integer values, and not a vector operation.
- The valid choices for T are given in the following table:

Opcode	Valid Values for T
<op>	8b, 16b, 4h, 8h, 2s, 4s, or 2d
f<op>	2s, 4s, 2d
(s|u)q<op>	8b, 16b, 4h, 8h, 2s, 4s, 2d

- <sop> can be uadd, sadd, usub, or ssub.
- If the modifier 2 is present, then the operation is performed using the upper 64 bits of the registers holding the narrower elements.
- The valid choices for Td/Ts are given in the following table:

Opcode	Valid Values for Td/Ts
<sop>l	8h/8b, 4s/4h, 2d/2s
<sop>l2	8h/16b, 4s/8h, 2d/4s
<sop>w	8h/8b, 4s/4h, 2d/2s
<sop>w2	8h/16b, 4s/8h, 2d/4s

10.6.1.2 Operations

Name	Effect	Description
{f}<op>	Vd[] ← Vn[]<op>Vm[]	The operation is applied to corresponding elements of Vn and Vm. The results are stored in the corresponding elements of Vd.
*x*q<op>	Vd[] ← ∫Vn[]<op>Vm[]⌉	The operation is applied to corresponding elements of Vn and Vm. The results are saturated then stored in the corresponding elements of Vd.
<sop>l	Vd[] ←≺ Vn[]<sop>Vm[] ≻	The operation is applied to corresponding elements of Vn and Vm. The results are zero or sign extended then stored in the corresponding elements of Vd.
<sop>w	Vd[] ← Vn[]<sop> ≺ Vm[] ≻	The elements of Vm are sign or zero extended, then the operation is applied with corresponding elements of Vn. The results are stored in the corresponding elements of Vd.
<sop>l2	Vd[] ←≺ Vn[]<sop>Vm[] ≻	The operation is applied to corresponding elements from the upper 64-bits of Vn and Vm. The results are zero or sign extended then stored in the corresponding elements of Vd.
<sop>w2	Vd[] ← Vn[]<sop> ≺ Vm[] ≻	The elements in the upper 64 bits of Vm are sign or zero extended, then the operation is applied with corresponding elements of Vn. The results are stored in the corresponding elements of Vd.

10.6.1.3 Examples

```
add     v1.16b,v6.16b,v8.16b // Add two vectors of 16 8-bit integers
sqadd   v2.8s,v7.8s,v9.8s    // Add two vectors of 8 16-bit signed
                             // integers, and saturate the results
```

10.6.2 *Vector add and subtract with narrowing*

These instructions add or subtract the corresponding elements of two vectors and narrow the resulting elements by taking the *upper* (most significant) half:

addhn   Vector add and narrow,
raddhn   Vector add, round, and narrow,
subhn   Vector subtract and narrow, and
rsubhn   Vector subtract, round, and narrow.

The results are stored in the corresponding elements of the destination register. Results can be optionally rounded instead of truncated.

10.6.2.1 Syntax

```
{r}<op>hn{2}    Vd.Td, Vn.Ts, Vm.Ts
```

- <op> is either add or sub.
- If r is present, then the result is rounded instead of truncated.
- If 2 is present, then the upper 64 bits of the destination vector are used.
- The valid choices for Td/Ts are given in the following table:

Opcode	Valid Values for Td/Ts
r<op>hn	8b/8h, 4h/4s, 2s/2d
r<op>hn2	16b/8h, 8h/4s, 4s/2d

10.6.2.2 Operations

Name	Effect	Description

{r}<op>hn	$shift \leftarrow size \div 2$
	if r is present **then**
	$\quad x \leftarrow \|Vn[]<op>Vm[]\|$
	$\quad Vd[] \leftarrow{>} x \gg shift \prec$
	else
	$\quad x \leftarrow Vn[]<op>Vm[]$
	$\quad Vd[] \leftarrow{>} x \gg shift \prec$
	end if
		The operation is applied to corresponding elements of Vn and Vm. The results are optionally rounded, then narrowed by taking the most significant half, and stored in the corresponding elements of Vd.

continued on next page

Name	Effect	Description

{r}<op>hn2	$shift \leftarrow size \div 2$
	if r is present **then**
	$\quad x \leftarrow \|Vn[]<op>Vm[]\|$
	$\quad Vd[] \leftarrow \succ x \gg shift \prec$
	else
	$\quad x \leftarrow Vn[]<op>Vm[]$
	$\quad Vd[] \leftarrow \succ x \gg shift \prec$
	end if
		The operation is applied to corresponding elements of Vn and Vm. The results are optionally rounded, then narrowed by taking the most significant half, and stored in the corresponding elements of the *upper* 64 bits of Vd.

10.6.2.3 Examples

```
addhn  v1.2s,v6.2d,v8.2d // Add and narrow
rsubhn v4.8b,v5.8h,v3.8h // Subtract round and narrow
```

10.6.3 Add or subtract and divide by two

These instructions add or subtract corresponding integer elements from two vectors, then shift the result right by one bit:

hadd   Vector halving add,

rhadd   Vector halving add and round, and

hsub   Vector halving subtract.

The results are stored in corresponding elements of the destination vector.

10.6.3.1 Syntax

```
(s|u){r}hadd.<type> Vd.T, Vn.T, Vm.T
(s|u)hsub.<type>    Vd.T, Vn.T, Vm.T
```

- If <r> is specified, then the result is rounded instead of truncated.
- T must be 8b, 16b, 4h, 8h, 2s, or 4s.

10.6.3.2 Operations

Name	Effect	Description
(s|u)hadd	$Vd[] \leftarrow (Vn[] + Vm[]) \gg 1$	The corresponding elements of Vn and Vm are added together. The results are shifted right one bit and stored in the corresponding elements of Vd.
(s|u)rhadd	$Vd[] \leftarrow \|Vn[] + Vm[]\| \gg 1$	The corresponding elements of Vn and Vm are added together. The Results are rounded, then shifted right one bit and stored in the corresponding elements of Vd.
(s|u)hsub	$Vd[] \leftarrow (Vn[] - Vm[]) \gg 1$	The elements of Vn are subtracted from the corresponding elements of Vm. Results are shifted right one bit and stored in the corresponding elements of Vd.

10.6.3.3 Examples

```
    srhadd    v1.16b,v6.16b,v8.16b // Add elements and divide by 2
    uhsub     v1.4h,v6.4h,v8.4h    // Subtract and divide by 2
```

10.6.4 Add elements pairwise

These instructions add vector elements pairwise:

addp Vector add pairwise,
addlp Vector add long pairwise, and
adalp Vector add and accumulate long pairwise.

The long versions can be used to prevent overflow.

10.6.4.1 Syntax

```
    addp        Vd.T, Vn.T, Vm.T
    faddp       Vd.Tf, Vn.Tf, Vm.Tf
    (s|u)addlp  Vd.Td, Vn.Ts
    (s|u)adalp  Vd.Td, Vn.Ts
```

- T must be 8b, 16b, 4h, 8h, 2s, 4s, or 2d.
- Tf must be 2s, 4s, or 2d.
- Td/Ts must be 4h/8b, 8h/16b, 2s/4h, 4s/8h, 1d/2s, or 2d/4s.

10.6.4.2 Operations

Name	Effect	Description

{f}addp	$n \leftarrow$ # of elements
	for $0 \leq i < (n \div 2)$ **do**
	$\quad$ Vd[i] $\leftarrow$ Vm[i] + Vm[$i + 1$]
	end for
	for $(n \div 2) \leq i < n$ **do**
	$\quad j \leftarrow i - (n \div 2)$
	$\quad$ Vd[i] $\leftarrow$ Vn[j] + Vn[$j + 1$]
	end for
		Add elements of two vectors pairwise and store the results in another vector.

addlp	$n \leftarrow$ # of elements
	for $0 \leq i < (n \div 2)$ by 2 **do**
	$\quad$ Vd[i] $\leftarrow$
	$\qquad \prec$ Vm[i] $\succ + \prec$ Vm[$i + 1$] $\succ$
	end for
		Add elements of a vector pairwise and store the results in another vector.

adalp	$n \leftarrow$ # of elements
	for $0 \leq i < (n \div 2)$ by 2 **do**
	$\quad$ Vd[i] $\leftarrow$ Vd[i]+
	$\qquad \prec$ Vm[i] $\succ + \prec$ Vm[$i + 1$] $\succ$
	end for
		Add elements of a vector pairwise and accumulate the results in another vector.

10.6.4.3 Examples

```
addp    v1.16b,v6.16b,v8.16b  // Add pairwise
saddlp  v3.2d,v4.4s     // Extend and add pairwise
uadalp  v2.4h,v6.8b     // Extend, add pairwise, and accumulate
```

10.6.5 Absolute difference

These instructions subtract the elements of one vector from another and store or accumulate the absolute value of the results:

abd　　　Vector integer absolute difference,
fabd　　Vector floating point absolute difference,
aba　　　Vector integer absolute difference and accumulate,

`faba`	Vector floating point absolute difference and accumulate,
`abal`	Vector absolute difference and accumulate long, and
`abdl`	Vector absolute difference long.

The long versions can be used to prevent overflow.

10.6.5.1 Syntax

```
(s|u)<op>        Vd.T, Vn.T, Vm.T
fabd             Vd.Tf, Vn.Tf, Vm.Tf
fabd             Fd, Fn, Fm
(s|u)<op>l{2}    Vd.Td, Vn.Ts, Vm.Ts
```

- `<op>` is either `aba` or `abd`.
- When a scalar register is specified (F is S or D), a scalar operation is performed instead of a vector operation.
- If 2 is present, then the upper 64 bits of the source registers are used.
- T must be 8b, 16b, 4h, 8h, 2s, or 4s.
- Tf must be 2s, 4s, or 2d.
- Td/Ts must be one of 4s/8h, 8h/16b, or 2d/4s.
- The valid choices for Td/Ts are given in the following table:

Opcode	Valid Types for Td/Ts
(s|u)<op>l	8h/8b, 4s/4h, 2d/2s
(s|u)<op>l2	8h/16b, 4s/8h, 2d/4s

10.6.5.2 Operations

Name	Effect	Description
`{f}abd`	Vd[] ← |Vn[] − Vm[]|	Subtract corresponding elements and take the absolute value.
`(s|u)aba`	Vd[] ← Vd[] + |Vn[] − Vm[]|	Subtract corresponding elements and take the absolute value. Accumulate the results.
`(s|u)abdl`	Vd[] ← |≺ Vn[] ≻ − ≺ Vm[] ≻|	Extend and subtract corresponding elements, then take the absolute value.
`(s|u)abal`	Vd[] ← Vd[]+ |≺ Vn[] ≻ − ≺ Vm[] ≻|	Extend and subtract corresponding elements, then take the absolute value and accumulate the results

10.6.5.3 Examples

```
uabd    v0.8b,v1.8b,v2.8b // unsigned absolute difference
fabd    v3.4s,v6.4s,v8.4s // floating point absolute difference
uaba    v0.8b,v1.8b,v2.8b // unsigned absolute difference and
                          // accumulate
sabdl   v0.2d,v1.2s,v2.2s // signed absolute difference-long result
sabdl2  v0.2d,v1.4s,v2.4s // signed absolute difference-use upper
                          // half of v1 and v2 - long result
```

10.6.6 Absolute value and negate

These operations compute the absolute value or negate each element in a vector:

abs   Vector absolute value,

neg   Vector negate,

fabs   Vector floating point absolute value, and

fneg   Vector floating point negate,

The saturating versions can be used to prevent overflow.

10.6.6.1 Syntax

```
{sq}<op>   Vd.T, Vn.T
f<op>      Vd.Tf, Vn.Tf
```

- <op> is either abs or neg.
- T may be 8b, 16b, 4h, 8h, 2s, 4s, or 2d.
- Tf may be 2s, 4s, or 2d.

10.6.6.2 Operations

Name	Effect	Description

{sq}abs	if sq is present **then**
	 $Vd[] \leftarrow \lfloor |Vm[]| \rceil$
	 else
	 $Vd[] \leftarrow |Vm[]|$
	 end if
		Copy absolute value of each element of Vm to the corresponding element of Vd, optionally saturating the result.

continued on next page

Name	Effect	Description

{sq}neg	**if** sq is present **then**
	 $Vd[] \leftarrow \int -Vm[] \rfloor$
	 else
	 $Vd[] \leftarrow -Vm[]$
	 end if
		Copy absolute value of each element of Vm to the corresponding element of Vd, optionally saturating the result.

fabs	$Vd[] \leftarrow \lvert Vm[] \rvert$	Copy absolute value of each element of Vm to the corresponding element of Vd.
fneg	$Vd[] \leftarrow -Vm[]$	Copy absolute value of each element of Vm to the corresponding element of Vd.

10.6.6.3 Examples

```
    abs    v1.2s,v6.2s    // Get absolute values
    sqneg  v3.16b,v4.16b  // Negate and saturate
```

10.6.7 Get maximum or minimum elements

The following four instructions select the maximum or minimum elements and store the results in the destination vector:

max   Vector integer maximum,

min   Vector integer minimum,

fmax   Vector floating point maximum,

fmin   Vector floating point minimum,

maxp   Vector integer pairwise maximum,

minp   Vector integer pairwise minimum,

fmaxp   Vector floating point pairwise maximum,

fminp   Vector floating point pairwise minimum,

fmaxnm   Vector floating point maxnum,

fminnm   Vector floating point minnum,

fmaxnmp   Vector floating point pairwise maxnum, and

fminnmp   Vector floating point pairwise minnum.

10.6.7.1 Syntax

```
(s|u)<op>{p}    Vd.T, Vn.T, Vm.T
f<op>{p}        Vd.Tf, Vn.Tf, Vm.Tf
```

- <op> is either max or min.
- T must be 8b, 16b, 4h, 8h, 2s, or 4s.
- Tf must be 2s, 4s, or 2d.
- If nm is present, then the result is as described in Chapter 9, Section 9.7.5, on page 311.

10.6.7.2 Operations

Name	Effect	Description

(s|u|f)max	$n \leftarrow$ # of elements
	for $0 \leq i < n$ **do**
	  if $Vn[i] > Vm[i]$ **then**
	    $Vd[i] \leftarrow Vn[i]$
	  else
	    $Vd[i] \leftarrow Vm[i]$
	  end if
	end for
		Compare corresponding elements and copy the greater of each pair into the corresponding element in the destination vector.

(s|u|f)maxp	$n \leftarrow$ # of elements
	for $0 \leq i < (n \div 2)$ **do**
	  if $Vm[i] > Vm[i+1]$ **then**
	    $Vd[i] \leftarrow Vm[i]$
	  else
	    $Vd[i] \leftarrow Vm[i+1]$
	  end if
	end for
	for $(n \div 2) \leq i < n$ **do**
	  if $Vn[i] > Vn[i+1]$ **then**
	    $Vd[i + (n \div 2)] \leftarrow Vn[i]$
	  else
	    $Vd[i + (n \div 2)] \leftarrow$
	    $Vn[i+1]$
	  end if
	end for
		Compare elements pairwise and copy the greater of each pair into an element in the destination vector.

continued on next page

Name	Effect	Description

(s|u|f)min	$n \leftarrow$ # of elements
	for $0 \leq i < n$ **do**
	  if Vn[i] < Vm[i] **then**
	    Vd[i] $\leftarrow$ Vn[i]
	  else
	    Vd[i] $\leftarrow$ Vm[i]
	  end if
	end for
		Compare corresponding elements and copy the lesser of each pair into the corresponding element in the destination vector.

(s|u|f)minp	$n \leftarrow$ # of elements
	for $0 \leq i < (n \div 2)$ **do**
	  if Vm[i] < Vm[$i + 1$] **then**
	    Vd[i] $\leftarrow$ Vm[i]
	  else
	    Vd[i] $\leftarrow$ Vm[$i + 1$]
	  end if
	end for
	for $(n \div 2) \leq i < n$ **do**
	  if Vn[i] < Vn[$i + 1$] **then**
	    Vd[$i + (n \div 2)$] $\leftarrow$ Vn[i]
	  else
	    Vd[$i + (n \div 2)$] $\leftarrow$
	    Vn[$i + 1$]
	  end if
	end for
		Compare elements pairwise and copy the lesser of each pair into an element in the destination vector.

10.6.7.3 Examples

```
  umin      v1.16b,v2.16b,v3.16b  // Get minimum values
  fmaxp     v0.4s,v0.4s,v5.4s     // Get maximum values pairwise
```

10.6.8 Count bits

These instructions can be used to count leading sign bits or zeros, or to count the number of bits that are set, for each element in a vector:

cls   Vector count leading sign bits,
clz   Vector count leading zero bits, and
cnt   Vector count set bits.

10.6.8.1 Syntax

```
cl(s|z)      Vd.T, Vm.T
cnt          Vd.Tn, Vm.Tn
```

- T must be 8b, 16b, 4h, 8h, 2s, or 4s.
- Tn must be 8b or 16b.

10.6.8.2 Operations

Name	Effect	Description

cls	$n \leftarrow$ # of elements
	for $0 \le i < n$) **do**
	   Vd[i] $\leftarrow$
	      leading_sign_bits(Vm[i])
	end for
		Count the number of consecutive bits that are the same as the sign bit for each element in Fm, and store the counts in the corresponding elements of Fd.

clz	$n \leftarrow$ # of elements
	for $0 \le i < n$) **do**
	   Vd[i] $\leftarrow$
	      leading_zero_bits(Vm[i])
	end for
		Count the number of leading zero bits for each element in Fm, and store the counts in the corresponding elements of Fd.

cnt	$n \leftarrow$ # of elements
	for $0 \le i < n$) **do**
	   Vd[i] $\leftarrow$
	      count_one_bits(Vm[i])
	end for
		Count the number of bits in Fm that are set to one, and store the counts in the corresponding elements of Fd.

10.6.8.3 Examples

```
cls    v1.4s,v6.4s  // Count leading sign bits
cnt    v0.8b,v4.8b  // Count bits that are 1
```

10.6.9 Scalar saturating operations

The following instructions perform basic saturating operations on scalars:

qadd   Scalar saturating add,
qsub   Scalar saturating subtract,
qdmulh   Scalar saturating multiply (high half), and
qshl   Scalar saturating shift left.

10.6.9.1 Syntax

```
(s|u)qadd       Fd, Fn, Fm
(s|u)qsub       Fd, Fn, Fm
sq{r}dmulh      <Fx>d, <Fx>n, <Fx>m
(s|u)q{r}shl    Fd, Fn, Fm
```

- F is b, h, s d.
- \<Fx\> is h or s.

10.6.9.2 Operations

Name	Effect	Description
*x*qadd	$Fd \leftarrow \lfloor Fn + Fm \rceil$	Add Fm to Fn, saturate, and store the result in Fd.
*x*qsub	$Fd \leftarrow \lfloor Fn + Fm \rceil$	Subtract Fm from Fn, saturate, and store the result in Fd.
sqrdmulh	$Fd \leftarrow \lfloor (Fn + Fm) \gg n \rceil$	Multiply Fm from Fn, saturate, optionally round, and store the upper half of the result in Fd.
*x*qrshl	$Fd \leftarrow \lfloor Fn \ll Fm \rceil$	Shift Fn left or right by Fm, optionally rounding, saturate, and store the result in Fd.

10.6.9.3 Examples

```
sqadd     b0,b1,b2    // Add two bytes
sqrdmulh  h1,h5,h6    // Multiply two 16 bit integers and
                      // store the upper 16 bits.
```

10.7 Multiplication and division

There is no integer divide instruction in Advanced SIMD. Integer division is accomplished with multiplication by the reciprocal, as was described in Chapter 7 and Chapter 8. For division by a constant, the constant reciprocal can be computed in advance, and simply loaded into a register. For division by a variable, special instructions are provided for computing the reciprocal.

10.7.1 Vector multiply and divide

These instructions are used to multiply the corresponding elements from two vectors:

mul	Vector integer multiply,
mla	Vector integer multiply accumulate,
mls	Vector integer multiply subtract,
fmul	Vector floating point multiply,
fdiv	Vector floating point divide,
fmla	Vector floating point multiply accumulate,
fmls	Vector floating point multiply subtract,
mull	Vector multiply long,
mlal	Vector multiply accumulate long,
mlsl	Vector multiply subtract long,
pmul	Vector polynomial multiply, and
pmull	Vector polynomial multiply long.

10.7.1.1 Syntax

```
    mul          Vd.T, Vn.T, Vm.T
    mla          Vd.T, Vn.T, Vm.T
    mls          Vd.T, Vn.T, Vm.T
   (s|u)mull{2}  Vd.Td, Vn.Ts, Vm.Ts
   (s|u)mlal{2}  Vd.Td, Vn.Ts, Vm.Ts
   (s|u)mlsl{2}  Vd.Td, Vn.Ts, Vm.Ts
    fmul{x}      Vd.Tf, Vn.Tf, Vm.Tf
    fdiv         Vd.Tf, Vn.Tf, Vm.Tf
    fmla         Vd.Tf, Vn.Tf, Vm.Tf
    fmls         Vd.Tf, Vn.Tf, Vm.Tf
    pmul         Vd.8b, Vn.8b, Vm.8b
    pmull{2}     Vd.8b, Vn.16b, Vm.16b
```

- T may be 8b, 16b, 4h, 8h, 2s, or 4s.
- Tf may be 2s, 4s, or 2d.
- T2 may be 8b or 16b.
- If x is present, then $0 \times \pm\infty \to \pm 2$ (vector).
- If 2 is present, then
  - Td/Ts may be 8h/16b, 4s/8h, or 2d/4s.
  - the upper 64 bits of the source vectors are used.
- If 2 is *not* present, then Td/Ts may be 8h/8b, 4s/4h, or 2d/2s.

10.7.1.2 Operations

Name	Effect	Description
{f}mul{x}	Vd[] ← Vn[] × Vm[]	Multiply corresponding elements from two vectors and store the results in a third vector.
{f}mla	Vd[] ← Vd[] + (Vn[] × Vm[])	Multiply corresponding elements from two vectors and add the results to a third vector.
{f}mls	Vd[] ← Vd[] − (Vn[] × Vm[])	Multiply corresponding elements from two vectors and subtract the results from a third vector.
fdiv	Vd[] ← Vd[] ÷ (Vn[] × Vm[])	Divide elements from Vn by corresponding elements from Vm and store the results in a third vector.
(s|u)mull	Vd[] ← Vn[] × Vm[]	Multiply corresponding elements from two vectors and store the results in a third vector.
(s|u)mlal	Vd[] ← Vd[] + (Vn[] × Vm[])	Multiply corresponding elements from two vectors and add the results in a third vector.
(s|u)mlsl	Vd[] ← Vd[] − (Vn[] × Vm[])	Multiply corresponding elements from two vectors and subtract the results from a third vector.
pmul{l}	Vd[] ← Vd[] − (Vn[] × Vm[])	Multiply corresponding elements from two vectors and store the results in a third vector. This instruction is particularly useful for binary field multiplication, which is used in cryptography.

10.7.1.3 Examples

```
mul     v1.8h,v6.8h,v8.8h   // Multiply elements
smlal   v0.4s,v4.4h,v5.4h   // Multiply-accumulate long
fmul    v1.4s,v6.4s,v7.4s   // Multiply elements
```

10.7.2 Multiply vector by element

These instructions are used to multiply each element in a vector by a scalar:

mul	Vector by scalar integer multiply,
mla	Vector by scalar integer multiply accumulate,
mls	Vector by scalar integer multiply subtract,
fmul	Vector by scalar floating point multiply,
fmla	Vector by scalar floating point multiply accumulate,
fmls	Vector by scalar floating point multiply subtract,
mull	Vector by scalar multiply long,
mlal	Vector by scalar multiply accumulate long, and
mlsl	Vector by scalar multiply subtract long.

10.7.2.1 Syntax

```
<op>            Vd.T, Vn.T, Vm.Ts[x]
(s|u)<op>l{2}   Vd.Ta, Vn.Tb, Vm.Ts[x]
fmul{x}         Vd.Tf, Vn.Tf, Vm.Ts2[x]
fmla            Vd.Tf, Vn.Tf, Vm.Ts2[x]
fmls            Vd.Tf, Vn.Tf, Vm.Ts2[x]
```

- <op> is either mul, mla, or mls.
- T/Ts must be 4h/h, 8h/h, 2s/s, or 4s/s.
- Tf/Ts2 must be 2s/s, 4s/s, or 2d/d.
- If x is present, then $0 \times \pm\infty \to \pm 2$ (vector).
- If 2 is present, then
  - Ta/Tb/Tc is 4s/8h/h or 2d/4s/s.
  - the upper 64 bits of the source vectors are used.
- If 2 is *not* present, then Ta/Tb/Tc is 4s/4h/h or 2d/2s/s.

10.7.2.2 Operations

Name	Effect	Description
{f}mul{x}	$Vd[] \leftarrow Vn[] \times Vm[x]$	Multiply all elements from a vector by one element of another vector and store the results in a third vector.

continued on next page

Name	Effect	Description
{f}mla	Vd[] ← Vd[] + (Vn[] × Vm[x])	Multiply all elements from a vector by one element of another vector and add the results to a third vector.
{f}mls	Vd[] ← Vd[] − (Vn[] × Vm[x])	Multiply all elements from a vector by one element of another vector and subtract the results from a third vector.
(s|u)mull	Vd[] ← Vn[] × Vm[x]	Multiply all elements from a vector by one element of another vector and store the results in a third vector.
(s|u)mlal	Vd[] ← Vd[] + (Vn[] × Vm[x])	Multiply all elements from a vector by one element of another vector and add the results to a third vector.
(s|u)mlsl	Vd[] ← Vd[] − (Vn[] × Vm[x])	Multiply all elements from a vector by one element of another vector and subtract the results from a third vector.

10.7.2.3 Examples

```
    mul     v1.4h,v6.4h,v7.h[1]   // Multiply elements
    umlal   v0.4s,v4.4h,v5.h[0]   // Multiply-accumulate long
    fmul    v1.4s,v6.4s,v7.s[1]   // Multiply elements
```

10.7.3 Saturating vector multiply and double

These instructions perform multiplication, double the results, and perform saturation:

sqdmull   Saturating Multiply Double,
sqdmlal   Saturating Multiply Double Accumulate, and
sqdmlsl   Saturating Multiply Double Subtract.

10.7.3.1 Syntax

```
    sqd<op>l{2}  Vd.Td, Vn.Ts, Vm.Ts
    sqd<op>l{2}  Vd.Td, Vn.Ta, Vm.Tb[x]
```

- <op> is either mul, mla, or mls.
- If 2 is present, then Td/Ts is 4s/8h or 2d/4s and the upper half of Vn and Vm are used. Otherwise, Td/Ts is 4s/4h or 2d/2s and the lower half of Vn and Vm are used.

- If 2 is present, then Td/Ta/Tb is 4s/8h/h or 2d/4s/s, and the upper half of Vn is used. Otherwise Td/Ta/Tb is 4s/4h/h or 2d/4s/s, and the lower half of Vn is used.
- If the third operand is a scalar ([x] is specified) and Tb is h, then Vm must be in the range v0-v15.

10.7.3.2 Operations

Name	Effect	Description

sqdmull	**if** second operand is scalar **then**
	 $Vd[] \leftarrow \lfloor Vn[] \times Vm[x] \times 2 \rceil$
	 else
	 $Vd[] \leftarrow \lfloor Vn[] \times Vm[] \times 2 \rceil$
	 end if
		Multiply elements, double the results, and store in the destination vector with saturation.

sqdmlal	**if** second operand is scalar **then**
	 $Vd[] \leftarrow$
	 $\lfloor Vd[] + Vn[] \times Vm[x] \times 2 \rceil$
	 else
	 $Vd[] \leftarrow$
	 $\lfloor Vd[] + Vn[] \times Vm[] \times 2 \rceil$
	 end if
		Multiply elements, double the results, and add to the destination vector with saturation.

sqdmlsl	**if** second operand is scalar **then**
	 $Vd[] \leftarrow$
	 $\lfloor Vd[] - Vn[] \times Vm[x] \times 2 \rceil$
	 else
	 $Vd[] \leftarrow$
	 $\lfloor Vd[] - Vn[] \times Vm[] \times 2 \rceil$
	 end if
		Multiply elements, double the results, and subtract from the destination vector with saturation.

10.7.3.3 Examples

```
sqdmull  v1.4s,v6.4h,v8.4h    // Multiply elements, double,
                              // saturate
sqdmlal2 v0.4s,v4.8h,v5.h[0]  // Multiply elements, double,
                              // saturate, accumulate
```

10.7.4 Saturating multiply and double (high)

These instructions perform multiplication, double the results, perform saturation, and store the high half of the results:

sqdmulh   Saturating Multiply Double (High), and

sqrdmulh

Saturating Multiply Double and Round (High).

10.7.4.1 Syntax

```
sq{r}dmulh  Vd.T, Vn.T, Vm.T
sq{r}dmulh  Vd.Td, Vn.Td, Vm.Ts[x]
```

- T must be 4h, 8h, 2s, or 4s.
- Td/Ts must be 4h/h, 8h/h, 2s/s, or 4s/s.
- If Ts is h, then Vm must be in the range v0-v15.

10.7.4.2 Operations

Name	Effect	Description

sqdmulh	$n \leftarrow$ # of elements in Ts
	if [x] is present **then**
	  Vd[] $\leftarrow$
	    Vn[] $\times$ Vm[x] $\times 2 \gg n$
	else
	  Vd[] $\leftarrow$
	    Vn[] $\times$ Vm[] $\times 2 \gg n$
	end if
		Multiply elements, double the results and store the high half in the destination vector with saturation.

sqrdmulh	$n \leftarrow$ # of elements in Ts
	if [x] is present **then**
	  Vd[] $\leftarrow$
	    $\|$Vn[] $\times$ Vm[x] $\times 2\| \gg n$
	else
	  Vd[] $\leftarrow$
	    $\|$Vn[] $\times$ Vm[] $\times 2\| \gg n$
	end if
		Multiply elements, double the results, round, and store the high half in the destination vector with saturation.

10.7.4.3 Examples

```
sqrdmulh v1.4h,v6.4h,v8.h[3]   // Multiply elements, double, round
                               // saturate, store high half
sqdmulh  v0.4s,v4.4s,v5.s[2]   // Multiply elements, double,
                               // accumulate high half, saturate
```

10.7.5 Estimate reciprocals

In general, multiplication is faster than division. In many cases of vector arithmetic, it is faster to calculate reciprocals and use multiplication. These instructions perform the initial estimates of the reciprocal values:

recpe   Reciprocal Estimate, and
rsqrte   Reciprocal Square Root Estimate.

These work on floating point and unsigned fixed point vectors. The estimates from this instruction are accurate to within about eight bits. If higher accuracy is desired, then the Newton-Raphson method can be used to improve the initial estimates. For more information, see the Reciprocal Step instructions on page 369.

10.7.5.1 Syntax

```
u<op>    Vd.Ta, Vn.Ta
f<op>    Vd.Tb, Vn.Tb
```

- <op> is either recpe or rsqrte.
- Ta must be 2s or 4s.
- Tb must be 2s, 4s, or 2d.

10.7.5.2 Operations

Name	Effect	Description

(u|f)recpe	$n \leftarrow$ # of elements
	for $0 \leq i < n$) **do**
	$\quad$ Vd[i] $\leftarrow \approx (1 \div \text{Vm}[i])$
	end for
		Find an approximate reciprocal of each element in a vector.

(u|f)rsqrte	$n \leftarrow$ # of elements
	for $0 \leq i < n$) **do**
	$\quad$ Vd[i] $\leftarrow \approx (1 \div \sqrt{\text{Vm}[i]})$
	end for
		Find an approximate reciprocal square root of each element in a vector.

10.7.5.3 Examples

```
urecpe   v1.4s,v6.4s   // Get initial reciprocal estimates
frecpe   v4.2d,v5.2d   // Get initial reciprocal estimates
```

10.7.6 Reciprocal step

These instructions are used to perform one Newton-Raphson step for improving the reciprocal estimates:

frecps   Reciprocal Step, and
frsqrts   Reciprocal Square Root Step.

For each element in the vector, the following equation can be used to improve the estimates of the reciprocals:

$$x_{n+1} = x_n(2 - dx_n).$$

Where x_n is the estimated reciprocal from the previous step, and d is the number for which the reciprocal is desired. This equation converges to $\frac{1}{d}$ if x_0 is obtained using **vrecpe** on d. The **vrecps** instruction computes

$$x'_{n+1} = 2 - dx_n,$$

so one additional multiplication is required to complete the update step. The initial estimate x_0 must be obtained using the **vrecpe** instruction.

For each element in the vector, the following equation can be used to improve the estimates of the reciprocals of the square roots:

$$x_{n+1} = x_n\frac{3 - dx_n^2}{2}.$$

Where x_n is the estimated reciprocal from the previous step, and d is the number for which the reciprocal is desired. This equation converges to $\frac{1}{\sqrt{d}}$ if x_0 is obtained using **vrsqrte** on d. The **vrsqrts** instruction computes

$$x'_{n+1} = \frac{3 - dx_n}{2},$$

so two additional multiplications are required to complete the update step. The initial estimate x_0 must be obtained using the **vrsqrte** instruction.

10.7.6.1 Syntax

```
f<op>       Vd.T, Vn.T, Vm.T
f<op>       Fd, Fn, Fm
```

* <op> is either **recps** or **rsqrts**.
* T must be 2s, 4s, or 2d.
* F is s or d.

10.7.6.2 Operations

Name	Effect	Description

vrecpe	$n \leftarrow$ # of elements
	for $0 \leq i < n$) **do**
	$\quad$ Vd[i] $\leftarrow 2 - $ Vn[i] $\times$ Vm[i]
	end for
		Perform most of the Newton Raphson reciprocal improvement step.

vrsqrte	$n \leftarrow$ # of elements
	for $0 \leq i < n$) **do**
	$\quad$ Vd[i] $\leftarrow$
	$\quad\quad (3 - $ Vn[i] $\times$ Vm[i]) $\div 2$
	end for
		Perform most of the Newton Raphson reciprocal square root improvement step.

10.7.6.3 Examples

```
1    // Divide elements of v0 by elements of v1 and store in v3
2    // Doing a loop and testing for convergence would be slow,
3    // so we will just do two improvement steps and hope it is
4    // close enough.
5    frecpe   v3.2d,v1.2d       // Get initial reciprocal estimates
6    frecps   v4.2d,v1.2d,v3.2d // Improve estimates
7    fmul     v3.2d,v3.2d,v4.2d // Finish improvement step
8    frecps   v4.2d,v1.2d,v3.2d // Improve estimates
9    fmul     v3.2d,v3.2d,v4.2d // Finish improvement step
10   fmul     v3.2d,v3.2d,v0.2d // Perform division
```

10.7.7 Multiply scalar by element

These instructions are used to multiply each element in a vector by a scalar:

fmul   Vector by scalar floating point multiply,
fmla   Vector by scalar floating point multiply accumulate,
fmls   Vector by scalar floating point multiply subtract.

10.7.7.1 Syntax

```
    fmul{x}        Fd, Fn, Vm.Ts[x]
    fmla           Fd, Fn, Vm.Ts[x]
    fmls           Fd, Fn, Vm.Ts[x]
```

• Ts must be s or d.

- If Ts is s, then Vm must be in the range s0 to s15.
- If x is present, then $0 \times \pm\infty \rightarrow \pm 2$ (vector).

10.7.7.2 Operations

Name	Effect	Description
fmul{x}	Fd ← Fn × Vm[x]	Multiply scalar register Fn by one element of register Vm and store the results in scalar register Fd.
fmla	Fd ← Fd + (Fn × Vm[x])	Multiply scalar register Fn by one element of register Vm vector and add the results to scalar register Fd.
fmls	Fd ← Fd − (Fn × Vm[x])	Multiply scalar register Fn by one element of register Vm vector and subtract the results from scalar register Fd.

10.7.7.3 Examples

```
fmul    d0,d1,v7.d[1]     // Multiply d1 by v7[i]
fmla    s0,s3,v5.s[3]     // Multiply-accumulate
```

10.7.8 Saturating multiply scalar by element and double

These instructions perform multiplication, double the results, and perform saturation:

sqdmull   Saturating Multiply Scalar by Element and Double,
sqdmlal   Saturating Multiply Scalar by Element, Double, and Accumulate, and
sqdmlsl   Saturating Multiply Scalar by Element, Double, and Subtract.

10.7.8.1 Syntax

```
sqd<op>l     Fd, Fn, Vm.Ts[x]
sq{r}dmulh   Fa, Fb, Vc.Tc[x]
```

- <op> is either mull, mlal, or mlsl.
- Fd/Fn/Ts must be s/h/h or d/s/s.
- If Ts is h, then Vm must be in the range v0-v15.

- F/Tc must be h/h or s/s.
- If Tc is h, then Vc must be in the range v0-v15.

10.7.8.2 Operations

Name	Effect	Description
sqdmull	$Fd \leftarrow \lfloor Fn \times Vm[x] \times 2 \rceil$	Multiply, double the results, saturate, and store in the destination register.
sqdmlal	$Fd \leftarrow \lfloor Fd + Fn \times Vm[x] \times 2 \rceil$	Multiply elements, double the results, add to the destination register, and saturate.
sqdmlsl	$Vd[] \leftarrow \lfloor Fd - Fn \times Vm[x] \times 2 \rceil$	Multiply elements, double the results, subtract from the destination register, and saturate..
sqdmulh	$Fa \leftarrow$ $Fb \times Vc[x] \times 2 \gg n$	Multiply, double the results, saturate, and store the high half in the destination register.
sqrdmulh	$Fa \leftarrow$ $\|Fb \times Vc[x] \times 2\| \gg n$	Multiply elements, double the results, round, saturate, and store the high half in the destination register.

10.7.8.3 Examples

```
sqdmull   d1,s6,v8.s[1]    // Multiply, double, saturate
sqdmlal   s0,h4,v5.h[0]    // Multiply double, saturate, accumulate
sqdmlal   s0,h4,h5         // Alias for previous instruction
```

10.8 Shift instructions

The Advanced SIMD shift instructions operate on vectors. Shifts are often used for multiplication and division by powers of two. The results of a left shift may be larger than the destination register, resulting in overflow. A shift right is equivalent to division. In some cases, it may be useful to round the result of a division, rather than truncating. Advanced SIMD provides versions of the shift instruction which perform saturation and/or rounding of the result.

10.8.1 Vector shift left by immediate

These instructions shift each element in a vector left by an immediate value:

shl   Unsigned Shift Left Immediate,
qshl   Saturating Signed or Unsigned Shift Left Immediate,
sqshlu   Saturating Signed Shift Left Immediate Unsigned, and
shll   Signed or Unsigned Shift Left Immediate Long.

Overflow conditions can be avoided by using the saturating version, or by using the long version, in which case the destination is twice the size of the source.

10.8.1.1 Syntax

```
shl          Vd.T, Vn.T, #shift
(s|u)qshl    Vd.T, Vn.T, #shift
sqshlu       Vd.T, Vn.T, #shift
(s|u)shll{2} Vd.Td, Vn.Ts, #shift
```

- UXTL Vd.Td, Vn.Ts is an alias for USHLL Vd.Td,Vn.Ts,#0.
- UXTL2 Vd.Td, Vn.Ts is an alias for USHLL2 Vd.Td,Vn.Ts,#0.
- SXTL Vd.Td, Vn.Ts is an alias for SSHLL Vd.Td,Vn.Ts,#0.
- SXTL2 Vd.Td, Vn.Ts is an alias for SSHLL2 Vd.Td,Vn.Ts,#0.
- T is 8b, 16b, 4h, 8h, 2s, 4s, or 2d.
- If 2 is present, then Td/Ts is 8h/16b, 4s/8h, or 2d/4s.
- If 2 is *not* present, then Td/Ts is 8h/8b, 4s/4h, or 2d/2s.
- shift is in the range 0 to $size(\text{T}) - 1$.
- If the instruction begins with u, then the elements are treated as unsigned integers.
- If s is present, then the elements are treated as signed integers.

10.8.1.2 Operations

Name	Effect	Description
shl	$Vd[] \leftarrow Vm[] \ll imm$	Each element of Vm is shifted left by the immediate value and stored in the corresponding element of Vd. Bits shifted past the end of an element are lost.

continued on next page

Name	Effect	Description
(s|u)shll	Vd[] ←⊰Vm[]⊱≪ *imm*	Each element of Vm is sign or zero extended (depending on \<type\>) then shifted left by the immediate value and stored in the corresponding element of Vd.
(s|u)qshl{u}	Vd[] ← ⌠Vm[] ≪ *imm* ⌡	Each element of Vm is shifted left by the immediate value and stored in the corresponding element of Vd. If the result of the shift is outside the range of the destination element, then the value is saturated. If **u** was specified, then the destination is treated as unsigned for saturation. Otherwise, it is treated as signed for saturation.

10.8.1.3 Examples

```
shl     v1.4h,v6.4h,#4      // shift each 16-bit word left
uqshl    v1.16b,v6.16b,#1  // Multiply each byte by two
```

10.8.2 Vector shift right by immediate

These instructions shift each element in a vector right by an immediate value:

shr   Shift Right Immediate,

rshr   Shift Right Immediate and Round,

shrn   Shift Right Immediate and Narrow,

rshrn Shift Right Immediate Round and Narrow,

sra   Shift Right and Accumulate Immediate, and

rsra   Shift Right Round and Accumulate Immediate.

10.8.2.1 Syntax

```
(s|u){r}shr       Vd.T, Vn.T, #shift
(s|u){r}sra       Vd.T, Vn.T, #shift
{r}shrn{2}        Vd.Td,Vn.Ts, #shift
```

- T is 8b, 16b, 4h, 8h, 2s, 4s, or 2d.
- If 2 is present, then Td/Ts is 16b/8h, 8h/4s, or 4s/2d.
- If 2 is *not* present, then Td/Ts is 8h/8b, 4s/4h, or 2d/2s.
- shift is in the range 0 to $size(T) - 1$ (or $size(Td) - 1$).

10.8.2.2 Operations

Name	Effect	Description

v{r}shr	**if** r is present **then**
	 $\quad Vd[] \leftarrow \|Vm[] \gg imm\|$
	 else
	 $\quad Vd[] \leftarrow Vm[] \gg imm$
	 end if
		Each element of Vm is shifted right by the immediate value, with zero extension, and stored in the corresponding element of Vd. Results can be rounded.

v{r}shrn	**if** r is present **then**
	 $\quad Vd[] \leftarrow$
	 $\qquad \succ \|Vm[] \gg imm\| \prec$
	 else
	 $\quad Vd[] \leftarrow \succ Vm[] \gg imm \prec$
	 end if
		Each element of Vm is shifted right by the immediate value, with zero extension, optionally rounded, then narrowed and stored in the corresponding element of Vd.

v{r}sra	**if** r is present **then**
	 $\quad Vd[] \leftarrow Vd[] + \|Vm[] \gg imm\|$
	 else
	 $\quad Vd[] \leftarrow Vd[] + Vm[] \gg imm$
	 end if
		Each element of Vm is shifted right by the immediate value, with sign or zero extension, and accumulated in the corresponding element of Vd. Results can be rounded.

10.8.2.3 Examples

```
ssra      v1.8h,v6.8h,#4    // shift each 32-bit integer
ursra     v2.4s,v8.4s,#2    // Divide by 4 with rounding
```

10.8.3 Vector saturating shift right by immediate

These instructions shift each element in a quad word vector right by an immediate value:

qshrn   Saturating Shift Right Narrow,
qrshrn   Saturating Rounding Shift Right Narrow,
sqshrun   Signed Saturating Shift Right Unsigned Narrow, and
sqrshrun   Signed Saturating Rounding Shift Right Unsigned Immediate.

10.8.3.1 Syntax

```
(s|u)q{r}shrn   Vd.Td, Vn.Ts, #shift
sq{r}shrun      Vd.Td2, Vn.Ts2, #shift
```

- If 2 is present, the Td/Ts is 16b/8h, 8h/4s, or 4s/2d
- If 2 is *not* present, the Td/Ts is 8b/8h, 4h/4s, or 2s/2d to elsize(Td).
- shift is in the range 1 to $size$(Td).

10.8.3.2 Operations

Name	Effect	Description

xq{r}shrn	if r is present **then**
	 Vd[] ←
	 ≻∫‖Vm[] ≫ imm‖⌊≺
	 else
	 Vd[] ←
	 ≻∫Vm[] ≫ imm⌊≺
	 end if
		Each element of Vm is shifted right with sign extension by the immediate value, optionally rounded, then saturated and narrowed, and stored in the corresponding element of Vd.

sq{r}shrun	if r is present **then**
	 Vd[] ←
	 ≻∫‖Vm[] ≫ imm‖⌊≺
	 else
	 Vd[] ←
	 ≻∫Vm[] ≫ imm⌊≺
	 end if
		Each element of Vm is shifted right with zero extension by the immediate value, optionally rounded, then saturated and narrowed, and stored in the corresponding element of Vd.

10.8.3.3 Examples

```
uqshrn    v1.4h,v6.4s,#4   // shift, saturate and narrow
sqrshrn   v1.8b,v6.8h,#4   // shift, round, saturate, and narrow
```

10.8.4 Shift left or right by variable

These instructions shift each element in a vector left or right, using the least significant byte of the corresponding element of a second vector as the shift amount:

shl	Shift Left or Right by Variable,
rshl	Shift Left or Right by Variable and Round,
qshl	Saturating Shift Left or Right by Variable, and
qrshl	Saturating Shift Left or Right by Variable and Round.

10.8.4.1 Syntax

```
(s|u){q}{r}shl    Vd.T, Vn.T, Vm.T
(s|u){r}shl       Dd, Dn, Dm
```

- T is 8b, 16b, 4h, 8h, 2s, 4s, or 2d.
- If double precision register s are specified (Dd, Dn, Dm), then the operation is a scalar operation instead of a vector operation.
- If the shift value is positive, the operation is a left shift.
- If shift value is negative, then it is a right shift.
- A shift value of zero is equivalent to a move.
- If the operation is a right shift, and r is specified, then the result is rounded rather than truncated.
- Results are saturated if q is specified.
- If q is present, then the results are saturated.
- If r is present, then right shifted values are rounded rather than truncated.

10.8.4.2 Operations

Name	Effect	Description

vshl	**if** q is present **then**
	 if r is present **then**
	  $Vd[] \leftarrow \lfloor \|Vn[] \ll Vm[]\| \rceil$
	 else
	  $Vd[] \leftarrow \lfloor Vn[] \ll Vm[] \rceil$
	 end if
	else
	 if r is present **then**
	  $Vd[] \leftarrow \|Vn[] \ll Vm[]\|$
	 else
	  $Vd[] \leftarrow Vn[] \ll Vm[]$
	 end if
	end if
		Each element of Vm is shifted left by the immediate value and stored in the corresponding element of Vd. Bits shifted past the end of an element are lost.

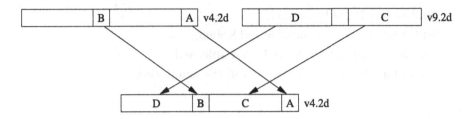

Figure 10.7: Effects of sli v4.2d,v9.2d,#6.

10.8.4.3 Examples

```
   sshl    v0.8b,v1.8b,v3.8b  // use elements in q3 to shift
                              // elements of q1
   urshl   v0.8h,v1.8h,v3.8h  // use elements in q3 to shift
                              // elements of q1 with rounding
```

10.8.5 Shift and insert

These instructions perform bitwise shifting of each element in a vector, then combine bits from the source with bits from the destination. Fig. 10.7 provides an example. The instructions are:

sli   Shift Left and Insert, and
sri   Shift Right and Insert.

10.8.5.1 Syntax

```
   s<dir>i    Vd.T, Vm.T, #shift // vector
   s<dir>i    Dd, Dm, #shift     // scalar
```

- T is 8b, 16b, 4h, 8h, 2s, 4s, or 2d.
- <dir> must be l for a left shift, or r for a right shift.
- shift is the amount that elements are to be shifted, and must be between zero and $size(T) - 1$ for sli, or between one and $size(T)$ for sri.

10.8.5.2 Operations

Name	Effect	Description
vsli	$mask \leftarrow (1 \ll imm + 1) - 1$ $Vd[] \leftarrow (mask \wedge Vd[])\vee$ $(Vm[] \ll imm)$	Each element of Vm is shifted left and combined with lower shift bits of the corresponding element of Vd.
vsri	$mask \leftarrow$ $\neg(1 \ll size - imm + 1) - 1$ $Vd[] \leftarrow (mask \wedge Vd[])\vee$ $(Vm[] \gg imm)$	Each element of Vm is shifted right and combined with upper shift bits of the corresponding element of ttVd.

10.8.5.3 Examples

```
sli    v1.8h,v6.8h,#4   // shift each element left and combine
sri    v0.2s,v4.2s,#4   // vector shift right and combine
```

10.8.6 Scalar shift left by immediate

These instructions shift each element in a vector left by an immediate value:

shl   Unsigned Shift Left Immediate,

qshl   Saturating Signed or Unsigned Shift Left Immediate,

sqshlu   Saturating Signed Shift Left Immediate Unsigned, and

shll   Signed or Unsigned Shift Left Immediate Long.

Overflow conditions can be avoided by using the saturating version, or by using the long version, in which case the destination is twice the size of the source.

10.8.6.1 Syntax

```
shl        Dd, Dn, #shift
uqshl      Fd, Fn, #shift
sqshl{u}   Fd, Fn, #shift
```

- F may be b, h, s, or d.
- If the instruction begins with u, then the scalars are treated as unsigned integers.
- If s is present, then the scalars are treated as signed integers.

10.8.6.2 Operations

Name	Effect	Description
shl	$Fd \leftarrow Fm \ll imm$	Shift Fm left by the immediate value and store result in Fd.
(uqshl	$Fd \leftarrow \lfloor Fm \ll imm \rceil$	Shift Fm left by the immediate value and store result in Fd. The results are saturated as unsigned integers.
(sqshl{u}	$Fd \leftarrow \lfloor Fm \ll imm \rceil$	Shift Fm left by the immediate value and store result in Fd If u was specified, then the destination is treated as unsigned for saturation. Otherwise, it is treated as signed for saturation.

10.8.6.3 Examples

```
    shl     d0,d1,#32
    uqshl   b0,b1,#3
```

10.8.7 Scalar shift right by immediate

These instructions shift each element in a vector right by an immediate value:

shr   Shift Right Immediate,

rshr   Shift Right Immediate and Round,

sra   Shift Right and Accumulate Immediate, and

rsra   Shift Right Round and Accumulate Immediate.

10.8.7.1 Syntax

```
    (s|u){r}shr     Dd, Dm, #shift
    (s|u){r}sra     Dd, Dm, #shift
```

- shift is in the range 0 to $size(T) - 1$ (or $size(Td) - 1$).

10.8.7.2 Operations

Name	Effect	Description

(s|u)rshr	**if** r is present **then**
	 $Dd \leftarrow \|Dm \gg imm\|$
	else
	 $Dd \leftarrow Dm \gg imm$
	end if
		Shift Dm right by the immediate value, optionally round, and store in Dd.

(s|u){r}sra	**if** r is present **then**
	 $Dd \leftarrow Dd + \|Dm \gg imm\|$
	else
	 $Dd \leftarrow Dd + (Dm \gg imm)$
	end if
		Shift Dm right by the immediate value, optionally round, and add result to Dd.

10.8.7.3 Examples

```
ssra    d0,d1,#4  // shift right and accumulate
urshr   d3,d2,#1  // Divide by 2 with rounding
```

10.8.8 Scalar saturating shift right by immediate

These instructions shift each element in a quad word vector right by an immediate value:

qshrn Saturating Shift Right Narrow,

qrshrn Saturating Rounding Shift Right Narrow,

sqshrun Signed Saturating Shift Right Unsigned Narrow, and

sqrshrun Signed Saturating Rounding Shift Right Unsigned Immediate.

10.8.8.1 Syntax

```
(s|u)q{r}shrn  <Fd>d, <Fn>n, #shift
sq{r}shrun     <Fd>d, <Fn>n, #shift
```

• <Fd>/<Fn> must be b/h, h/s, or s/d.

10.8.8.2 Operations

Name	Effect	Description
xq{r}shrn	**if** r is present **then** $$Fd \leftarrow \succ\!\!\int \|\|Fn \gg imm\|\| \rceil \prec$$ **else** $$Fd \leftarrow \succ\!\!\int Fn \gg imm \rceil \prec$$ **end if**	Shift Fn right, optionally round, then saturate and narrow. Store result in **Fd**.
sq{r}shrun	**if** r is present **then** $$Fd \leftarrow \succ\!\!\int \|\|Fn \gg imm\|\| \rceil \prec$$ **else** $$Fd \leftarrow \succ\!\!\int Fn \gg imm \rceil \prec$$ **end if**	Shift Fn right, optionally round, then saturate and narrow as un-signed. Store result in **Fd**.

10.8.8.3 Examples

```
    uqshrn      s0,d1,#4  // shift, saturate, and narrow
    sqrshrn     h0,s2,#4  // shift, round, saturate, and narrow
```

10.9 Unary arithmetic

Advanced SIMD provides several unary operations for integer and floating point values. It provides instructions for bitwise complement, negation, reversing bits an elements, and other operations.

10.9.1 Vector unary arithmetic

not   Vector 1's Complement,
qadd   Vector Saturating Accumulate,
fsqrt   Vector Floating Point Square Root,
rbit   Vector bit reverse,
rev   Reverse Elements.

10.9.1.1 Syntax

```
    not         Vd.T, Vn.T
    rbit        Vd.T, Vn.T
```

```
rev16      Vd.T, Vn.T
(us|su)qadd  Vd.Ta, Vn.Ta
fsqrt      Vd.Tb, Vn.Tb
rev32      Vd.Tc, Vn.Tc
rev64      Vd.Td, Vn.Td
```

- T is 8b or 16b.
- Ta is 8b, 16b, 4h, 8h, 2s, 4s, or 2d.
- Tb is 2s, 4s, or 2d.
- Tc is 8b, 16b, 4h, or 8h.
- Td is 8b, 16b, 4h, 8h, 2s, or 4s.

10.9.1.2 Operations

Name	Effect	Description
not	$Vd \leftarrow \neg Vn$	Store 1's complement of Vn in Vd.
suqadd	$Vd[i] \leftarrow \lfloor Vd[i] + Vm[i] \rceil$	Signed integer saturating accumulate of unsigned value.
usqadd	$Vd[i] \leftarrow \lfloor Vd[i] + Vm[i] \rceil$	Unsigned integer saturating accumulate of signed value.
rbit	$Vd[i] \leftarrow \lfloor bitrev(Vm[i]) \rceil$	Reverse bits in each element of Vm.
rev	$Vd[i] \leftarrow \lfloor rev(Vm[i]) \rceil$	Reverse elements of Vm.

10.9.1.3 Examples

Fig. 10.8 provides some illustrated examples of the `rev` instructions.

```
1  not      v3.8b,v4.8b
2  usqadd   v7.4h,v12.4h
3  rev32    v3.8h,v4.8h
4  rev64    v8.4s,v9.4s
5  rev64    v5.8h,v7.8h
```

10.9.2 Scalar unary arithmetic

abs   Integer absolute value,
neg   Integer absolute value,
qadd Vector Saturating Accumulate,
fsqrt Vector Floating Point Square Root,

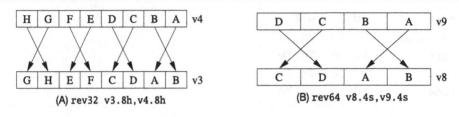

(A) rev32 v3.8h,v4.8h   (B) rev64 v8.4s,v9.4s

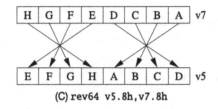

(C) rev64 v5.8h,v7.8h

Figure 10.8: Examples of the rev instruction.

rbit   Vector bit reverse, and

rev   Reverse Elements.

10.9.2.1 Syntax

```
    abs          Dd, Dn
    neg          Dd, Dn
    sqabs        Fd, Fn
    sqneg        Fd, Fn
    (su|us)qadd  Fd, Fn
```

• F is b, h, s, or d.

10.9.2.2 Operations

Name	Effect	Description
abs	Dd ← |Dn|	Store absolute value of Dn in Dd.
neg	Dd ← −Dn	Store 2's complement of Dn in Dd.
sqabs	Fd ← ⌈|Fn|⌋	Store saturated absolute value of Fn in Fd.
sqneg	Fd ← ⌈−Fn⌋	Store saturated 2's complement of Fn in Fd.

continued on next page

Name	Effect	Description
suqadd	$Fd \leftarrow \lceil Fd + Fm \rfloor$	Signed integer saturating accumulate of unsigned value.
usqadd	$Fd \leftarrow \lceil Fd + \lceil Fm \rfloor$	Unsigned integer saturating accumulate of signed value.

10.9.2.3 Examples

```
abs.s16     d0,d2
sqneg       s1,s4
```

10.10 Vector reduce instructions

These instructions operate across all lanes in a vector, and produce a scalar.

10.10.1 Reduce across lanes

Advanced SIMD provides instructions for summing the elements in a vector, and for getting the maximum or minimum value from a vector. These instructions are:

addv   Integer Sum Elements to Scalar,
maxv   Integer Maximum Element to Scalar,
minv   Integer Minimum Element to Scalar,
fmaxv   Floating Point Maximum Element to Scalar, and
fminv   Floating Point Minimum Element to Scalar.

There are long versions of the addv instruction which will prevent overflow.

10.10.1.1 Syntax

```
addv          Fd, Vn.T
(s|u)addlv    Fa, Vn.Tn
(s|u)maxv     Fd, Vn.T
(s|u)minv     Fd, Vn.T
fmax{nm}v     Sd, Vn.4s
fmin{nm}v     Sd, Vn.4s
```

- F/T may be b/8b, b/16b, h/4h, h/8h, s/2s, or s/4s.
- Fa/Tn may be h/8b, h/16b, s/4h, s/8h, d/2s, or d/4s.

- If nm is present, then comparison between a nan and a numerical value will return the numerical value.

10.10.1.2 Operations

Name	Effect	Description

$x$addlv	$Fd \leftarrow 0$
	$n \leftarrow$ # of elements
	for $0 \leq i < n$ **do**
	    $Fd \leftarrow Fd + Vn[i]$
	end for
		Sum the elements of Vn and store the result in Fd.

$x$maxv	$Fd \leftarrow Vn[0]$
	$n \leftarrow$ # of elements
	for $1 \leq i < n$ **do**
	    if $Vn[i] > Fd$ **then**
	        $Fd \leftarrow Vn[i]$
	    end if
	end for
		Store maximum value from Vn in Fd.

$x$minv	$Fd \leftarrow Vn[0]$
	$n \leftarrow$ # of elements
	for $0 \leq i < n$ **do**
	    if $Vn[i] < Fd$ **then**
	        $Fd \leftarrow Vn[i]$
	    end if
	end for
		Store minimum value from Vn in Fd.

10.10.1.3 Examples

```
addv   s3,v2.4s
umaxv  b0,v4.16b
```

10.10.2 Reduce pairwise

The pairwise instructions are similar to the vector reduce instructions, but always operate on two elements of the source vector. These instructions are:

addp   Integer Sum Elements to Scalar,
faddp   Integer Maximum Element to Scalar,
fmaxp   Floating Point Maximum Element to Scalar, and

fminp Floating Point Minimum Element to Scalar.

There are long versions of the **addv** instruction which will prevent overflow.

10.10.2.1 Syntax

```
addp        Dd, Vn.2d
faddp       Fd, Vn.T
fmax{nm}p   Fd, Vn.T
fmin{nm}p   Fd, Vn.T
```

- F/T must be either s/2s or d/2d.
- If nm is present, then comparison between a nan and a numerical value will return the numerical value.

10.10.2.2 Operations

Name	Effect	Description
addp	$Dd \leftarrow Vn[0] + Vn[1]$	Sum the two elements of Vn and store the result in Fd.
faddp	$Fd \leftarrow Vn[0] + Vn[1]$	Sum the two elements of Vn and store the result in Fd.

*x*maxv	**if** $Vn[0] > Vn[1]$ **then**
	   $Fd \leftarrow Vn[0]$
	 else
	   $Fd \leftarrow Vn[1]$
	 end if
		Store maximum value from Vn in Fd.

*x*minv	**if** $Vn[0] < Vn[1]$ **then**
	   $Fd \leftarrow Vn[0]$
	 else
	   $Fd \leftarrow Vn[1]$
	 end if
		Store minimum value from Vn in Fd.

10.10.2.3 Examples

```
addp   d3,v2.2d
fmaxp  s0,v4.2s
```

10.11 Comparison operations

Advanced SIMD provides instructions to perform comparisons between vectors. Since there are multiple pairs of items to be compared, the comparison instructions set one element in a result vector for each pair of items. After the comparison operation, each element of the result vector will have every bit set to zero (for false) or one (for true). Note that if the elements of the result vector are interpreted as signed two's-complement numbers, then the value 0 represents false and the value -1 represents true. Note that summing the elements of the result vector (as signed integers) will give the two's complement of the number of comparisons which were true.

10.11.1 Vector compare mask

The following instructions perform comparisons of all of the corresponding elements of two vectors in parallel:

cm   Vector integer compare mask, and

fcm   Vector floating point compare mask.

10.11.1.1 Syntax

```
cm<op>     Vd.T, Vn.T, Vm.T
cm<op2>    Vd.T, Vn.T, #0
fcm<op2>   Vd.Tf, Vn.Tf, Vm.Tf
fcm<op2>   Vd.Tf, Vn.Tf, #0
```

- <op> is one of eq, hs, ge, hi, gt, ls, le, lo, or lt.
- <op2> is one of eq, ge, gt, le, or lt.
- T is 8b, 16b, 4h, 8h, 2s, 4s, or 2d.
- Tf is 2s, 4s, or 2d.
- If the third operand is #0, then it is treated as a vector of the correct size in which every element is zero.
- V can be d or q.

10.11.1.2 Operations

Name	Effect	Description

{f}cm<op>	**for** $i \in vector\_length$ **do**
	    if Vn[i]<op>Vm[i]
	    then
	        Vd[i] ← 111...
	    else
	        Vd[i] ← 000...
	    end if
	end for
		Compare each scalar in Vn to the corresponding scalar in Vm. Set the corresponding scalar in Vd to all ones if <op> is true, and all zeros if <op> is not true.

10.11.1.3 Examples

```
    cmeq   v0.8b,v1.8b,v2.8b     // 8 8-bit comparisons
    cwge   v0.4h,v1.4h,v2.4h     // 4 16-bit signed comparisons
    cwgt   v0.8h,v1.8h,v2.8h     // 8 16-bit unsigned comparisons
    fcmle  v0.2s,v1.2s,v2.2s     // 2 single precision comparisons
    fcmlt  v0.4s,v1.4s,v2.4s     // 4 single precision comparisons
    cmeq   v0.16b,v1.16b,#0      // 16 8-bit comparisons
    cmge   v0.8b,v1.8b,#0        // 8 8-bit signed comparisons
    fcmgt  v0.2s,v1.2s,#0        // 2 single precision comparisons
```

10.11.2 Vector absolute compare mask

The following instruction performs comparisons between the absolute values of all of the corresponding elements of two vectors in parallel:

fac   Vector floating point absolute compare mask.

10.11.2.1 Syntax

```
    fac<op>  Vd.T, Vn.T, Vm.T
```

- <op> is one of ge, gt, le, or lt.
- T is 2s, 4s, or 2d.

10.11.2.2 Operations

Name	Effect	Description

fac\<op\>	**for** $i \in vector\_length$ **do**
	  if $\|Vn[i]\|<op>\|Vm[i]\|$
	  then
	    $Vd[i] \leftarrow 111\ldots$
	  else
	    $Vd[i] \leftarrow 000\ldots$
	  end if
	end for
		Compare each scalar in Vn to the corresponding scalar in Vm. Set the corresponding scalar in Vd to all ones if \<op\> is true, and all zeros if \<op\> is not true.

10.11.2.3 Examples

```
facgt   v0.2s,v1.2s,v2.2s   // 2 single precision comparisons
facge   v0.4s,v1.4s,v2.4s   // 4 single precision comparisons
```

10.11.3 Vector test bits mask

Advanced SIMD provides the following vector version of the ARM `tst` instruction:

cmtst   Vector test bits compare mask.

10.11.3.1 Syntax

```
cmtst      Vd.T, Vn.T, Vm.T
```

- T is 8b, 16b, 4h, 8h, 2s, 4s, or 2d.

10.11.3.2 Operations

Name	Effect	Description

cmtst	**for** $i \in vector\_length$ **do**
	  if $(Vm[i] \wedge Vn[i]) \neq 0$ **then**
	    $Vd[i] \leftarrow 111\ldots$
	  else
	    $Vd[i] \leftarrow 000\ldots$
	  end if
	end for
		Perform logical AND between each scalar in Vn and the corresponding scalar in Vm. Set the corresponding scalar in Vd to all ones if the result is not zero, and all zeros otherwise.

10.11.3.3 Examples

```
    cmtst    v0.8b,v1.8b,v2.8b
    cmtst    v0.4h,v1.4h,v2.4h
```

10.11.4 Scalar compare mask

The following instructions perform comparisons of the specified scalar registers:

cm   Scalar integer compare mask, and

fcm   Scalar floating point compare mask,

If the comparison is true, then all bits in the destination register are set to one. Otherwise, all bits in the destination register are set to zero.

10.11.4.1 Syntax

```
    cm<op>      Dd, Dn, Dm
    cm<op2>     Dd, Dn, #0
    fcm<op2>    Fd, Fn, Fm
    fcm<op2>    Fd, Fn, #0
```

- The integer comparison can only operate on 64-bit integers.
- F can be either s for single precision, or d for double precision floating point.
- <op> is one of eq, hs, ge, hi, gt, ls, le, lo, or lt.
- <op2> is one of eq, ge, gt, le, or lt.

10.11.4.2 Operations

Name	Effect	Description

cm<op>	**if** Dn<op>Dm
	then
	   Dd ← 111...
	else
	   Dd ← 000...
	end if
		Compare 64-bit integer Dn to the 64-bit integer in Dm. Set Dd to all ones if <op> is true, and all zeros if <op> is false.

continued on next page

Name	Effect	Description

fcm\<op\>	**if** Fn\<op\>Fm
	   then
	      Fd ← 111...
	   else
	      Fd ← 000...
	   end if
		Compare floating point value in Fn to floating point value in Fm. Set Dd to all ones if \<op\> is true, and all zeros if \<op\> is false.

10.11.4.3 Examples

```
cmeq  d0,d1,d2
fcmle s3,s4,#0
fcmgt d2,d3,d9
```

10.11.5 Scalar absolute compare mask

The following instruction performs comparisons between the absolute values of two scalars:

fac   Scalar floating point absolute compare mask.

10.11.5.1 Syntax

```
fac<op>  Fd, Fn, Fm
```

- \<op\> is one of ge, gt, le, or lt.
- T is 2s, 4s, or 2d.

10.11.5.2 Operations

Name	Effect	Description

fac\<op\>	**if** |Fn|\<op\>|Fm|
	   then
	      Fd ← 111...
	   else
	      Fd ← 000...
	   end if
		Compare each scalar in Vn to the corresponding scalar in Vm. Set the corresponding scalar in Vd to all ones if \<op\> is true, and all zeros if \<op\> is not true.

10.11.5.3 Examples

```
facgt  d3,d1,d2
facge  s0,s5,s8
```

10.11.6 Scalar test bits mask

The scalar test bits instruction performs a logical AND operation between two source registers. If the result is not zero, then every bit in the result register is set to one. Otherwise, every bit in the result register is set to zero. The instruction is:

cmtst   Scalar test bits compare mask.

10.11.6.1 Syntax

```
cmtst     Dd, Dn, Dm
```

• Only 64-bit comparisons can be performed.

10.11.6.2 Operations

Name	Effect	Description

cmtst	**if** $(Dm \wedge Dn) \neq 0$ **then**
	 $\quad$ Dd $\leftarrow 111\ldots$
	 else
	 $\quad$ Dd $\leftarrow 000\ldots$
	 end if
		Perform logical AND between scalar Fn scalar in Vm. Set Vd to all ones if the result is not zero, and all zeros otherwise.

10.11.6.3 Examples

```
cmtst   d6,d5,d2
cmtst   d0,d1,d3
```

10.12 Performance mathematics: a final look at sine

In Chapter 9, two versions of the sine function were given. Those implementations used scalar FP/NEON instructions for single-precision and double-precision. Those previous implementations are already faster than the implementations provided by GCC, However, it may be possible to gain a little more performance by taking advantage of the Advanced SIMD architecture.

10.12.1 Single precision

Listing 10.1 Advanced SIMD implementation of the $\sin x$ function using single precision.

```
1   // sin_N_f implements the sine function using NEON single
2   // precision floating point. It computes sine by summing
3   // the first 8 terms of the Taylor series.
4   // -------------------------------------------------------------
5           .data
6           // The following is a table of constants used in the
7           // Taylor series approximation for sine
8           .align  8               // Align to cache (256-byte boundary)
9   ctab:   .word   0x3F800000      //  1.000000000000000  1/1!
10          .word   0xBE2AAAAB      // -0.166666671633720 -1/3!
11          .word   0x3C088889      //  0.008333333767951  1/5!
12          .word   0xB9500D01      // -0.000198412701138 -1/7!
13          .word   0x3638EF1D      //  0.000002755731884  1/9!
14          .word   0xB2D7322B      // -0.000000025052108 -1/11!
15          .word   0x2F309231      //  0.000000000160590  1/13!
16          .word   0xAB573F9F      // -0.000000000000765 -1/15!
17  // -------------------------------------------------------------
18          .text
19          .align 2
20          .global sin_N_f
21  sin_N_f:// Use vectors of length 4
22          // 1) Load the entire coefficient table into v16,v17
23          // 2) Calculate powers of x as vectors in v0,v1
24          // 3) Multiply powers of x by coefficients
25          // 4) Sum the results
26          fmul    s1,s0,s0         // Put x^2 in s1
27          ldr     x0,=ctab         // load pointer to table
28          fmul    s2,s1,s1         // Put x^4 in s2
29          fmul    s4,s0,s1         // Put x^3 in s4
30          ld1     {v16.4s},[x0],#16 // load first half of table
31          mov     v2.2s[1],v2.2s[0] // make vector of x^4,x^4
32          mov     v0.2s[1],v4.2s[0] // make vector of x^3,x
33          ld1     {v17.4s},[x0]    // load second half of table
34          fmul    s5,s2,s2         // put x^8 in s5
35          fmul    v3.2s,v0.2s,v2.2s // make vector of x^7, x^5
36          dup     v2.4s,v5.4s[0]   // make vector of x^8,x^8,x^8,x^8
37          mov     v0.2d[1],v3.2d[0] // make vector of x^7,x^5,x^3,x
38          fmul    v1.4s,v0.4s,v2.4s // make vector of x^15,x^13,x^11,x^9
39          fmul    v0.4s,v0.4s,v16.4s // calculate first four terms
40          fmla    v0.4s,v1.4s,v17.4s // accumulate last four terms
41          faddp   v0.4s,v0.4s,v0.4s // add terms pairwise
42          faddp   s0,v0.2s         // add terms pairwise
43          ret
```

Listing 10.1 shows a single precision floating point implementation of the sine function, using Advanced SIMD vector instructions. It performs the same operations as the previous implementations of the sine function, but performs many of the calculations in parallel. This implementation is slightly faster than the previous version. In addition to being faster, it also uses nine terms of the Taylor series, so it should be more accurate as well.

10.12.2 Double precision

Listing 10.2 Advanced SIMD implementation of the $\sin x$ function using double precision.

```
// sin_N_d implements the sine function using Advanced SIMD
// double precision vector instructions. It computes sine by
// summing the first 11 terms of the Taylor series.
// ---------------------------------------------------------------
        .data
        // The following is a table of constant coefficients used in
        // the Taylor series approximation for sine
        .align  8               // Align to cache (256-byte boundary)
ctab:   .dword  0xBFC5555555555555 // -0.1666666666666666574 -1/3!
        .dword  0x3F81111111111111 //  0.0083333333333333332  1/5!
        .dword  0xBF2A01A01A01A01A // -0.0001984126984126984 -1/7!
        .dword  0x3EC71DE3A556C734 //  0.0000027557319223986  1/9!
        .dword  0xBE5AE64567F544E4 // -0.0000000250521083854 -1/11!
        .dword  0x3DE6124613A86D09 //  0.0000000001605904384  1/13!
        .dword  0xBD6AE7F3E733B81F // -0.0000000000007647164 -1/15!
        .dword  0x3CE952C77030AD4A //  0.0000000000000028115  1/17!
        .dword  0xBC62F49B46814157 // -0.0000000000000000082 -1/19!
        .dword  0x3BD71B8EF6DCF572 //  0.0000000000000000000  1/21!
// ---------------------------------------------------------------
        .text
        .align 2
        .global sin_N_d
sin_N_d:// Use vectors of length 2
        // 1) Load the entire coefficient table into v16-v20
        // 2) Create vectors for powers of x as needed
        // 3) Multiply powers of x by coefficients
        // 4) Accumulate the results
        ldr     x0,=ctab        // load pointer to table
        fmul    d1,d0,d0        // Put x^2 in d1
        ld1     {v16.2d},[x0],#16// load first part
        fmul    d2,d1,d1        // Put x^4 in d2
        ld1     {v17.2d},[x0],#16// load second part of table
        fmul    d4,d0,d1        // Put x^3 in d4
        ld1     {v18.2d},[x0],#16// load third part of table
        mov     v2.2d[1],v2.2d[0] // make vector of x^4
```

```
36    fmul    d3,d4,d1        // Put x^5 in d3
37    ld1     {v19.2d},[x0],#16// load fourth part of table
38    mov     v4.2d[1],v3.2d[0]  // Copy x^5 to v4[1]
39    ld1     {v20.2d},[x0],#16// load fifth part of table
40    // We have a vector containing x^3 and x^5 in v4,
41    // Start creating remaining powers, two at a time,
42    // multiplying the powers by the coefficients, and
43    // accumulating the results.
44    fmul    v5.2d,v4.2d,v2.2d  // get x^7 and x^9
45    fmul    v4.2d,v4.2d,v16.2d // get terms 2 and 3
46    fmul    v6.2d,v5.2d,v2.2d  // get x^11 and x^13
47    fmla    v4.2d,v5.2d,v17.2d // accumulate terms 4 and 5
48    fmul    v7.2d,v6.2d,v2.2d  // get x^14 and x^15
49    fmla    v4.2d,v6.2d,v18.2d // accumulate terms 6 and 7
50    fmul    v5.2d,v7.2d,v2.2d  // get x^16 and x^17
51    fmla    v4.2d,v7.2d,v19.2d // accumulate terms 8 and 9
52    fmla    v4.2d,v5.2d,v20.2d // accumulate terms 11 and 13
53    faddp   d2,v4.2d       // add terms 2-7 together
54    fadd    d0,d0,d2       // add result to term 1
55    ret
```

Listing 10.2 shows a double precision floating point implementation of the sine function. It also uses Advanced SIMD vector instructions. Both of the implementations in this chapter are faster than the corresponding implementations in Chapter 9 because they use a large number of registers, do not contain loops, and are written carefully to use vector instructions, ordered so that multiple instructions can be at different stages in the pipeline at the same time. This technique of gaining performance is known as *loop unrolling*. In addition to being faster, the vector implementations use more terms of the Taylor series, so they may also be more accurate.

10.12.3 Performance comparison

Table 10.1 compares the implementations from Listing 10.1 and Listing 10.2 with the FP/-NEON implementations from Chapter 9 and the sine function provided by GCC. When compiler optimization is not used, the single precision scalar FP/NEON implementation achieves a speedup of about 3.41, and the Advanced SIMD implementation achieves a speedup of about 4.21 compared to the GCC implementation. The double precision scalar FP/NEON implementation achieves a speedup of about 2.22, and the Advanced SIMD achieves a speedup of about 2.41 compared to the GCC implementation.

When the compiler optimization is used (-Ofast), the single precision scalar FP/NEON implementation achieves a speedup of about 3.04, and the Advanced SIMD implementation

Table 10.1: **Performance of sine function with various implementations.**

Optimization	Implementation	CPU seconds
None	Single Precision C	6.958
	Single Precision FP/NEON scalar Assembly	2.038
	Single Precision Advanced SIMD Assembly	1.654
	Double Precision C	6.682
	Double Precision FP/NEON scalar Assembly	3.014
	Double Precision Advanced SIMD Assembly	2.778
Full	Single Precision C	4.409
	Single Precision FP/NEON scalar Assembly	1.449
	Single Precision Advanced SIMD Assembly	1.142
	Double Precision C	5.758
	Double Precision FP/NEON scalar Assembly	2.134
	Double Precision Advanced SIMD Assembly	1.941

achieves a speedup of about 3.86 compared to the GCC implementation. The double precision scalar FP/NEON implementation achieves a speedup of about 2.70, and the Advanced SIMD implementation achieves a speedup of about 2.97 compared to the GCC implementation. The single precision Advanced SIMD version was 1.27 times as fast as the FP/NEON scalar version, and the double precision Advanced SIMD implementation was 1.10 times as fast as the FP/NEON scalar implementation.

Although the FP/NEON versions of the sine functions were already much faster than the C standard library, re-writing them using Advanced SIMD resulted in further performance improvement. The take-away lesson is that a programmer can improve performance by writing some functions in assembly that are specifically targeted to run on a specific platform. To achieve optimal or near-optimal performance, it is important for the programmer to be aware of advanced features available on the hardware platform that is being used.

10.13 Alphabetized list of advanced SIMD instructions

Name	Page	Operation
aba	354	Vector integer absolute difference and accumulate
abal	355	Vector absolute difference and accumulate long
abd	354	Vector integer absolute difference
abdl	355	Vector absolute difference long
abs	383	Integer absolute value

continued on next page

Name	Page	Operation
abs	356	Vector absolute value
adalp	353	Vector add and accumulate long pairwise
add	348	Vector integer add
addhn	351	Vector add and narrow
addl	348	Vector add long
addlp	353	Vector add long pairwise
addp	386	Integer Sum Elements to Scalar
addp	353	Vector add pairwise
addv	385	Integer Sum Elements to Scalar
addw	349	Vector add wide
and	346	Vector bitwise AND
bic	347	Vector Bit clear immediate
bic	346	Vector bit clear
bif	346	Vector insert if false
bit	346	Vector insert if true
bsl	346	Vector bitwise select
cls	359	Vector count leading sign bits
clz	359	Vector count leading zero bits
cm	391	Scalar integer compare mask
cm	388	Vector integer compare mask
cmtst	393	Scalar test bits compare mask
cmtst	390	Vector test bits compare mask
cnt	359	Vector count set bits
cvtf	343	Vector convert integer or fixed point to floating point
dup	333	Duplicate Scalar
eor	346	Vector bitwise Exclusive-OR
ext	339	Byte Extract
faba	355	Vector floating point absolute difference and accumulate
fabd	354	Vector floating point absolute difference
fabs	356	Vector floating point absolute value
fac	392	Scalar floating point absolute compare mask
fac	389	Vector floating point absolute compare mask

continued on next page

Name	Page	Operation
fadd	348	Vector floating point add
faddp	386	Integer Maximum Element to Scalar
fcm	391	Scalar floating point compare mask
fcm	388	Vector floating point compare mask
fcvt	343	Vector convert floating point to integer or fixed point
fcvtl	344	Vector convert from half to single precision
fcvtn	344	Vector convert from single to half precision
fcvtxn	344	Vector convert from double to single precision
fdiv	362	Vector floating point divide
fmadd	310	Fused Multiply Accumulate
fmax	357	Vector floating point maximum
fmaxnm	357	Vector floating point maxnum
fmaxnmp	357	Vector floating point pairwise maxnum
fmaxp	386	Floating Point Maximum Element to Scalar
fmaxp	357	Vector floating point pairwise maximum
fmaxv	385	Floating Point Maximum Element to Scalar
fmin	357	Vector floating point minimum
fminnm	357	Vector floating point minnum
fminnmp	357	Vector floating point pairwise minnum
fminp	387	Floating Point Minimum Element to Scalar
fminp	357	Vector floating point pairwise minimum
fminv	385	Floating Point Minimum Element to Scalar
fmla	364	Vector by scalar floating point multiply accumulate
fmla	370	Vector by scalar floating point multiply accumulate
fmla	362	Vector floating point multiply accumulate
fmls	364	Vector by scalar floating point multiply subtract
fmls	370	Vector by scalar floating point multiply subtract
fmls	362	Vector floating point multiply subtract
fmov	336	Vector Floating Point Move Immediate
fmsub	310	Fused Multiply Subtract
fmul	364	Vector by scalar floating point multiply
fmul	370	Vector by scalar floating point multiply

continued on next page

Name	Page	Operation
fmul	362	Vector floating point multiply
fneg	356	Vector floating point negate
fnmadd	310	Fused Multiply Accumulate and Negate
fnmsub	310	Fused Multiply Subtract and Negate
frecps	369	Reciprocal Step
frint	345	Round Floating Point to Integer
frsqrts	369	Reciprocal Square Root Step
fsqrt	382	Vector Floating Point Square Root
fsqrt	383	Vector Floating Point Square Root
fsub	349	Vector floating point subtract
hadd	352	Vector halving add
hsub	352	Vector halving subtract
ld<n>	329	Load Multiple Structured Data
ld<n>	327	Load Structured Data
ld<n>r	331	Load Copies of Structured Data
max	357	Vector integer maximum
maxp	357	Vector integer pairwise maximum
maxv	385	Integer Maximum Element to Scalar
min	357	Vector integer minimum
minp	357	Vector integer pairwise minimum
minv	385	Integer Minimum Element to Scalar
mla	364	Vector by scalar integer multiply accumulate
mla	362	Vector integer multiply accumulate
mlal	364	Vector by scalar multiply accumulate long
mlal	362	Vector multiply accumulate long
mls	364	Vector by scalar integer multiply subtract
mls	362	Vector integer multiply subtract
mlsl	364	Vector by scalar multiply subtract long
mlsl	362	Vector multiply subtract long
mov	334	Copy element into vector
movi	336	Vector Move Immediate
mul	364	Vector by scalar integer multiply

continued on next page

Name	Page	Operation
mul	362	Vector integer multiply
mull	364	Vector by scalar multiply long
mull	362	Vector multiply long
mvni	336	Vector Move NOT Immediate
neg	383	Integer absolute value
neg	356	Vector negate
not	382	Vector 1's Complement
orn	346	Vector bitwise NOR
orr	346	Vector bitwise OR
orr	347	Vector bitwise OR immediate
pmul	362	Vector polynomial multiply
pmull	362	Vector polynomial multiply long
qadd	360	Scalar saturating add
qadd	382	Vector Saturating Accumulate
qadd	383	Vector Saturating Accumulate
qadd	348	Vector saturating add
qdmulh	360	Scalar saturating multiply (high half)
qrshl	377	Saturating Shift Left or Right by Variable and Round
qrshrn	376	Saturating Rounding Shift Right Narrow
qrshrn	381	Saturating Rounding Shift Right Narrow
qshl	377	Saturating Shift Left or Right by Variable
qshl	373	Saturating Signed or Unsigned Shift Left Immediate
qshl	379	Saturating Signed or Unsigned Shift Left Immediate
qshl	360	Scalar saturating shift left
qshrn	376	Saturating Shift Right Narrow
qshrn	381	Saturating Shift Right Narrow
qsub	360	Scalar saturating subtract
qsub	349	Vector saturating subtract
raddhn	351	Vector add, round, and narrow
rbit	382	Vector bit reverse
rbit	384	Vector bit reverse
recpe	368	Reciprocal Estimate

continued on next page

Name	Page	Operation
rev	382	Reverse Elements
rev	384	Reverse Elements
rhadd	352	Vector halving add and round
rshl	377	Shift Left or Right by Variable and Round
rshr	374	Shift Right Immediate and Round
rshr	380	Shift Right Immediate and Round
rshrn	374	Shift Right Immediate Round and Narrow
rsqrte	368	Reciprocal Square Root Estimate
rsra	374	Shift Right Round and Accumulate Immediate
rsra	380	Shift Right Round and Accumulate Immediate
rsubhn	351	Vector subtract, round, and narrow
shl	377	Shift Left or Right by Variable
shl	373	Unsigned Shift Left Immediate
shl	379	Unsigned Shift Left Immediate
shll	373	Signed or Unsigned Shift Left Immediate Long
shll	379	Signed or Unsigned Shift Left Immediate Long
shr	374	Shift Right Immediate
shr	380	Shift Right Immediate
shrn	374	Shift Right Immediate and Narrow
sli	378	Shift Left and Insert
smov	334	Copy signed integer element from vector to AARCH64 register
sqdmlal	365	Saturating Multiply Double Accumulate
sqdmlal	371	Saturating Multiply Scalar by Element, Double, and Accumulate
sqdmlsl	365	Saturating Multiply Double Subtract
sqdmlsl	371	Saturating Multiply Scalar by Element, Double, and Subtract
sqdmulh	366	Saturating Multiply Double (High)
sqdmull	365	Saturating Multiply Double
sqdmull	371	Saturating Multiply Scalar by Element and Double
sqrdmulh	367	Saturating Multiply Double and Round (High)
sqrshrun	376	Signed Saturating Rounding Shift Right Unsigned Immediate
sqrshrun	381	Signed Saturating Rounding Shift Right Unsigned Immediate
sqshlu	373	Saturating Signed Shift Left Immediate Unsigned

continued on next page

Name	Page	Operation
sqshlu	379	Saturating Signed Shift Left Immediate Unsigned
sqshrun	376	Signed Saturating Shift Right Unsigned Narrow
sqshrun	381	Signed Saturating Shift Right Unsigned Narrow
sra	374	Shift Right and Accumulate Immediate
sra	380	Shift Right and Accumulate Immediate
sri	378	Shift Right and Insert
st<n>	329	Store Multiple Structured Data
st<n>	327	Store Structured Data
sub	349	Vector integer subtract
subhn	351	Vector subtract and narrow
subl	349	Vector subtract long
subw	349	Vector subtract wide
tbl	341	Table Lookup
tbx	341	Table Lookup with Extend
trn	337	Transpose Matrix
umov	334	Copy unsigned integer element from vector to AARCH64 register
uzp	339	Unzip Vectors
zip	339	Zip Vectors

10.14 Advanced SIMD intrinsics

The C compiler may provide C (and C++) programs direct access to the Advanced SIMD instructions through the Advanced SIMD intrinsics library. The intrinsics are a large set of functions that are built into the compiler. Most of the intrinsics functions map to one Advanced SIMD instruction. There are additional functions provided for typecasting (reinterpreting) SIMD vectors, so that the C compiler does not complain about mismatched types. It is usually shorter and more efficient to write the Advanced SIMD code directly as assembly language functions and link them to the C code. However only those who know assembly language are capable of doing that.

10.15 Chapter summary

Advanced SIMD can dramatically improve performance of algorithms that can take advantage of data parallelism. However, compiler support for automatically vectorizing and using

Advanced SIMD instructions is still immature. Advanced SIMD intrinsics allow C and C++ programmers to access these instructions, by making them look like C functions. It is usually just as easy and more concise to write AArch64 assembly code as it is to use the intrinsics functions. A careful assembly language programmer can usually beat the compiler, sometimes by a wide margin.

Exercises

10.1.   What is the advantage of using IEEE half-precision? What is the disadvantage?

10.2.   Advanced SIMD achieved relatively modest performance gains on the sine function, when compared to FP/NEON.

    a. Why?

    b. List some tasks for which Advanced SIMD could significantly outperform scalar FP/NEON.

10.3.   There are some limitations on the size of the structure that can be loaded or stored using the ld\<n\> and st\<n\> instructions. What are the limitations?

10.4.   The sine function in Listing 10.2 uses a technique known as "loop unrolling" to achieve higher performance. Name at least three reasons why this code is more efficient than using a loop?

10.5.   Reimplement the fixed-point sine function from Listing 8.7 using Advanced SIMD instructions. Hint: you should not need to use a loop. Compare the performance of your Advanced SIMD implementation with the performance of the original implementation.

10.6.   Reimplement Exercise 9.10. using Advanced SIMD instructions.

10.7.   Fixed point operations may be faster than floating point operations. Modify your code from the previous example so that it uses the following definitions for points and transformation matrices:

```
typedef int point[3];      // Point is an array of S(15,16)
typedef int matrix[4][4];  // Matrix is a 2-D array of S(15,16)
```

Use saturating instructions and/or any other techniques necessary to prevent overflow. Compare the performance of the two implementations.

## CHAPTER 11

Devices

As mentioned in Chapter 1, a computer system consists of three main parts: the CPU, memory, and devices. The typical computing system has many devices of various types for performing certain functions. Some devices, such as data caches, are closely coupled to the CPU, and are typically controlled by executing special CPU instructions that can only be accessed in Assembly language. However, most of the devices on a typical system are accessed and controlled through the system data bus. These devices appear to the programmer to be ordinary memory locations.

The hardware in the system bus decodes the addresses coming from the CPU, and some addresses correspond to devices rather than memory. Fig. 11.1 shows the memory layout for a typical system. The exact locations of the devices and memory are chosen by the system hardware designers. From the programmer's standpoint, writing data to certain memory addresses results in the data being transferred to a device rather than being stored in memory. The programmer must read documentation on the hardware design to determine exactly where the devices are located in memory.

There are devices that allow data to be read or written from external sources, devices that can measure time, devices for moving data from one location in memory to another, devices for modifying the addresses of memory regions, and devices for even more esoteric purposes. Some devices are capable of sending signals to the CPU, to indicate that they need attention, while others simply wait for the CPU to check on their status.

A modern computer system, such as the Raspberry Pi, has dozens or even hundreds of devices. Programmers write device driver software for each device. A device driver provides a few standard function calls for each device, so that it can be used easily. The specific set of function calls depends on the type of device and the design of the operating system. Operating system designers strive to define a small set of device types, and define a standard software interface for each type in order to make devices interchangeable.

11.1 Accessing devices directly under Linux

Devices are typically controlled by writing specific values to the device's internal *device registers*. For the AArch64 processor, access to most device registers is accomplished using the load and store instructions. Each device is assigned a *base address* in memory. This ad-

405

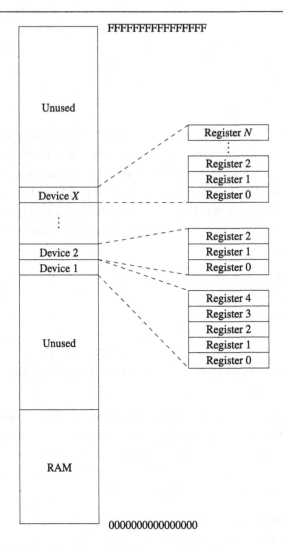

Figure 11.1: Typical hardware address mapping for memory and devices.

dress corresponds with the first register inside the device. The device may also have other registers that are accessible at some pre-defined *offset address* from the base address. Some registers are read-only, some are write-only, and some are read-write. To use the device, the programmer must read from, and write appropriate data to, the correct device registers. For every device, there is a programmer's model and documentation explaining what each register in the device does. Some devices are well-designed, easy to use, and well-documented. Some devices are not, and the programmer must work harder to write software to use them.

The Raspberry Pi 3 and some versions of the Raspberry Pi 2 use the Broadcom BCM2837 system-on-chip. This single chip contains the CPU and several other devices, to form a complete computer system on one chip. The Raspberry Pi 3B+ uses the Broadcom BCM2837B0, which provides 5G wireless Internet, and a few other advanced features, but also contains all of the base hardware in the BCM2837. On all of these Raspberry Pi models, the devices are mapped to a memory region beginning at hardware address **0x00000000F3000000**.

Linux is a powerful, multi-user, multitasking operating system. The Linux kernel manages all of the devices, and protects them from direct access by user programs. User programs are intended to access devices by making system calls. The kernel accesses the devices on behalf of the user programs, ensuring that an errant user program cannot misuse the devices and other resources on the system. Attempting to directly access the registers in any device will result in an exception. The kernel will take over and kill the offending process.

However, our programs will need direct access to the device registers. Linux allows user programs to gain direct access through the mmap() system call. Listing 11.1 shows how four devices can be mapped into the memory space of a user program on a Raspberry Pi. In most cases, the user program will need administrator privileges in order to perform the mapping. The operating system does not usually give permission for ordinary users to access devices directly. However Linux does provide the ability to change permissions on /dev/mem or for user programs to run with elevated privileges.

Listing 11.1 Function to map devices into the user program memory on a Raspberry Pi.

```
/* Raspberry Pi devices
   ----------------------------------------------------------------
   This file provides a function "IO_init" that will map some
   devices into the user program's memory space.  Pointers to
   the devices are stored in global variables, and the user
   program can then use those pointers to access the device
   registers.

   Another function, "IO_close" is provided to unmap the devices.
   ----------------------------------------------------------------*/
        // These are the addresses for the I/O devices (after
        // the firmware boot code has remapped them).
        .equ    PERI_BASE,  0x3F000000 // start of all devices
        // Base Physical Address of the GPIO registers
        .equ    GPIO_BASE, (PERI_BASE + 0x200000)
        // Base Physical Address of the PWM registers
        .equ    PWM_BASE,  (PERI_BASE + 0x20C000)
        // Base Physical Address of the UART 0 device
        .equ    UART0_BASE,(PERI_BASE + 0x201000)
        // Base Physical Address of the Clock/timer registers
```

```
21          .equ    CLK_BASE,  (PERI_BASE + 0x101000)
22          // some constants from /usr/include/mman.h
23          .equ    MAP_FAILED,-1
24          .equ    MAP_SHARED, 1
25          .equ    PROT_READ,  1
26          .equ    PROT_WRITE, 2
27          // some constants from /usr/include/fcntl.h
28          .equ    O_RDONLY,   00000000
29          .equ    O_WRONLY,   00000001
30          .equ    O_RDWR,     00000002
31          .equ    O_CREAT,    00000100
32          .equ    O_EXCL,     00000200
33          .equ    O_NOCTTY,   00000400
34          .equ    O_TRUNC,    00001000
35          .equ    O_APPEND,   00002000
36          .equ    O_NONBLOCK, 00004000
37          .equ    O_NDELAY,   O_NONBLOCK
38          .equ    O_SYNC,     00010000
39          .equ    O_FSYNC,    O_SYNC
40          .equ    O_ASYNC,    00020000
41          // Size of a memory block (may depend on how kerenl
42          // was compiled).
43          .equ    BLOCK_SIZE,(4*1024)
44
45          // The following global variables will hold the addresses
46          // of the devices that can be accessed directly after
47          // IO_init has been called.
48          .data
49          .global gpiobase
50  gpiobase:        .dword  0
51          .global pwmbase
52  pwmbase :        .dword  0
53          .global uart0base
54  uart0base:       .dword  0
55          .global clkbase
56  clkbase :        .dword  0
57
58          // strings used for printing messages
59          .section .rodata
60  memdev:          .asciz  "/dev/mem"
61  successstr:      .asciz  "Successfully opened /dev/mem\n"
62  mappedstr:       .asciz  "Mapped %s device at 0x%01611X\n"
63  openfailed:      .asciz  "IO_init: failed to open /dev/mem: "
64  mapfailedmsg:    .asciz  "IO_init: mmap of %s failed: "
65  gpiostr:         .asciz  "GPIO"
66  pwmstr:          .asciz  "PWM"
67  uart0str:        .asciz  "UART0"
68  clkstr:          .asciz  "CLK"
69
```

```
70          .text
71  // --------------------------------------------------------------
72  // IO_init() maps devices into memory space and stores their
73  // addresses in global variables.
74  // --------------------------------------------------------------
75          .global IO_init
76  IO_init:
77          stp     x29,x30,[sp, #-16]! // Push FP and LR to stack
78          // Try to open /dev/mem
79          adr     x0,memdev       // load address of "/dev/mem"
80          mov     w1,#(O_RDWR+O_SYNC) // set up flags
81          bl      open            // call the open syscall
82          cmp     w0,#0           // check result
83          bge     init_opened     // if open failed,
84          adr     x0,openfailed   // print message and exit
85          bl      printf
86          bl      __errno_location
87          ldr     w0, [x0]
88          bl      strerror
89          bl      perror
90          mov     x0,#0           // return 0 for failure
91          b       init_exit
92  init_opened:
93          // Open succeeded. Now map the devices
94          mov     x29,x0          // move file descriptor to x29
95          adr     x0,successstr   // print message
96          bl      printf
97          // Map the GPIO device
98          mov     x0,x29          // copy file descriptor
99          ldr     x1,=GPIO_BASE   // address of device in memory
100         bl      trymap
101         cmp     w0,#MAP_FAILED
102         bne     map_ok
103         adr     x1,gpiostr      // if failed, load pointer to string,
104         b       map_failed_exit // print message, and return
105 map_ok: adr     x1,gpiobase     // load address of global pointer
106         str     x0,[x1]         // store pointer in gpiobase
107         mov     x2,x1
108         ldr     x2,[x2]
109         adr     x0,mappedstr    // print success message
110         adr     x1,gpiostr
111         bl      printf
112         // Map the PWM device
113         mov     x0,x29          // move file descriptor to x29
114         ldr     x1,=PWM_BASE    // address of device in memory
115         bl      trymap
116         cmp     w0,#MAP_FAILED
117         bne     pwm_ok
118         adr     x1,pwmstr       // if failed, load pointer to string,
```

```
119        b         map_failed_exit // print message, and return
120 pwm_ok: adr       x1,pwmbase      // load address of global pointer
121        str       x0,[x1]         // store pointer in pwmbase
122        mov       x2,x1
123        ldr       x2,[x2]
124        adr       x0,mappedstr    // print success message
125        adr       x1,pwmstr
126        bl        printf
127        // Map the UART0 device
128        mov       x0,x29          // move file descriptor to r4
129        ldr       x1,=UART0_BASE  // address of device in memory
130        bl        trymap
131        cmp       w0,#MAP_FAILED
132        bne       uart_ok
133        adr       x1,uart0str     // if failed, load pointer to string,
134        b         map_failed_exit // print message, and return
135 uart_ok:adr      x1,uart0base    // load address of pointer
136        str       x0,[x1]         // store pointer in uart0base
137        mov       x2,x1
138        ldr       x2,[x2]
139        adr       x0,mappedstr    // print success message
140        adr       x1,uart0str
141        bl        printf
142        // Map the clock manager device
143        mov       x0,x29          // move file descriptor to r4
144        ldr       x1,=CLK_BASE    // address of device in memory
145        bl        trymap
146        cmp       w0,#MAP_FAILED
147        bne       clk_ok
148        adr       x1,clkstr       // if failed, load pointer to string,
149        b         map_failed_exit // print message, and return
150 clk_ok: adr      x1,clkbase      // load address of pointer
151        str       x0,[x1]         // store pointer in clkbase
152        mov       x2,x1
153        ldr       x2,[x2]
154        adr       x0,mappedstr    // print success message
155        adr       x1,clkstr
156        bl        printf
157        // All mmaps have succeded.
158        // Close file and return 1 for success
159        mov       w5,#1
160        b         init_close
161 map_failed_exit:
162        // At least one mmap failed. Print error,
163        // unmap everthing and return
164        adr       x0,mapfailedmsg
165        bl        printf
166        bl        __errno_location
167        ldr       x0, [x0]
```

```
168          bl        strerror
169          bl        perror
170          bl        IO_close
171          mov       x0,#0
172   init_close:
173          mov       x0,x29          // close /dev/mem
174          bl        close
175   init_exit:
176          ldp       x29,x30,[sp],#16 // pop FP and LR from stack
177          ret
178   // ----------------------------------------------------------
179   // trymap(int fd, unsigned offset) Calls mmap.
180   trymap:
181          stp       x29,x30,[sp, #-16]! // Push FP and LR to stack
182          mov       x6,x1           // copy address to x5
183          mov       x7,#0xFFF       // set up a mask for aligning
184          and       x29,x6,x7       // get offset from page boundary
185          bic       x5,x6,x7        // align phys addr to page boundary
186          mov       x4,x0           // put fd in x4
187          mov       x3,#MAP_SHARED
188          mov       x2,#(PROT_READ + PROT_WRITE)
189          mov       x1,#BLOCK_SIZE
190          mov       x0,x6           // request offset as virt address
191          bl        mmap
192          cmp       x0,#-1
193          beq       mapex
194          add       x0,x0,x29       // add offset from page boundary
195   mapex:  ldp       x29,x30,[sp],#16 // pop FP and LR from stack
196          ret
197   // ----------------------------------------------------------
198   // IO_close unmaps all of the devices
199          .global IO_close
200   IO_close:
201          stp       x27,x28,[sp, #-16]! // Push x27,x28 to stack
202          stp       x29,x30,[sp, #-16]! // Push FP and LR to stack
203
204          adr       x27,gpiobase    // get address of first pointer
205          mov       x28,#4          // there ar 4 pointers
206   IO_closeloop:
207          ldr       x0,[x27]                // load address of device
208          mov       x1,#BLOCK_SIZE
209          cmp       x0,#0
210          ble       closed
211          bl        munmap          // unmap it
212          mov       x0,#0
213          str       x0,[x27],#8     // store and increment
214          subs      x28,x28,#1
215          bgt       IO_closeloop
216   closed: ldp       x29,x30,[sp],#16 // pop FP and LR from stack
```

```
217          ldp      x27,x28,[sp],#16 // pop x27 and x28
218          ret
```

11.2 General purpose digital input/output

One type of device commonly found on embedded systems, is the General Purpose I/O
(GPIO) device. Although there are many variations on this device provided by different manufacturers, they all provide similar capabilities. The device provides some set of input and/or
output bits, that allow signals to be transferred to or from the outside world.

11.2.1 Typical GPIO device

Each bit of input or output in a GPIO device is generally referred to as a pin, and a group of
pins is referred to as a GPIO port. Ports commonly support 8, 16, or 32 bits of input and/or
output, although other sizes are also found in some systems. Some GPIO devices support multiple ports, and some systems have multiple GPIO devices.

A system with a GPIO device usually has some type of connector or wires that allow external
inputs or outputs to be connected to the GPIO device. For example, the IBM PC has a type of
GPIO device that was originally intended no allow users to connect a parallel printer. In that
platform, the GPIO device is commonly referred to as the parallel printer port.

Some GPIO devices, such as the one on the IBM PC, are arranged as sets of pins that can be
switched as a group to either input or output. In many modern GPIO devices, each pin can be
set up independently to read (for input) or drive (for output) different input and output voltages. On some devices, the amount of drive current available can be configured. Some GPIO
devices include the ability to configure built-in pull up and/or pull down resistors. On most
older GPIO devices, the input and output voltages are typically limited to the supply voltage
of the GPIO device, and the device may be damaged by greater voltages. Newer GPIO devices generally can tolerate 5 Volts on inputs, regardless of the supply voltage of the device.

GPIO devices are very common in systems that are intended to be used for embedded applications. For most GPIO devices:

- individual pins or groups of pins can be configured,
- pins can be configured to be input or output,
- pins can be disabled so that they are neither input nor output,
- input values can be read by the CPU (typically high=1, low=0),
- output values can be read or written by the CPU, and
- input pins can be configured to generate interrupt requests.

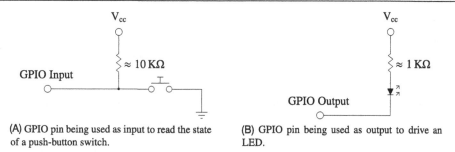

(A) GPIO pin being used as input to read the state of a push-button switch.

(B) GPIO pin being used as output to drive an LED.

Figure 11.2: GPIO pins being used for input and output.

Some GPIO devices may also have more advanced features, such as the ability to use Direct Memory Access (DMA) to send data without requiring the CPU to move each byte or word.

Fig. 11.2 shows two common ways to use GPIO pins. Fig. 11.2A shows a GPIO pin that has been configured for input, and connected to a push-button switch. When the switch is open, the pull-up resistor pulls the voltage on the pin to a high state. When the switch is closed, the pin is pulled to a low state and some current flows through the pull-up resistor to ground. Typically, the pull-up resistor would be around 10K Ohm. The specific value is not critical, but it must be high enough to limit the current to a small amount when the switch is closed.

Fig. 11.2B shows a GPIO pin that is configured as an output and is being used to drive an LED. When a 1 is output on the pin, it is at the same voltage as Vcc, and no current flows. The LED is off. When a 0 is output on the pin, current is drawn through the resistor and the led, and through the pin to ground. This causes the LED to be illuminated. Selection of the resistor is not critical, but it must be small enough to light the LED without allowing enough current to destroy either the LED or the GPIO circuitry. This is typically around 1K Ohm. Note that, in general, GPIO pins can sink more current than they can source, so it is most common to connect LEDs and other devices in the way shown.

11.2.2 Raspberry Pi GPIO

The Broadcom BCM28xx system-on-chip (SOC), which is used in the Raspberry Pi, contains 54 GPIO pins which are split into two banks. The GPIO pins on these systems are named using the following format: GPIOx, where x is a number between 0 and 53. The GPIO pins are highly configurable. Each pin can be used either for general purpose I/O, or can be configured to serve up to six pre-defined alternate functions. Configuring a GPIO pin for an alternate function usually allows some other device within the BCM28xx to use the pin. For example GPIO4 can be used

- for general purpose I/O,
- to send the signal generated by General Purpose Clock 0 to external devices,
- to send bit one of the Secondary Address Bus to external devices, or
- to receive JTAG data for programming the firmware of the device.

The last 8 GPIO pins, GPIO46–GPIO53 have no alternate functions, and are used only for GPIO.

In addition to the alternate function, all GPIO pins can be configured individually as input or output. When configured as input, a pin can also be configured to detect when the signal changes, and to send an interrupt to the CPU. Each input pin also has internal pull-up and pull-down resistors, which can be enabled or disabled by the programmer.

The GPIO pins on the BCM28xx SOC are very flexible and are quite complex, but are well designed and not difficult to program, once the programmer understands how the pins operate, and what the various registers do. There are 41 registers that control the GPIO pins. The base address for the GPIO device is 3F200000. The 41 registers and their offsets from the base address are shown in Table 11.1.

11.2.2.1 Setting the GPIO pin function

The first six 32-bit registers in the device are used to select the alternate functions for each of the 54 GPIO pins. The function of each of the pins is controlled by a group of three bits in one of these registers. The mapping is very regular. Bits 0-2 of GPIOFSEL0 control the function of GPIO pin 0. Bits 3-5 of GPIOFSEL0 control the function of GPIO pin 1, and so on, up to bits 27-29 of GPIOFSEL0, which control the function of GPIO pin 9. The next pin, pin 10, is controlled by bits 0-2 of GPIOFSEL1. The pins are assigned in sequence through the remaining bits, until bits 27-29 control GPIO pin 19. The remaining four GPIOFSEL registers control the remaining GPIO pins. Note that bits 30 and 31 of all of the GPIOFSEL registers are not used, and most of the bits in GPIOFSEL5 are not assigned to any pin. The meaning of each combination of the three bits is shown in Table 11.2. Note that the encoding is not as simple as one might expect.

The procedure for setting the function of a GPIO pin is as follows:

- Determine which GPIOFSEL register controls the desired pin.
- Determine which bits of the GPIOFSEL register are used.
- Determine what the bit pattern should be.
- Read the GPIOFSEL register.
- Clear the correct bits using the bic instruction.
- Set them to the correct pattern using the orr instruction.

Table 11.1: Raspberry Pi GPIO register map.

Offset	Name	Description	Size	R/W
00_{16}	GPFSEL0	GPIO Function Select 0	32	R/W
04_{16}	GPFSEL1	GPIO Function Select 1	32	R/W
08_{16}	GPFSEL2	GPIO Function Select 2	32	R/W
$0C_{16}$	GPFSEL3	GPIO Function Select 3	32	R/W
10_{16}	GPFSEL4	GPIO Function Select 4	32	R/W
14_{16}	GPFSEL5	GPIO Function Select 5	32	R/W
$1C_{16}$	GPSET0	GPIO Pin Output Set 0	32	W
20_{16}	GPSET1	GPIO Pin Output Set 1	32	W
28_{16}	GPCLR0	GPIO Pin Output Clear 0	32	W
$2C_{16}$	GPCLR1	GPIO Pin Output Clear 1	32	W
34_{16}	GPLEV0	GPIO Pin Level 0	32	R
38_{16}	GPLEV1	GPIO Pin Level 1	32	R
40_{16}	GPEDS0	GPIO Pin Event Detect Status 0	32	R/W
44_{16}	GPEDS1	GPIO Pin Event Detect Status 1	32	R/W
$4C_{16}$	GPREN0	GPIO Pin Rising Edge Detect Enable 0	32	R/W
50_{16}	GPREN1	GPIO Pin Rising Edge Detect Enable 1	32	R/W
58_{16}	GPFEN0	GPIO Pin Falling Edge Detect Enable 0	32	R/W
$5C_{16}$	GPFEN1	GPIO Pin Falling Edge Detect Enable 1	32	R/W
64_{16}	GPHEN0	GPIO Pin High Detect Enable 0	32	R/W
68_{16}	GPHEN1	GPIO Pin High Detect Enable 1	32	R/W
70_{16}	GPLEN0	GPIO Pin Low Detect Enable 0	32	R/W
74_{16}	GPLEN1	GPIO Pin Low Detect Enable 1	32	R/W
$7C_{16}$	GPAREN0	GPIO Pin Async. Rising Edge Detect 0	32	R/W
80_{16}	GPAREN1	GPIO Pin Async. Rising Edge Detect 1	32	R/W
88_{16}	GPAFEN0	GPIO Pin Async. Falling Edge Detect 0	32	R/W
$8C_{16}$	GPAFEN1	GPIO Pin Async. Falling Edge Detect 1	32	R/W
94_{16}	GPPUD	GPIO Pin Pull-up/down Enable	32	R/W
98_{16}	GPPUDCLK0	GPIO Pin Pull-up/down Enable Clock 0	32	R/W
$9C_{16}$	GPPUDCLK1	GPIO Pin Pull-up/down Enable Clock 1	32	R/W

For example, Listing 11.2 shows the sequence of code would be used to set GPIO pin 26 to alternate function 1.

11.2.2.2 Setting GPIO output pins

To use a GPIO pin for output, the function select bits for that pin must be set to 001. Once that is done, the output can be driven high or low by using the GPSET and GPCLR registers. GPIO pin 0 is set to a high output by writing a 1 to bit 0 of GPSET0, and it is set to low out-

Table 11.2: GPIO pin function select bits.

MSB-LSB	Function
000	Pin is an input
001	Pin is an output
100	Pin performs alternate function 0
101	Pin performs alternate function 1
110	Pin performs alternate function 2
111	Pin performs alternate function 3
011	Pin performs alternate function 4
010	Pin performs alternate function 5

Listing 11.2 AArch64 Assembly code to set GPIO pin 26 to alternate function 1.

```
     .equ  GPIOFSEL2, 0x3F200008
     .equ  mask1, (0b111 << (3*7))
     .equ  mask2, (0b101 << (3*7))
       :
     ldr    x0,=GPIOFSEL2
     ldr    x1,[x0]
     bic    x1,x1,#mask1
     mov    x2,#mask2
     orr    x1,x1,x2
     str    x1,[x0]
```

put by writing a 1 to bit 0 of GPCLR0. GPIO pin 1 is similarly controlled by bit 1 in GPSET0 and GPCLR0. Each of the GPIO pins numbered 0 through 31 is assigned one bit in GPSET0 and one bit in GPCLR0. GPIO pin 32 is assigned to bit 0 of GPSET1 and GPCLR1, GPIO pin 33 is assigned to bit 1 of GPSET1 and GPCLR1, and so on. Since there are only 54 GPIO pins, bits 22-31 of GPSET1 and GPCLR1 are not used. The programmer can set or clear several outputs simultaneously by writing the appropriate bits in the GPSET and GPCLR registers.

11.2.2.3 Reading GPIO input pins

To use a GPIO pin for input, the function select bits for that pin must be set to 000. Once that is done, the input can be read at any time by reading the appropriate GPLEV register and examining the bit that corresponds with the input pin. GPIO pin 0 is read as bit 0 of GPLEV0, GPIO pin 1 is similarly read as bit 1 of GPLEV1. Each of the GPIO pins numbered 0 through 31 is assigned one bit in GPLEV0. GPIO pin 32 is assigned to bit 0 of GPLEV1, GPIO pin 33 is assigned to bit 1 of GPLEV1, and so on. Since there are only 54 GPIO pins, bits 22-31 of

Table 11.3: GPPUD control codes.

Code	Function
00	Disable pull-up and pull-down
01	Enable pull-down
10	Enable pull-up

GPLEV1 are not used. The programmer can read the status of several inputs simultaneously by reading one of the GPLEV registers and examining the bits corresponding to the appropriate pins.

11.2.2.4 Enabling internal pull-up or pull-down

Input pins can be configured with internal pull-up or pull-down resistors. This can simplify the design of the system. For instance, Fig. 11.2A, shows a push-button switch connected to an input, with an external pull-up resistor. That resistor is unnecessary if the internal pull-up for that pin is enabled.

Enabling the pull-up or pull-down is a two step process. The first step is to configure the type of change to be made, and the second step is to perform that change on the selected pin(s). The first step is accomplished by writing to the GPPUD register. The valid binary control codes are shown in Table 11.3.

Once the GPPUD register is configured, the selected operation can be performed on multiple pins by writing to one or both of the GPPUDCLK registers. GPIO pins are assigned to bits in these two registers in the same way as the pins are assigned in the GPLEV, GPSET, and GPCLR registers. Writing 1 to bit 0 of GPPUDCLK0 will configure the pull-up or pull-down for GPIO pin 0, according to the control code that is currently in the GPPUD register.

11.2.2.5 Detecting GPIO events

The GPEDS registers are used for detecting events that have occurred on the GPIO pins. For instance a pin may have transitioned from low to high, and back to low. If the CPU does not read the GPLEV register often enough, then such an event could be missed. The GPEDS registers can be configured to capture such events so that the CPU can detect that they occurred.

GPIO pins are assigned to bits in these two registers in the same way as the pins are assigned in the GPLEV, GPSET, and GPCLR registers. If bit 1 of GPEDS0 is set, then that indicates that an event has occurred on GPIO pin 0. Writing a 0 to that bit will clear the bit and allow the event detector to detect another event. Each pin can be configured to detect specific types of events by writing to the GPREN, GPHEN, GPLEN, GPAREN, and GPAFEN registers. For more information, refer to the *BCM2835 ARM Peripherals* manual.

Figure 11.3: The Raspberry Pi expansion header location.

11.2.2.6 GPIO pins available on the Raspberry Pi

The Raspberry Pi provides access to several of the 54 GPIO pins, through the expansion header. The expansion header is a group of physical pins sticking up in the corner of the Raspberry Pi board. Fig. 11.3 shows where the header is located on the Raspberry Pi. Wires can be connected these pins and then the GPIO device can be programmed to send and/or receive digital information. Fig. 11.4 shows which signals are attached to the various pins. Some of the pins are available to provide power and ground to the external devices.

Table 11.4 shows some useful alternate functions available on each pin of the Raspberry Pi expansion header. Many of the alternate functions available on these pins are not really useful. Those functions have been left out of the table. The most useful alternate functions are probably GPIO 14 and 15, which can be used for serial communication, and GPIO 18, which can be used for pulse width modulation. Pulse width modulation is covered in Section 11.3.2, and serial communication is covered in Section 11.5. The Serial Peripheral Interface (SPI) functions

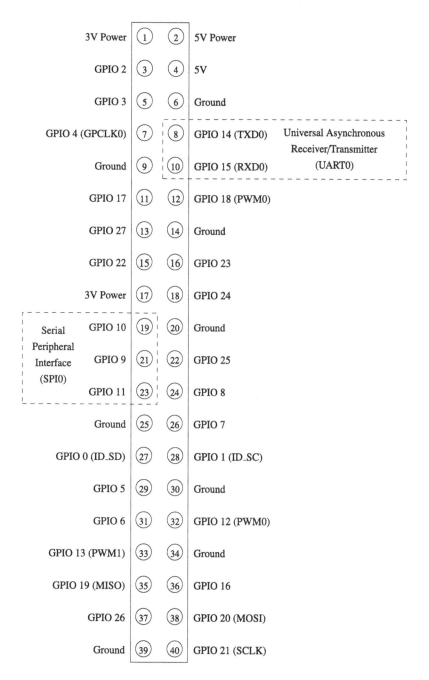

Figure 11.4: The Raspberry Pi extended expansion header pin assignments.

Table 11.4: Raspberry Pi expansion
header useful alternate functions.

Pin	Alternate Function	
	0	5
GPIO 0	SDA0	
GPIO 1	SCL0	
GPIO 2	SDA1	
GPIO 3	SCL1	
GPIO 4	GPCLK0	
GPIO 5	GPCLK1	
GPIO 6	GPCLK2	
GPIO 7	SPI0_CE1_N	
GPIO 8	SPI0_CE0_N	
GPIO 9	SPI0_MISO	
GPIO 10	SPI0_MOSI	
GPIO 11	SPI0_SCLK	
GPIO 12	PWM0	
GPIO 13	PWM1	
GPIO 14	TXD0	TXD1
GPIO 15	RXD0	RXD1
GPIO 16		CTS1
GPIO 17		RTS1
GPIO 18	PCM_CLK	PWM0
GPIO 19	PCM_FS	PWM1
GPIO 20	PCM_DIN	GPCLK0
GPIO 21	PCM_DOUT	GPCLK1

could also be useful for connecting the Raspberry Pi to other devices which support SPI. Also, the SDA and SCL functions could be used to communicate with I^{2}C devices.

11.3 Pulse modulation

The GPIO device provides a method for sending digital signals to external devices. This can be useful to control devices that have basically two states: on and off. In some situations, it is useful to have the ability to turn a device on at varying levels. For instance, it could be useful to control a motor at any required speed, or control the brightness of a light source. One way that this can be accomplished is through pulse modulation.

The basic idea is that the computer sends a stream of pulses to the device. The device acts as a low-pass filter, which averages the digital pulses into an analog voltage. By varying the

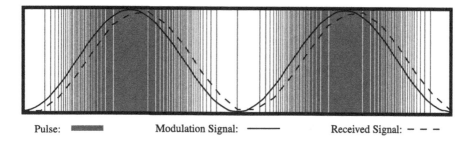

Figure 11.5: Pulse Density Modulation.

percentage of time that the pulses are high versus low, the computer can control how much average energy is sent to the device. The percentage of time that the pulses are high versus low is known as the *duty cycle*. Varying the duty cycle is referred to as *modulation*. There are two major types of pulse modulation: pulse density modulation (PDM) and pulse width modulation (PWM). Most pulse modulation devices are configured in three steps as follows:

1. The base frequency of the clock that drives the PWM device is configured. This step is usually optional.
2. The mode of operation for the pulse modulation device is configured by writing to one or more configuration registers in the pulse modulation device.
3. The cycle time is set by writing a "range" value into a register in the pulse modulation device. This value is usually set as a multiple of the base clock cycle time.

Once the device is configured, the duty cycle can be changed easily by writing to one or more registers in the pulse modulation device.

11.3.1 Pulse density modulation

With Pulse Density Modulation (PDM), also known as Pulse Frequency Modulation (PFM), the duration of the positive pulses does not change, but the time between them (the pulse density) is modulated. When using PDM devices, the programmer typically sets the device cycle time t_c in a register, then uses another register to specify the number of pulses d that are to be sent during a device cycle. The number of pulses is typically referred to as the duty cycle and must be chosen such that $0 \leq d \leq t_c$. For instance, if $t_c = 1024$, then the device cycle time is 1024 times the cycle time of the clock that drives the device. If $d = 512$, then the device will send 512 pulses, evenly spaced, during the device cycle. Each pulse will have the same duration as the base clock. The device will continue to output this pulse pattern until d is changed.

Fig. 11.5 shows a signal that is being sent using PDM, and the resulting set of pulses. Each pulse transfers a fixed amount of energy to the device. When the pulses arrive at the device,

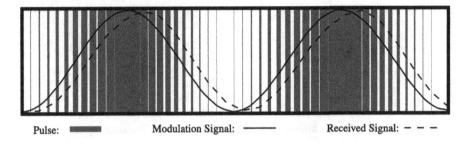

Pulse: ▬▬▬   Modulation Signal: ——   Received Signal: — — —

Figure 11.6: Pulse Width Modulation.

they are effectively filtered using a low pass filter. The resulting received signal is also shown. Notice that the received signal has a delay, or phase shift, caused by the low-pass filtering. This approach is suitable for controlling certain types of devices, such as lights and speakers.

However, when driving such devices directly with the digital pulses, care must be taken that the minimum frequency of pulses remains above the threshold that can be detected by human senses. For instance, when driving a speaker, the minimum pulse frequency must be high enough that the individual pulses cannot be distinguished by the human ear. This minimum frequency is around 40 KHz. Likewise, when driving an LED directly, the minimum frequency must be high enough that the eye cannot detect the individual pulses, because they will be seen as a flickering effect. That minimum frequency is around 70 Hz. To reduce or alleviate this problem, designers may add a low-pass filter between the PWM device and the device that is being driven.

11.3.2 Pulse width modulation

In pulse width modulation, the frequency of the pulses remains fixed, but the duration of the positive pulse (the pulse width) is modulated. When using PWM devices, the programmer typically sets the device cycle time t_c in a register, then uses another register to specify the number of base clock cycles, d, for which the output should be high. The percentage $\frac{d}{t_c} \times$ 100 is typically referred to as the duty cycle and d must be chosen such that $0 \leq d \leq t_c$. For instance, if $t_c = 1024$, then the device cycle time is 1024 times the cycle time of the clock that drives the device. If $d = 512$, then the device will output a high signal for 512 clock cycles, then output a low signal for 512 clock cycles. It will continue to repeat this pattern of pulses until d is changed.

Fig. 11.6 shows a signal that is being sent using pulse width modulation. The pulses are also shown. Each pulse transfers some energy to the device. The width of each pulse determines how much energy is transferred. When the pulses arrive at the device, they are effectively filtered using a low pass filter. The resulting received signal is shown by the dashed line. As with PDM, the received signal has a delay, or phase shift, caused by the low-pass filtering.

One advantage of PWM over PDM is that the digital circuit is not as complex. Another advantage of PWM over PDM is that the frequency of the pulses does not vary, so it is easier for the programmer to set the base frequency high enough that the individual pulses cannot be detected by human senses. Also, when driving motors it is usually necessary to match the pulse frequency to the size and type of motor. Mismatching the frequency can cause loss of efficiency, as well as overheating of the motor and drive electronics. In severe cases, this can cause premature failure of the motor and/or drive electronics. With PWM, it is easier for the programmer to control the base frequency, and thereby avoid those problems.

11.3.3 Raspberry Pi PWM device

The Broadcom BCM28xx system-on-chip includes a device that can create two PWM signals. One of the signals (PWM0) can be routed through GPIO pin 18 (alternate function 5), where it is available on the Raspberry Pi expansion header at pin 12. PWM0 can also be routed through GPIO pin 40. On the Raspberry Pi, pin 40 it is sent through a low-pass filter, and then to the Raspberry Pi audio output port as the right stereo channel. The other signal (PWM1) can be routed through GPIO pin 45. From there, it is sent through a low-pass filter, and then to the Raspberry Pi audio output port as the left stereo channel. So, both PWM channels are accessible, but PWM1 is only accessible through the audio output port, after it has been low-pass filtered. The raw PWM0 signal is available through the Raspberry Pi expansion header at pin 12.

There are three modes of operation for the BCM28xx PWM device:

1. pulse density modulation mode,
2. pulse width modulation mode, and
3. serial transmission mode.

The following paragraphs explain how the device can be used in basic PWM mode. This is the most simple and straightforward mode for this device. Information on how to use the PDM and serial transmission modes, the FIFO, and DMA is available in the *BCM2835 ARM Peripherals* manual.

Table 11.5:   Raspberry Pi PWM register map.

Offset	Name	Description	Size	R/W
00_{16}	PWMCTL	PWM Control	32	R/W
04_{16}	PWMSTA	PWM FIFO Status	32	R/W
08_{16}	PWMDMAC	PWM DMA Configuration	32	R/W
10_{16}	PWMRNG1	PWM Channel 1 Range	32	R/W
14_{16}	PWMDAT1	PWM Channel 1 Data	32	R/W
18_{16}	PWMFIF1	PWM FIFO Input	32	R/W
20_{16}	PWMRNG2	PWM Channel 2 Range	32	R/W
24_{16}	PWMDAT2	PWM Channel 2 Data	32	R/W

The base address of the PWM device is 0x3F20C000 and it contains eight registers. Table 11.5 shows the offset, name, and a short description for each of the registers. The mode of operation is selected for each channel independently by writing appropriate bits in the PWM-CTL register. The base clock frequency is controlled by the clock manager device, which is explained in Section 11.4.1. By default, the system startup code sets the base clock for the PWM device to 100 MHz.

Table 11.6 shows the names and short descriptions of the bits in the PWMCTL register. There are eight bits used for controlling channel 1 and eight bits for controlling channel 2. PWENn is the master enable bit for channel n. Setting that bit to 0 disables the PWM channel, while setting it to 1 enables the channel. MODEn is used to select whether the channel is in serial transmission mode or in the PDM/PWM mode. If MODEn is set to 0, then MSENn is used to choose whether channel n is in PDM mode or PWM mode. If MODEn is set to 1, then RPTLn, SBITn, USEFn, and CLRFn are used to manage the operation of the FIFO for channel n. POLAn is used to enable or disable inversion of the output signal for channel n.

The PWMRNGn registers are used to define the base period for the corresponding channel. In PDM mode, evenly distributed pulses are sent within a period of length defined by this register. The number of pulses sent during the base period is controlled by writing to the corresponding PWMDATn register. In PWM mode, this register defines the base frequency for the pulses, and the duty cycle is controlled by writing to the corresponding PWMDATn register. Example 19 gives an overview of the steps needed to configure PWM0 for use in pulse width modulation mode.

Table 11.6: Raspberry Pi PWM control register bits.

Bit	Name	Description	Values
0	PWEN1	Channel 1 Enable	0: Channel is disabled 1: Channel is enabled
1	MODE1	Channel 1 Mode	0: PDM or PWM mode 1: Serial mode
2	RPTL1	Channel 1 Repeat Last	0: Transmission stops when FIFO empty 1: Last data is sent repeatedly
3	SBIT1	Channel 1 Silence Bit	0: Output goes low when not transmitting 1: Output goes high when not transmitting
4	POLA1	Channel 1 Polarity	0: 0 is low voltage and 1 is high voltage 1: 1 is low voltage and 0 is high voltage
5	USEF1	Channel 1 Use FIFO	0: Data register is used 1: FIFO is used
6	CLRF1	Channel 1 Clear FIFO	Write 0: No effect Write 1: Causes FIFO to be emptied
7	MSEN1	Channel 1 PWM Enable	0: PDM mode 1: PWM mode
8	PWEN2	Channel 2 Enable	0: Channel is disabled 1: Channel is enabled
9	MODE2	Channel 2 Mode	0: PDM or PWM mode 1: Serial mode
10	RPTL2	Channel 2 Repeat Last	0: Transmission stops when FIFO empty 1: Last data is sent repeatedly
11	SBIT2	Channel 2 Silence Bit	0: Output goes low when not transmitting 1: Output goes high when not transmitting
12	POLA2	Channel 2 Polarity	0: 0 is low voltage and 1 is high voltage 1: 1 is low voltage and 0 is high voltage
13	USEF2	Channel 2 Use FIFO	0: Data register is used 1: FIFO is used
14	Unused	Reserved	
16	MSEN2	Channel 2 PWM Enable	0: PDM mode 1: PWM mode
16–31	Unused	Reserved	

In serial mode, serialized data is transmitted within the period. The data to be sent serially can be sent through the PWMDAT*n* register of through the PWMFIF1 register. If the value in PWMRNG*n* is less than 32, only the first PWMRNG*n* bits of serial data are sent. If it is larger than 32, then zero bits are added at the end of the data.

Example 19. Example of determining clock values on the Raspberry Pi.

Suppose we wish to use PWM0 to perform pulse width modulation, with a base frequency of 100 KHz, and the ability to control the duty cycle with a resolution of 0.1%. The steps would be as follows:

1. Verify that the clock manager device is configured to send a 100 MHz clock to the pulse modulator device through PWM_CLK.
2. To obtain a frequency of 100 KHz from a 100 MHz clock, it is necessary to divide by 1000. Therefore the second step is to store 1000 in the PWMRNG1 register.
3. Before enabling the PWM channel, it is prudent to initialize the duty cycle. The safest initial value is 0%, or completely off. This is accomplished by writing zero to the PWMDAT1 register.
4. Enable PWM channel 1 to operate in PWM mode by setting bit zero of PWM-CTL to 1, bit one of PWMCTL to 0, bit five of PWMCTL to 0, and bit seven of PWMCTL to 1.

Once this initialization is performed, we can set or change the duty cycle at any time by writing a value between 0 and 1000 to the PWMDAT1 register.

11.4 Common system devices

There are some classes of device that are found in almost every system, including the smallest embedded systems. Such common devices include hardware for managing the clock signals sent to other devices, and serial communications (typically RS232). Most mid-sized or large systems also include devices for managing virtual memory, managing the cache, driving a display, interfacing with keyboard and mouse, accessing disk and other storage devices, and networking. Small embedded systems may have devices for converting analog signals to digital, and vice-versa, pulse width modulation, and other purposes. Some systems, such as the Raspberry Pi, and pcDuino, have all or most of the devices of large systems, as well as most of the devices found on embedded systems. In this chapter, we look at two devices found on almost every system.

11.4.1 Clock management device

Very simple computer systems can be driven by a single clock. Most devices, including the CPU, are designed as state machines. The clock device sends a square-wave signal at a fixed frequency to all devices that need it. The clock signal tells the devices when to transition from

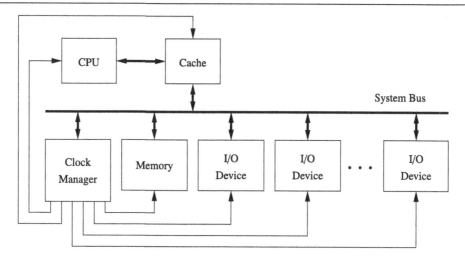

Figure 11.7: Typical system with a clock management device.

the current state to the next state. Without the clock signal, none of the devices would do anything.

More complex computers may contain devices which need to run at different rates. This requires the system to have separate clock signals for each device (or group of devices). System designers often solve this problem by adding a clock manager device to the system. This device allows the programmer to configure the clock signals that are sent to the other devices in the system. Fig. 11.7 shows a typical system. The clock manager, just like any other device, is configured by the CPU writing data to its registers using the system bus.

11.4.2 Raspberry Pi clock manager

The BCM28xx system-on-chip contains an ARM CPU and several devices. Some of the devices need their own clock to drive their operation at the correct frequency. Some devices such as serial communications receivers and transmitters, need configurable clocks so that the programmer has control over the speed of the device. To provide this flexibility and allow the programmer to have control over the clocks for each device, the BCM28xx includes a clock manager device, which can be used to configure the clock signals driving the other devices in the system.

The Raspberry Pi has a 19.2 MHz oscillator which can be used as a base frequency for any of the clocks. The BCM28xx also has three phase-locked-loop circuits that boost the oscillator to higher frequencies. Table 11.7 shows the frequencies that are available from various sources.

Table 11.7: Clock sources available for the clocks provided by the
clock manager.

Number	Name	Frequency	Note
0	GND	0 Hz	Clock is stopped
1	oscillator	19.2 MHz	
2	testdebug0	Unknown	Used for system testing
3	testdebug1	Unknown	Used for system testing
4	PLLA	650 MHz	May not be available
5	PLLC	200 MHz	May not be available
6	PLLD	500 MHz	
7	HDMI auxiliary	Unknown	
8-15	GND	0 Hz	Clock is stopped

Table 11.8: Some registers in the clock manager device.

Offset	Name	Description
070_{16}	CM_GP0_CTL	GPIO Clock 0 (GPCLK0) Control
074_{16}	CM_GP0_DIV	GPIO Clock 0 (GPCLK0) Divisor
078_{16}	CM_GP1_CTL	GPIO Clock 1 (GPCLK1) Control
$07c_{16}$	CM_GP1_DIV	GPIO Clock 1 (GPCLK1) Divisor
080_{16}	CM_GP2_CTL	GPIO Clock 2 (GPCLK2) Control
084_{16}	CM_GP2_DIV	GPIO Clock 2 (GPCLK2) Divisor
098_{16}	CM_PCM_CTL	Pulse Code Modulator Clock (PCM_CLK) Control
$09c_{16}$	CM_PCM_DIV	Pulse Code Modulator Clock (PCM_CLK) Divisor
$0a0_{16}$	CM_PWM_CTL	Pulse Modulator Device Clock (PWM_CLK) Control
$0a4_{16}$	CM_PWM_DIV	Pulse Modulator Device Clock (PWM_CLK) Divisor
$0f0_{16}$	CM_UART_CTL	Serial Communications Clock (UART_CLK) Control
$0f4_{16}$	CM_UART_DIV	Serial Communications Clock (UART_CLK) Divisor

Each device clock can be driven by one of the PLLs, the external 19.2 MHz oscillator, a signal from the HDMI port, or either of two test/debug inputs.

Among the clocks controlled by the clock manager device are the core clock (CM_VPU), the system timer clock (PM_TIME) which controls the speed of the system timer, the GPIO clocks which are documented in the Raspberry Pi peripheral documentation, the pulse modulator device clocks, and the serial communications clocks. It is generally not a good idea to modify the settings of any of the clocks without good reason.

The base address of the clock manager device is 0x3f101000. Some of the clock manager registers are shown in Table 11.8. Each clock is managed by two registers: a control register and a divisor. The control register is used to enable or disable a clock, to select which source oscillator drives the clock, and to select an optional multi-stage noise shaping (MASH) filter

Table 11.9: Bit fields in the clock manager control registers.

Bit	Name	Description
3–0	SRC	Clock source chosen from Table 11.7.
4	ENAB	Writing a 0 causes the clock to shut down. The clock will not stop immediately. The BUSY bit will be 1 while the clock is shutting down. When the BUSY bit becomes 0, the clock has stopped and it is safe to re-configure it. Writing a 1 to this bit causes the clock to start.
5	KILL	Writing a 1 to this bit will stop and reset the clock. This does not shut down the clock cleanly, and could cause a glitch in the clock output.
6	–	Unused.
7	BUSY	A 1 in this bit indicates that the clock is running.
8	FLIP	Writing a 1 to this bit will invert the clock output. Do not change this bit while the clock is running.
10–9	MASH	Controls how the clock source is divided. 00: Integer division 01: 1-stage MASH division 10: 2-stage MASH division 11: 3-stage MASH division Do not change this while the clock is running.
23–11	–	Unused.
31–24	PASSWD	This field must be set to $5A_{16}$ every time the clock control register is written to.

level. MASH filtering is useful for reducing the perceived noise when a clock is being used to generate an audio signal. In most cases, MASH filtering should not be used.

Table 11.9 shows the meaning of the bits in the control registers for each of the clocks. Likewise, Table 11.10 shows the layout for the bits in the clock divisor registers for each of the clocks. The procedure for configuring one of the clocks is:

1. Read the desired clock control register.
2. Clear bit 4 in the word that was read, then OR it with $5A000000_{16}$ and store the result back to the desired clock control register.
3. Repeatedly read the desired clock control register, until bit 7 becomes 0.
4. Calculate the divisor required and store it into the desired clock divisor register.
5. Create a word to configure and start the clock. Begin with $5A000000_{16}$, and set bits 3–0 to select the desired clock source. Set bits 10–9 to select the type of division, and set bit four to 1 to enable the clock.
6. Store the control word into the desired clock control register.

Selection of the divisor depends on which clock source is used, what type of division is selected, and the desired output of the clock being configured. For example, to set the PWM

Table 11.10: Bit fields in the clock manager divisor registers.

Bit	Name	Description
11–0	DIVF	Fractional part of divisor Do not change this while the clock is running.
23–12	DIVI	Integer part of divisor. Do not change this while the clock is running.
31–24	PASSWD	This field must be set to $5A_{16}$ every time the clock divisor register is written to.

clock to 100 KHz, the 19.20 MHz clock can be used. Dividing that clock by 192 will provide a 100 KHz clock. To accomplish this, it is necessary to stop the PWM clock as described, store the value $5A0C0000_{16}$ in the PWM clock divisor register, and then start the clock by writing $5A000011_{16}$ into the PWM clock control register.

11.5 Serial communications

There are basically two methods for transferring data between two digital devices: parallel and serial. Parallel connections use multiple wires to carry several bits at one time. Parallel connections typically include extra wires to carry timing information. Parallel communications are used for transferring large amounts of data over very short distances. However, this approach becomes very expensive when data must be transferred more than a few meters. Serial, on the other hand, uses a single wire to transfer the data bits one at a time. When compared to parallel transfer, the speed of serial transfer typically suffers. However, because it uses much fewer wires, the distance may be greatly extended, reliability improved, and cost vastly reduced.

11.5.1 UART

One of the oldest and most common devices for communications between computers and peripheral devices is the Universal Asynchronous Receiver/Transmitter, or UART. The word "universal" indicates that the device is highly configurable and flexible. UARTs allow a receiver and transmitter to communicate without a synchronizing signal.

The logic signal produced by the digital UART typically oscillates between zero volts for a low level and five volts for a high level, and the amount of current that the UART can supply is limited. For transmitting the data over long distances, the signals may go through a level-shifting or amplification stage. The circuit used to accomplish this is typically called a *line driver*. This circuit boosts the signal provided by the UART and also protects the delicate digital outputs from short circuits and signal spikes. Various standards, such as RS-232, RS-422 and RS-485 define the voltages that the line driver uses. For example, the RS-232 standard

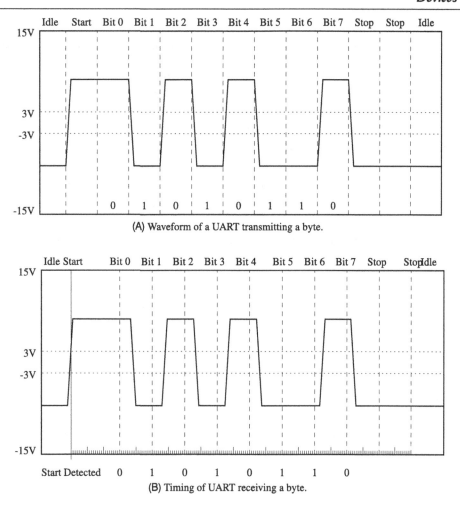

(A) Waveform of a UART transmitting a byte.

(B) Timing of UART receiving a byte.

Figure 11.8: Transmitter and receiver timings for two UARTS. The receiver clock is running slightly slower than the transmitter clock, but data is still received correctly.

specifies that valid signals are in the range of +3 to +15 volts, or −3 to −15 volts. The standards also specify the maximum time that is allowable when shifting from a high signal to a low signal and vice-versa, the amount of current that the device must be capable of sourcing and sinking, and other relevant design criteria.

The UART transmits data by sending each bit sequentially. The receiving UART re-assembles the bits into the original data. Fig. 11.8 shows how the transmitting UART converts a byte of data into a serial signal, and how the receiving UART samples the signal to recover the original data. Serializing the transmission and re-assembly of the data is accomplished using shift registers. The receiver and transmitter each have their own clocks, and are configured so that the clocks run at the same speed (or close to the same speed). In this case, the receiver's clock is running slightly slower than the transmitter's clock, but the data is still received correctly.

To transfer a group of bits, called a *data frame*, the transmitter typically first sends a *start bit*. Most UARTs can be configured to transfer between four and eight data bits in each group. The transmitting and receiving UARTs must be configured to use the same number of data bits. After each group of data bits, the transmitter will return the signal to the low state and keep it there for some minimum period. This period is usually the time that it would take to send two bits of data, and is referred to as the two *stop bits*. The stop bits allow the receiver to have some time to process the received byte and prepare for the next start bit. Fig. 11.8A shows what a typical RS-232 signal would look like when transferring the value 56_{16} (the ASCII 'V' character). The UART enters the idle state only if there is not another byte immediately ready to send. If the transmitter has another byte to send, then the start bit can begin at the end of the second stop bit.

Note that it is impossible to ensure that the receiver and transmitter have clocks which are running at exactly the same speed, unless they use the same clock signal. Fig. 11.8B shows how the receiver can reassemble the original data, even with a slightly different clock rate. When the start bit is detected by the receiver, it prepares to receive the data bits, which will be sent by the transmitter at an expected rate (within some tolerance). The receive circuitry of most UARTs is driven by a clock that runs 16 times as fast as the baud rate. The receive circuitry uses its faster clock to latch each bit in the middle of its expected time period. In Fig. 11.8B, the receiver clock is running slower than the transmitter clock. By the end of the data frame, the sample time is very far from the center of the bit, but the correct value is received. If the clocks differed by much more, or if more than eight data bits were sent, then it is very likely that incorrect data would be received. Thus, as long as their clocks are synchronized within some tolerance (which is dependent on the number of data bits and the baud rate), the data will be received correctly.

The RS-232 standard allows point-to-point communication between two devices for limited distances. With the RS-232 standard, simple one-way communications can be accomplished using only two wires: One to carry the serial bits, and another to provide a common ground. For bi-directional communication, three wires are required. In addition, the RS-232 standard specifies optional hand-shaking signals, which the UARTs can use to signal their readiness to transmit or receive data. The RS-422 and RS-485 standards allow multiple devices to be connected using only two wires.

The first UART device to enjoy widespread use was the 8250. The original version had 12 registers for configuration, sending, and receiving data. The most important registers are the ones that allow the programmer to set the transmit and receive bit rates, or baud. One baud is one bit per second. The baud is set by storing a 16 bit divisor in two of the registers in the

UART. The chip is driven by an external clock, and the divisor is used to reduce the frequency of the external clock to a frequency that is appropriate for serial communication. For example, if the external clock runs at 1 MHz, and the required baud is 1200, then the divisor must be $833.\overline{3} \approx 833$. Note that the divisor can only be an integer, so the device cannot achieve exactly 1200 baud. However, as explained previously, the sending and receiving devices do not have to agree precisely on the baud. During the transmission and reception of a byte, 1200.48 baud is close enough that the bits will be received correctly even if the other end is running slightly below 1200 baud. In the 8250, there was only one 8-bit register for sending data and only one 8-bit register for receiving data. The UART could send an interrupt to the CPU after each byte was transmitted or received. When receiving, the CPU had to respond to the interrupt very quickly. If the current byte was not read quickly enough by the CPU, it would be overwritten by the subsequent incoming byte. When transmitting, the CPU needed to respond quickly to interrupts to provide the next byte to be sent, or the transmission rate would suffer.

The next generation of UART device was the 16550A. This device is the model for most UART devices today. It features 16-byte input and output buffers, and the ability to trigger interrupts when a buffer is partially full or partially empty. This allows the CPU to move several bytes of data at a time, and results in much lower CPU overhead and much higher data transmission and reception rates. The 16550A also supports much higher baud rates than the 8250.

11.5.2 Raspberry Pi UART0

The BCM28xx system-on-chip provides two UART devices: UART0 and UART1. UART 1 is part of the I^2C device, and is not recommended for use as a UART. UART0 is a PL011 UART, which is based on the industry standard 16550A UART. The major differences are that the PL011 allows greater flexibility in configuring the interrupt trigger levels, the registers appear in different locations, and the locations of bits in some of the registers is different. So, although it operates very much like a 16550A, things have been moved to different locations. The transmit and receive lines can be routed through GPIO pin 14 and GPIO pin 15, respectively. UART0 has 18 registers, starting at its base address of $3f201000_{16}$. Table 11.11 shows the name, location, and a brief description for each of the registers.

UART_DR The UART Data Register is used to send and receive data. Data is sent or received one byte at a time. Writing to this register will add a byte to the transmit FIFO. Although the register is 32-bits, only the 8 least significant bits are used in transmission, and 12 least significant bits are used for reception. If the FIFO is empty, then the UART will begin transmitting the byte immediately. If the FIFO is full, then the last byte in the FIFO will be overwritten with the new byte that is written to the Data Register. When

Table 11.11:   Raspberry Pi UART0 register map.

Offset	Name	Description
00_{16}	UART_DR	Data Register
04_{16}	UART_RSRECR	Receive Status Register/Error Clear Register
18_{16}	UART_ FR	Flag register
20_{16}	UART_ILPR	not in use
24_{16}	UART_IBRD	Integer Baud rate divisor
28_{16}	UART_FBRD	Fractional Baud rate divisor
$2c_{16}$	UART_LCRH	Line Control register
30_{16}	UART_CR	Control register
34_{16}	UART_IFLS	Interrupt FIFO Level Select Register
38_{16}	UART_IMSC	Interrupt Mask Set Clear Register
$3c_{16}$	UART_RIS	Raw Interrupt Status Register
40_{16}	UART_MIS	Masked Interrupt Status Register
44_{16}	UART_ICR	Interrupt Clear Register
48_{16}	UART_DMACR	DMA Control Register
80_{16}	UART_ITCR	Test Control register
84_{16}	UART_ITIP	Integration test input reg
88_{16}	UART_ITOP	Integration test output reg
$8c_{16}$	UART_TDR	Test Data reg

Table 11.12:   Raspberry Pi UART data register.

Bit	Name	Description	Values
7–0	DATA	Data	Read: Last data received Write: Data byte to transmit
8	FE	Framing error	0: No error 1: The received character did not have a valid stop bit
9	PE	Parity error	0: No error 1: The received character did not have the correct parity, as set in the EPS and SPS bits of the Line Control Register (UART_LCRH)
10	BE	Break error	0: No error 1: A break condition was detected. The data input line was held low for longer than the time it would take to receive a complete byte, including the start and stop bits.
11	OE	Overrun error	0: No error 1: Data was not read quickly enough, and one or more bytes were overwritten in the input buffer
31–12	–	Not used	Write as zero, read as don't care

Table 11.13: Raspberry Pi UART Receive Status Register/Error Clear Register.

Bit	Name	Description	Values
0	FE	Framing error	0: No error 1: The received character did not have a valid stop bit
1	PE	Parity error	0: No error 1: The received character did not have the correct parity, as set in the EPS and SPS bits of the Line Control Register (UART_LCRH)
2	BE	Break error	0: No error 1: A break condition was detected. The data input line was held low for longer than the time it would take to receive a complete byte, including the start and stop bits.
3	OE	Overrun error	0: No error 1: Data was not read quickly enough, and one or more bytes were overwritten in the input buffer
31–4		Not used	Write as zero, read as don't care

this register is read, it returns the byte at the top of the receive FIFO, along with four additional status bits to indicate if any errors were encountered. Table 11.12 specifies the names and use of the bits in the UART Data Register.

UART_RSRECR   The UART Receive Status Register/Error Clear Register is used to check the status of the byte most recently read from the UART Data Register, and to check for overrun conditions at any time. The status information for overrun is set immediately when an overrun condition occurs. The Receive Status Register/Error Clear Register provides the same four status bits as the Data Register (but in bits 3–0 rather than bit 11–8). The received data character must be read first from the Data Register, before reading the error status associated with that data character from this register. Since the Data Register also contains these four bits, this register may not be required, depending on how the software is written. Table 11.13 describes the bits in this register.

UART_FR   The UART Flag Register can be read to determine the status of the UART. The bits in this register are used mainly when sending and receiving data using the FIFOs. When several bytes need to be sent, the TXFF flag should be checked to ensure that the transmit FIFO is not full before each byte is written to the data register. Wen receiving data, the RXFE bit can be used to determine whether or not there is more data to be read from the FIFO. Table 11.14 describes the flags in this register.

UART_ILPR   This is the IrDA register, which is supported by some PL011 UARTs. IrDA stands for the Infrared Data Association, which is a group of companies that cooperate to provide specifications for a complete set of protocols for wireless infrared communications. The name "IrDA" also refers to that set of protocols. IrDA is not implemented on the Raspberry Pi UART. Writing to this register has no effect and reading returns 0.

Table 11.14:   Raspberry Pi UART Flags Register bits.

Bit	Name	Description	Values
0	CTS	Clear To Send	0: Sender indicates they are ready to receive 1: Sender is NOT ready to receive
1	DSR	Data Set Ready	Not implemented: Write as zero, read as don't care
2	DCD	Data Carrier Detect	Not implemented: Write as zero, read as don't care
3	BUSY	UART is busy	0: UART is not transmitting data 1: UART is transmitting a byte
4	RXFE	Receive FIFO Empty	0: Receive FIFO contains bytes that have been received 1: Receive FIFO is empty
5	TXFF	Transmit FIFO is Full	0: There is room for at least one more byte in the transmit FIFO 1: Transmit FIFO is full – do not write to the data register at this time
6	RXFF	Receive FIFO is Full	0: There is no more room in the receive FIFO 1: There is still some space in the receive FIFO
7	TXFE	Transmit FIFO is Empty	0: There are no bytes waiting to be transmitted 1: There is at least one byte waiting to be transmitted
8	RI	Ring Indicator	Not implemented: Write as zero, read as don't care
31–9		Not used	Write as zero, read as don't care

Table 11.15:   Raspberry Pi UART integer baud rate divisor.

Bit	Name	Description	Values
15–0	IBRD	Integer Baud Rate Divisor	See Eq. (11.1)
31–16		Not used	Write as zero, read as don't care

Table 11.16:   Raspberry Pi UART fractional baud rate divisor.

Bit	Name	Description	Values
15–0	FBRD	Fractional Baud Rate Divisor	See Eq. (11.1)
31–16		Not used	Write as zero, read as don't care

UART_IBRD and UART_FBRD   UART_FBRD is the fractional part of the baud rate divisor value, and UART_IBRD is the integer part. The baud rate divisor is calculated as follows:

$$BAUDDIV = \frac{UARTCLK}{16 \times Baudrate} \tag{11.1}$$

where $UARTCLK$ is the frequency of the UART_CLK that is configured in the Clock Manager device. The default value is 3 MHz. $BAUDDIV$ is stored in two registers.

Table 11.17: Raspberry Pi UART line control register bits.

Bit	Name	Description	Values
0	BRK	Send Break	0: Normal operation 1: After the current character is sent, take the TXD output to a low level and keep it there
1	PEN	Parity Enable	0: Parity checking and generation is disabled 1: Generate and send parity bit and check parity on received data
2	EPS	Even Parity Select	0: Odd parity 1: Even parity
3	STP2	Two Stop Bits	0: Send one stop bit for each data word 1: Send two stop bits for each data word
4	FEN	FIFO Enable	0: Transmit and Receive FIFOs are disabled 1: Transmit and Receive FIFOs are enabled
6–5	WLEN	Word Length	00: 5 bits per data word 01: 6 bits per data word 10: 7 bits per data word 11: 8 bits per data word
31–7		Not used	Write as zero, read as don't care

UART_IBRD holds the integer part and UART_FBRD holds the fractional part. Thus $BAUDDIV$ should be calculated as a U(16,6) fixed point number. The contents of the UART_IBRD and UART_FBRD registers may be written at any time, but the change will not have any effect until transmission or reception of the current character is complete. Table 11.15 shows the arrangement of the integer baud rate divisor register, and Table 11.16 shows the arrangement of the fractional baud rate divisor register.

UART_LCRH UART_LCRH is the line control register. It is used to configure the communication parameters. This register must not be changed until the UART is disabled by writing zero to bit 0 of UART_CR, and the BUSY flag in UART_FR is clear. Table 11.17 shows the layout of the line control register.

UART_CR The UART Control Register is used for configuring, enabling, and disabling the UART. Table 11.18 shows the layout of the control register. To enable transmission, the TXE bit and UARTEN bit must be set to 1. To enable reception, the RXE bit and UARTEN bit, must be set to 1. In general, the following steps should be used to configure or re-configure the UART:

1. Disable the UART.
2. Wait for the end of transmission or reception of the current character.
3. Flush the transmit FIFO by setting the FEN bit to 0 in the Line Control Register.
4. Reprogram the Control Register.
5. Enable the UART.

Table 11.18: Raspberry Pi UART Control Register bits.

Bit	Name	Description	Values
0	UARTEN	UART Enable	0: UART disabled 1: UART enabled.
1	SIREN	Not used	Write as zero, read as don't care
2	SIRLP	Not used	Write as zero, read as don't care
3–6		Not used	Write as zero, read as don't care
7	LBE	Loopback Enable	0: Loopback disabled 1: Loopback enabled. Transmitted data is also fed back to the receiver.
8	TXE	Transmit enable	0: Transmitter is disabled 1: Transmitter is enabled
9	RXE	Receive enable	0: Receiver is disabled 1: Receiver is enabled
10	DTR	Not used	Write as zero, read as don't care
11	RTS	Complement of nUARTRTS	
12	OUT1	Not used	Write as zero, read as don't care
13	OUT2	Not used	Write as zero, read as don't care
14	RTSEN	RTS Enable	0: Hardware RTS disabled. 1: Hardware RTS Enabled
15	CTSEN	CTS Enable	0: Hardware CTS disabled. 1: Hardware CTS Enabled
16–31		Not used	Write as zero, read as don't care

Interrupt Control   The UART can signal the CPU by asserting an interrupt when certain conditions occur. This will be covered in more detail in Chapter 12. For now, it is enough to know that there are five additional registers which are used to configure and use the interrupt mechanism.

UART_IFLS defines the FIFO level that triggers the assertion of the interrupt signal. One interrupt is generated when the FIFO reaches the specified level. The CPU must clear the interrupt before another can be generated.

UART_IMSC is the interrupt mask set/clear register. It is used to enable or disable specific interrupts. This register determines which of the possible interrupt conditions are allowed to generate an interrupt to the CPU.

UART_RIS is the raw interrupt status register. It can be read to raw status of interrupts conditions, before any masking is performed.

UART_MIS is the masked interrupt status register. It contains the masked status of the interrupts. This is the register that the operating system should use to determine the cause of a UART interrupt.

UART_ICR is the interrupt clear register. Writing to it clears the interrupt conditions. The operating system should use this register to clear interrupts before returning from the interrupt service routine.

UART_DMACR The DMA control register is used to configure the UART to access memory directly, so that the CPU does not have to move each byte of data to or from the UART. DMA will be explained in more detail in Chapter 12.

Additional Registers The remaining registers: UART_ITCR, UART_ITIP, and UART_ITOP, are either unimplemented or are used for testing the UART. These registers should not be used.

11.5.3 Basic programming for the Raspberry Pi UART

Listing 11.3 shows four basic functions for initializing the UART, changing the baud rate, sending a character, and receiving a character using UART0 on the Raspberry Pi. Note that a large part of the code simply defines the location and offset for all of the registers (and bits) that can be used to control the UART.

Listing 11.3 Assembly functions for using the Raspberry Pi UART.

```
        // offsets to the UART registers
        .equ    UART_DR,        0x00 // data register
        .equ    UART_RSRECR,    0x04 // Recieve Status/Error clear
        .equ    UART_FR,        0x18 // flag register
        .equ    UART_ILPR,      0x20 // not used
        .equ    UART_IBRD,      0x24 // integer baud rate divisor
        .equ    UART_FBRD,      0x28 // fractional baud rate divisor
        .equ    UART_LCRH,      0x2C // line control register
        .equ    UART_CR,        0x30 // control register
        .equ    UART_IFLS,      0x34 // interrupt FIFO level select
        .equ    UART_IMSC,      0x38 // Interrupt mask set clear
        .equ    UART_RIS,       0x3C // raw interrupt status
        .equ    UART_MIS,       0x40 // masked interrupt status
        .equ    UART_ICR,       0x44 // interrupt clear register
        .equ    UART_DMACR,     0x48 // DMA control register
        .equ    UART_ITCR,      0x80 // test control register
        .equ    UART_ITIP,      0x84 // integration test input
        .equ    UART_ITOP,      0x88 // integration test output
        .equ    UART_TDR,       0x8C // test data register

        // error condition bits when reading the DR (data register)
        .equ    UART_OE,        (1<<11) // overrun error bit
        .equ    UART_BE,        (1<<10) // break error bit
        .equ    UART_PE,        (1<<9)  // parity error bit
        .equ    UART_FE,        (1<<8 ) // framing error bit
```

```
27
28          // Bits for the FR (flags register)
29          .equ    UART_RI,        (1<<8) // Unsupported
30          .equ    UART_TXFE,      (1<<7) // Transmit FIFO empty
31          .equ    UART_RXFF,      (1<<6) // Receive FIFO full
32          .equ    UART_TXFF,      (1<<5) // Transmit FIFO full
33          .equ    UART_RXFE,      (1<<4) // Receive FIFO empty
34          .equ    UART_BUSY,      (1<<3) // UART is busy xmitting
35          .equ    UART_DCD,       (1<<2) // Unsupported
36          .equ    UART_DSR,       (1<<1) // Unsupported
37          .equ    UART_CTS,       (1<<0) // Clear to send
38
39          // Bits for the LCRH (line control register)
40          .equ    UART_SPS,       (1<<7) // enable stick parity
41          .equ    UART_WLEN1,     (1<<6) // MSB of word length
42          .equ    UART_WLEN0,     (1<<5) // LSB of word length
43          .equ    UART_FEN,       (1<<4) // Enable FIFOs
44          .equ    UART_STP2,      (1<<3) // Use 2 stop bits
45          .equ    UART_EPS,       (1<<2) // Even parity select
46          .equ    UART_PEN,       (1<<1) // Enable parity
47          .equ    UART_BRK,       (1<<0) // Send break
48
49          // Bits for the CR (control register)
50          .equ    UART_CTSEN,     (1<<15) // Enable CTS
51          .equ    UART_RTSEN,     (1<<14) // Enable RTS
52          .equ    UART_OUT2,      (1<<13) // Unsupported
53          .equ    UART_OUT1,      (1<<12) // Unsupported
54          .equ    UART_RTS,       (1<<11) // Request to send
55          .equ    UART_DTR,       (1<<10) // Unsupported
56          .equ    UART_RXE,       (1<<9)  // Enable receiver
57          .equ    UART_TXE,       (1<<8)  // Enable transmitter
58          .equ    UART_LBE,       (1<<7)  // Enable loopback
59          .equ    UART_SIRLP,     (1<<2)  // Unsupported
60          .equ    UART_SIREN,     (1<<1)  // Unsupported
61          .equ    UART_UARTEN,    (1<<0)  // Enable UART
62
63          .text
64          .align  2
65 // ------------------------------------------------------------
66          .global UART_put_byte
67 UART_put_byte:
68          adr     x1,uartbase     // load address of pointer
69          ldr     x1,[x1]         // load base address of UART
70 putlp:   ldr     x2,[x1,#UART_FR] // read the flag resister
71          tst     x2,#UART_TXFF   // check if transmit FIFO is full
72          bne     putlp           // loop while transmit FIFO is full
73          str     x0,[x1,#UART_DR] // write the char to the FIFO
74          ret
75
```

```
 76    // ------------------------------------------------------------
 77            .global UART_get_byte
 78    UART_get_byte:
 79            adr     x1,uartbase       // load address of pointer
 80            ldr     x1,[x1]           // load base address of UART
 81    getlp:  ldr     x2,[x1,#UART_FR]  // read the flag resister
 82            tst     x2,#UART_RXFE     // check if receive FIFO is empty
 83            bne     getlp             // loop while receive FIFO is empty
 84            ldr     w0,[x1,#UART_DR]  // read the char from the FIFO
 85            tst     w0,#UART_OE       // check for overrun error
 86            bne     get_ok1
 87            // handle receive overrun error here - does nothing now
 88    get_ok1:
 89            tst     w0,#UART_BE       // check for break error
 90            bne     get_ok2
 91            // handle receive break error here - does nothing now
 92    get_ok2:
 93            tst     w0,#UART_PE       // check for parity error
 94            bne     get_ok3
 95            // handle receive parity error here - does nothing now
 96    get_ok3:
 97            tst     w0,#UART_FE       // check for framing error
 98            bne     get_ok4
 99            // handle receive framing error here - does nothing now
100    get_ok4:
101            ret     // return the character that was received
102
103    // ------------------------------------------------------------
104    // UART init will set default values:
105    // 115200 baud, no parity, 2 stop bits, 8 data bits
106            .global UART_init
107    UART_init:
108            adr     x1,uartbase       // load address of pointer
109            ldr     x1,[x1]           // load base address of UART
110            // set baud rate divisor
111            // (3MHz / ( 115200 * 16 )) = 1.62760416667
112            // = 1.101000 in binary
113            mov     w0,#1
114            str     w0,[x1,#UART_IBRD]
115            mov     w0,#0x28
116            str     w0,[x1,#UART_FBRD]
117            // set parity, word length, enable FIFOS
118            .equ BITS, (UART_WLEN1|UART_WLEN0|UART_FEN|UART_STP2)
119            mov     w0,#BITS
120            str     w0,[x1,#UART_LCRH]
121            // mask all UART interrupts
122            mov     w0,#0
123            str     w0,[x1,#UART_IMSC]
124            // enable receiver and transmitter and enable the uart
```

```
125        .equ FINALBITS, (UART_RXE|UART_TXE|UART_UARTEN)
126        ldr    w0,=FINALBITS
127        str    w0,[x1,#UART_CR]
128        // return
129        ret
130
131 // -----------------------------------------------------------
132 // UART_set_baud will change the baud rate to whatever is in w0
133 // The baud rate divisor is calculated as follows: Baud rate
134 // divisor BAUDDIV = (FUARTCLK/(16 Baud rate)) where FUARTCLK
135 // is the UART reference clock frequency. The BAUDDIV
136 // is comprised of the integer value IBRD and the
137 // fractional value FBRD. NOTE: The contents of the
138 // IBRD and FBRD registers are not updated until
139 // transmission or reception of the current character
140 // is complete.
141        .global UART_set_baud
142 UART_set_baud:
143        // set baud rate divisor using formula:
144        // (3000000.0 / ( W0 * 16 ))  ASSUMING 3Mhz clock
145        lsl    x1,x0,#4        // x1 <- desired baud * 16
146        ldr    w0,=(3000000<<6)// Load 3 MHz as a U(26,6) in w0
147        bl     divide          // divide clk freq by (baud*16)
148        asr    x1,x0,#6        // put integer divisor into x1
149        and    w0,w0,#0x3F     // put fractional divisor into w0
150        adr    x2,uartbase     // load base address of UART
151        ldr    x2,[x2]         // load base address of UART
152        str    w1,[x2,#UART_IBRD] // set integer divisor
153        str    w0,[x2,#UART_FBRD] // set fractional divisor
154        ret
```

11.6 Chapter summary

All input and output is accomplished by using devices. There are many types of device, and each device has its own set of registers which are used to control the device. The programmer must understand the operation of the device and the use of each register in order to use the device at a low level. Computer system manufacturers usually can provide documentation providing the necessary information for low-level programming. The quality of the documentation can vary greatly, and a general understanding of various types of devices can help in deciphering poor or incomplete documentation.

There are two major tasks where programming devices at the register level is required: operating system drivers and very small embedded systems. Operating systems provide an abstract view of each device and this allows programmers to use them more easily. However, someone must write that driver, and that person must have intimate knowledge of the device. On very small systems, there may not be a driver available. In that case, the device must be accessed directly. Even when an operating system provides a driver, sometimes it is necessary or desirable for the programmer to access the device directly. For example, some devices may provide modes of operation or capabilities that are not supported by the operating system driver. Linux provides a mechanism which allows the programmer to map a physical device into the program's memory space, and thereby gain access to the raw device registers.

Pulse modulation is a group of methods for generating analog signals using digital equipment. Pulse modulation is commonly used in control systems to regulate the power sent to motors and other devices. Pulse modulation techniques can have very low power loss compared to other methods of controlling analog devices, and the circuitry required is relatively simple.

The cycle frequency must be programmed to match the application. Typically, 10 Hz is adequate for controlling an electric heating element, while 120 Hz would be more appropriate for controlling an incandescent light bulb. Large electric motors may be controlled with a cycle frequency as low as 100 Hz, while smaller motors may need frequencies around 10,000 Hz. It can take some experimentation to find the best frequency for any given application.

Most modern computer systems have some type of Universal Asynchronous Receiver/Transmitter. These are serial communications devices, and are meant to provide communications with other systems using RS-232 (most commonly) or some other standard serial protocol. Modern systems often have a large number of other devices as well. Each device may need it's own clock source, to drive it at the correct frequency for its operation. The clock sources for all of the devices are often controlled by yet another device: the clock manager.

Although two systems may have different UARTs, these devices perform the same basic functions. The specifics about how they are programmed will vary from one system to another. However, there is always enough similarity between devices of the same class that a programmer who is familiar with one specific device can easily learn to program another similar device. The more experience a programmer has, the less time it takes to learn how to control a new device.

Exercises

11.1. Explain the relationships and differences between device registers, memory locations, and CPU registers.

11.2. Why is it necessary to map the device into user program memory before accessing it under Linux? Would this step be necessary under all operating systems or in the case where there is no operating system and our code is running on the "bare metal?"

11.3. What is the purpose of a GPIO device?

11.4. Draw a circuit diagram showing how to connect a push-button switch to GPIO 23 and an LED to GPIO 27 on the Raspberry Pi.

11.5. Assuming the system is wired according to the previous exercise, write two functions. One function must initialize the GPIO pins, and the other function must read the state of the switch and turn the LED on if the button is pressed, and off if the button is not pressed.

11.6. Write the code necessary to route the output from PWM0 to GPIO 18 on a Raspberry Pi.

11.7. Write ARM assembly programs to configure PWM0 and the GPIO device to send a signal out on Raspberry Pi header pin 12 with:
   a. period of 1 ms and duty cycle of 25%, and
   b. frequency of 150 Hz and duty cycle of 63%.

11.8. Write a function for setting the PWM clock on the Raspberry Pi to 2 MHz.

11.9. The UART_GET_BYTE function in Listing 11.3 contains skeleton code for handling errors, does not actually do anything when errors occur. Describe at least two ways that the errors could be handled.

## CHAPTER 12

Running without an operating system

The previous chapters assumed that the software would be running in user mode under an operating system. Sometimes, it is necessary to write assembly code to run on "bare metal," which simply means: without an operating system. For example, when we write an operating system kernel, it must run on bare metal and a significant part of the code (especially during the boot process) must be written in assembly language. Coding on bare metal is useful to deeply understand how the hardware works and what happens in the lowest levels of an operating system. There are some significant differences between code that is meant to run under an operating system and code that is meant to run on bare metal.

The operating system takes care of many details for the programmer. For instance, it sets up the stack, text, and data sections, initializes static variables, provides an interface to input and output devices and gives the programmer an abstracted view of the machine. When accessing data on a disk drive, the programmer uses the file abstraction. The underlying hardware only knows about blocks of data. The operating system provides the data structures and operations which allow the programmer to think of data in terms of files and streams of bytes. A user program may be scattered in physical memory, but the hardware memory management unit, managed by the operating system, allows the programmer to view memory as a simple memory map (such as shown in Fig. 1.7). The programmer uses *system calls* to access the abstractions provided by the operating system. On bare metal, there are no abstractions, unless the programmer creates them.

However, there are some software packages to help bare-metal programmers. For example, Newlib is a C standard library intended for use in bare-metal programs. Its major features are that:

- it implements the hardware-independent parts of the standard C library,
- for I/O, it relies on only a few low-level functions that must be implemented specifically for the target hardware, and
- many target machines are already supported in the Newlib source code.

To support a new machine, the programmer only has to write a few low-level functions in C and/or Assembly, which will initialize the system and perform low-level I/O on the target hardware.

ARM 64-Bit Assembly Language
https://doi.org/10.1016/B978-0-12-819221-4.00019-5
445

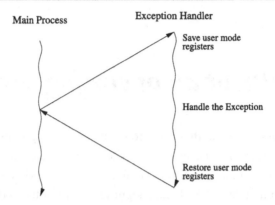

Figure 12.1: Basic exception processing.

12.1 Exception processing

Many early computers were not capable of protecting the operating system from user programs. That problem was solved mostly by building CPUs that support multiple "levels of privilege" for running programs. Almost all modern CPUs have the ability to operate in at least two modes:

User mode   is the mode that normal user programs use when running under an operating system, and

Privileged mode   is reserved for operating system code or exception handling. There are operations that can be performed in privileges mode which cannot be performed in user mode.

Many bare metal programs consist of a single thread of execution running in user mode to perform some task. This main program is occasionally interrupted by the occurrence of some exception. The exception is processed, and then control returns to the main thread. The main thread usually runs with user privileges, and exception processing is done in privileged mode.

Fig. 12.1 shows the sequence of events when an exception occurs in such a system. The main program typically would be running with the CPU in user mode. When the exception occurs, the CPU uses the *vector table* to find and execute the appropriate exception handler. Different CPU architectures have different methods for implementing vector tables, but all modern CPUs support some type of vector table. The exception handler must save any registers that it is going to use, execute the code required to handle the exception, then restore the registers. When it returns to the user mode process, everything will be as it was before the exception occurred. The user mode program continues executing as if the exception never occurred.

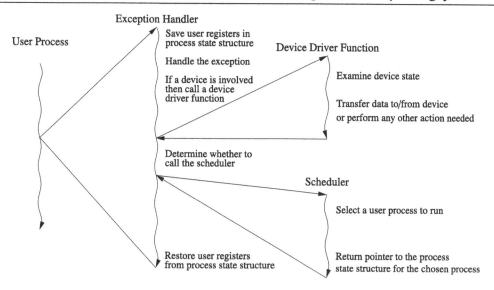

Figure 12.2: Exception processing with multiple user processes.

More complex systems may have multiple tasks, threads of execution, or user processes running concurrently. In a single-processor system, only one task, thread, or user process can actually be executing at any given instant, but when an exception occurs, the exception handler may change the currently active task, thread, or user process. This is the basis for all modern multi-processing systems. Fig. 12.2 shows how an exception may be processed on such a system. It is common on multi-processing systems for a timer device to be used to generate periodic interrupts, which allows the currently active task, thread, or user process to be changed at a fixed frequency.

Privileged mode is entered automatically by the hardware when certain exceptional circumstances occur. For example, when a hardware device needs attention, it can signal the processor by causing an interrupt. When this occurs, the processor immediately sets the appropriate exception level and begins executing the IRQ exception handler function. The exception handlers (or their addresses) are stored in a *vector table* at a known location in memory. When an exception occurs, the CPU automatically uses the vector table to find and execute the appropriate exception handler function.

12.2 AArch64 execution and exception states

The AArch64 processor provides two major modes of operation, referred to as *execution states*. They are 32-bit AArch32 state, and 64-bit AArch64 state. Both of these execution

Table 12.1:   The ARM User and System Registers.

	ELO	EL1	EL2	EL3
Stack Pointer	SP_ELO	SP_EL1	SP_EL2	SP_EL3
Exception Link Register		ELR_EL1	ELR_EL2	ELR_EL3
Saved PSTATE		SPSR_EL1	SPSR_EL2	SPSR_EL3

states provide privileged modes and a user mode. The AArch32 execution state allows the processor to execute code written for the ARMv7 and older processors.

12.2.1 AArch64 exception levels

In the AArch64 execution state, there are three privileged modes and one user mode. These are referred to as *exception levels*. The higher the exception level, the more privilege the code has. Typically, the system uses the exception levels as follows:

EL0   User applications,
EL1   OS kernel and other privileged code,
EL2   Hypervisor (support for virtual machines), and
EL3   Secure monitor (manage security contexts).

The major difference between EL0 and the higher levels is that code executing in EL0 cannot access system registers. EL1 can access most system registers, EL2 has additional privileges, and EL3 has all privileges. The only way that the processor can change from one exception level to a higher level is when an exception occurs. The only way that the processor can move to a lower exception level is by executing an exception return instruction. When changing the exception level, it is also possible to switch between AArch64 and AArch32 execution state. The processor also supports two security states: Secure and non-secure. EL3 is meant to manage the security state, and EL2 is meant to provide virtual machine capabilities. In many situations, only EL0 and EL1 are required, and some processors may not provide EL2 and/or EL3. On power-up and on reset, the processor enters the highest available exception level.

Each exception level has its own stack pointer, link register, and saved process state register (SPSR). Table 12.1 shows the names of these *banked* registers. When the exception level changes, the corresponding link register and stack pointer become active, and "replace" the user stack pointer and link register. Also, when an exception occurs, the current PSTATE register is copied into the SPSR for the exception level that is being entered. When an exception occurs:

1. The current PSTATE is copied to SPSR_ELn where n is the exception level being entered,
2. the PSTATE register is updated,

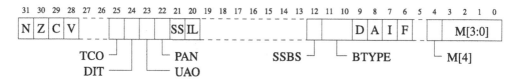

Figure 12.3: The PSTATE register.

3. the exception level stays the same or gets higher,
4. the return address is stored to ELR_ELn,
5. the program counter (PC) is set to the appropriate vector address, and
6. if it is a synchronous or SError exception, ESR_ELn is updated with the cause of exception.

To return from an exception, the `eret` instruction is executed. This instruction:

1. restores the PSTATE from SPSR_ELn, and
2. restores the program counter (PC) from ELR_ELn.

Software executing in an exception level higher than EL0 can select whether to use the default SP_ELx stack pointer or the SP_EL0 stack pointer by executing an

```
msr SPSel, #Imm1
```

instruction. This allows privileged code to access and modify the user mode stack. For example, when an exception occurs, registers can be saved on the user mode stack, then restored before going back to user mode.

12.2.2 System control and status registers

The PSTATE flags were introduced in Chapter 3. The PSTATE register also contains other fields that are used by system (or bare metal) code. Fig. 12.3 shows all of the fields in the PSTATE register. These fields are briefly described below:

TCO Tag Check Override is available in ARMv8.5 and above with the MemTag extension.
DIT Data Independent Timing is available in ARMv8.4 and above with the DIT extension.
UAO User Access Override is available in ARMv8.2 and above with the UAO extension.
PAN Privileged Access Never is available in ARMv8.1 and above with the PAN extension.
SS Software Step.
IL Illegal Execution state.

Table 12.2:   Mode bits in the PSTATE register.

M[3:0]	EL	SP
0000	EL0	SP_EL0
0100	EL1	SP_EL0
0101	EL1	SP_EL1
1000	EL2	SP_EL0
1001	EL2	SP_EL2
1100	EL3	SP_EL0
1101	EL3	SP_EL3

SSBS   Speculative Store Bypass is available in ARMv8.0 and above with the SSBS extension.

BTYPE   Branch Type Indicator is available in ARMv8.5 and above with the BTI extension.

D   Debug exception mask.

A   SError interrupt mask.

I   IRQ interrupt mask.

F   FIQ interrupt mask.

M[4]   Execution state (Mode bit 4). A zero in this bit indicates AArch64 state. One indicates AArch32 state.

M[3:0]   AArch64 Exception level and selected Stack Pointer. The values for this field are shown in Table 12.2

The most important fields are the Mode bits, which can be examined to determine which stack pointer is currently selected, which execution state the processor is in, and the current exception level. Some of the fields can also be changed by code running in privileged mode. Table 12.2 shows valid values for the Mode bits.

For exception levels other than EL0, the execution state (AArch32 or AArch64) is determined by one or more control register configuration bits. These bits can be set only in execution states higher than EL0. For EL0, the execution state is determined as part of the exception return to EL0, under the control of the exception level that the execution is returning from. Higher exception levels can change the execution state, but EL0 cannot.

There are some additional configuration and status registers available in higher exception levels. They include:

SCTLR_ELn   The System Control Register controls architectural features. For example the Memory Management Unit (MMU), caches, and memory alignment checking.

ACTLR_ELn   The Auxiliary Control Register controls processor specific features.

SCR_EL3   The Secure Configuration Register controls secure state and trapping of exceptions to EL3.

HCR_EL2   The Hypervisor Configuration Register controls virtualization settings, and trapping of exceptions to EL2.

MIDR_EL1   The Main ID Register describes the type of processor the code is running on.

MPIDR_EL1   The Multiprocessor Affinity Register contains the core and cluster ID of the core/cluster that the code is running on, in multi-core/cluster systems.

CTR_EL0   The Cache Type register contains information about the integrated caches.

12.3 AArch64 vector table

In AArch64, exceptions are divided into two major types:

Synchronous   exceptions are synchronized with the CPU clock. This type of exception includes data aborts from MMU, permission/alignment failures for load and store operations, service call instructions, and illegal instructions.

Asynchronous   exceptions may occur at any time, and may not be synchronized with the CPU clock. These exceptions include interrupt requests (IRQ and FIQ) from external devices, or SError (System Error) interrupts.

When an exception is processed, the exception level can stay the same or go to a higher exception level. If an exception occurs in EL0, then the new exception level must be higher than EL0. Synchronous exceptions are normally handled in the current exception level, except for EL0. Asynchronous exceptions can be routed to a higher exception level. Preparing to handle exceptions requires:

- setting up the vector table, and
- configuring asynchronous exception routing and masking.

Each exception must be handled by a dedicated function. Each function in the vector table may have up to 32 instructions. If the exception handler is more than 32 instructions, then a branch instruction is used to jump to a longer exception handler. Each exception level except for EL0 has its own vector table. The addresses of the vector tables are stored in three registers: VBAR_EL3, VBAR_EL2, and VBAR_EL1. Each vector table has sixteen entries, and each entry is 128 bytes long, giving room for 32 instructions. Vector tables must be aligned to a 2 kilobyte boundary.

Fig. 12.4 shows how each vector table is structured. Note that the figure has the lowest address at the bottom. When writing the assembly code, the bottom entry in the table comes

0x780	SError/vSError	
0x700	FIQ/vFIQ	Lower EL using AArch32
0x680	IRQ/vIRQ	
0x600	Synchronous	
0x580	SError/vSError	
0x500	FIQ/vFIQ	Lower EL using AArch64
0x480	IRQ/vIRQ	
0x400	Synchronous	
0x380	SError/vSError	
0x300	FIQ/vFIQ	Current EL with SPn
0x280	IRQ/vIRQ	
0x200	Synchronous	
0x180	SError/vSError	
0x100	FIQ/vFIQ	Current EL with SP0
0x080	IRQ/vIRQ	
0x000	Synchronous	

Figure 12.4: AArch64 vector table structure.

first, so that it gets the lowest address (at offset **0x000** from the start of the table. It is followed by the handler that is to be placed at **0x080**, relative to the start of the table, and so on.

12.3.1 Creating the vector tables

Listing 12.1 shows 16 functions making up one vector table. Each exception level, (EL1, EL2, and EL3) may have it's own vector table, for a total of 48 exception handler functions. Note that each exception handler ends with an **eret** (exception return) instruction. Up to 31 instructions can be inserted before the exception return instruction. If more instructions are needed, then they can be placed in a subroutine, and a **bl** instruction can be inserted in the vector table code.

Listing 12.1 Stubs for the exception handlers.

```
// FILE: handlers.s
        .text
        .align  2
// Typical exception vector table code.
//------------------------------------------------
        .balign 0x800  // align to a 2 kilobyte boundary
        .global Vector_table_el1
Vector_table_el1:
```

```
 9   curr_el_sp0_sync:
10           // The exception handler for the synchronous
11           // exception from the current EL using SP0.
12           eret    // return from exception
13
14           .balign 0x80    // align to next 128 byte boundary
15   curr_el_sp0_irq:
16           // The exception handler for the IRQ exception
17           // from the current EL using SP0.
18           eret    // return from exception
19
20           .balign 0x80    // align to next 128 byte boundary
21   curr_el_sp0_fiq:
22           // The exception handler for the FIQ exception
23           // from the current EL using SP0.
24           eret    // return from exception
25
26           .balign 0x80    // align to next 128 byte boundary
27   curr_el_sp0_serror:
28           // The exception handler for the system error
29           // exception from the current EL using SP0.
30           eret    // return from exception
31
32           .balign 0x80    // align to next 128 byte boundary
33   curr_el_spx_sync:
34           // The exception handler for the synchronous
35           // exception from the current EL using the
36           // current SP.
37           eret    // return from exception
38
39           .balign 0x80    // align to next 128 byte boundary
40   curr_el_spx_irq:
41           // The exception handler for IRQ exception
42           // from the current EL using the current SP.
43           eret    // return from exception
44
45           .balign 0x80    // align to next 128 byte boundary
46   curr_el_spx_fiq:
47           // The exception handler for the FIQ exception
48           // from the current EL using the current SP.
49           eret    // return from exception
50
51           .balign 0x80    // align to next 128 byte boundary
52   curr_el_spx_serror:
53           // The exception handler for the system error
54           // exception from the current EL using the
55           // current SP.
56           eret    // return from exception
57
```

```
58        .balign 0x80    // align to next 128 byte boundary
59  lower_el_aarch64_sync:
60        // The exception handler for the synchronous
61        // exception from a lower EL (AArch64).
62        eret    // return from exception
63
64        .balign 0x80    // align to next 128 byte boundary
65  lower_el_aarch64_irq:
66        // The exception handler for the IRQ exception
67        // from a lower EL (AArch64).
68        eret    // return from exception
69
70        .balign 0x80    // align to next 128 byte boundary
71  lower_el_aarch64_fiq:
72        // The exception handler for the FIQ exception
73        // from a lower EL (AArch64).
74        eret    // return from exception
75
76        .balign 0x80    // align to next 128 byte boundary
77  lower_el_aarch64_serror:
78        // The exception handler for the system error
79        // exception from a lower EL(AArch64).
80        eret    // return from exception
81
82        .balign 0x80    // align to next 128 byte boundary
83  lower_el_aarch32_sync:
84        // The exception handler for the synchronous
85        // exception from a lower EL(AArch32).
86        eret    // return from exception
87
88        .balign 0x80    // align to next 128 byte boundary
89  lower_el_aarch32_irq:
90        // The exception handler for the IRQ exception
91        // from a lower EL (AArch32).
92        eret    // return from exception
93
94        .balign 0x80    // align to next 128 byte boundary
95  lower_el_aarch32_fiq:
96        // The exception handler for the FIQ exception
97        // from a lower EL (AArch32).
98        eret    // return from exception
99
100       .balign 0x80    // align to next 128 byte boundary
101 lower_el_aarch32_serror:
102       // The exception handler for the system error
103       // exception from a lower EL(AArch32).
104       eret    // return from exception
```

12.3.2 Using the vector tables

Listing 12.2 shows how the processor is initialized to use the vector tables. This listing assumes that three vector tables have been created, each of which follows the pattern shown in Listing 12.1. The three vector table pointer registers are initialized by loading the address of a vector table into a general purpose register, then transferring the address to a vector table register.

Listing 12.2 Code fragment to initialize the exception vector tables.

```
// This code initializes the vector table pointers
// Initialize VBAR_EL3.
ldr     x1, =vector_table_el3
msr     VBAR_EL3, x1
// Initialize VBAR_EL2.
ldr     x1, =vector_table_el2
msr     VBAR_EL2, x1
// Initialize VBAR_EL1.
ldr     x1, =vector_table_el1
msr     VBAR_EL1, x1
```

12.3.3 Configuring asynchronous exceptions

Asynchronous exceptions, or "hardware interrupts" may occur at any time, and are used by hardware devices to signal that they require service. On the AArch64 processor, there are three bits in the PSTATE register which affect asynchronous exception processing:

I: when set to one, normal hardware interrupts are disabled,
F: when set to one, fast hardware interrupts are disabled, and
A: when set to one, system error interrupts are disabled.

Programs running in EL0 cannot modify these bits. Therefore, the operating system is guaranteed to gain control of the CPU whenever an interrupt occurs. The user program cannot disable interrupts and continue to run. Most operating systems use a hardware timer to generate periodic interrupts, thus the operating system is able to regain control of the CPU every few milliseconds. This prevents a malicious or defective user program from taking control of the CPU.

On startup or reset, the asynchronous exceptions are disabled. Before they can be enabled, routing rules must be set, and external devices must be configured to signal interrupts. Section 12.6.2 provides an overview of configuring the interrupt controller on the Raspberry Pi.

Listing 12.3 Code to route SError, IRQ and FIQ to EL3.

```
mrs    x0, SCR_EL3
orr    x0, x0, #(1<<3)  // Set the EA bit.
orr    x0, x0, #(1<<1)  // Set the IRQ bit.
orr    x0, x0, #(1<<2)  // Set the FIQ bit.
msr    SCR_EL3, x0
```

Listing 12.4 Code to route SError, IRQ and FIQ to EL2.

```
mrs    x0, HCR_EL2
orr    x0, x0, #(1<<5)  // Set the AMO bit.
orr    x0, x0, #(1<<4)  // Set the IMO bit.
orr    x0, x0, #(1<<3)  // set the FMO bit.
msr    HCR_EL2, x0
```

Listing 12.5 Code to enable asynchronous exceptions.

```
// Enable SError, IRQ and FIQ
msr    DAIFClr, #0x7
```

By default, all asynchronous exceptions are routed to EL1. SCR_EL3 specifies exceptions to be routed to EL3 Listing 12.3 shows how to route all three asynchronous exceptions to EL3, and HCR_EL2 specifies exceptions to be routed to EL2. Listing 12.4 shows how to route all three asynchronous exceptions to EL2. For both control registers, separate bits to control routing of IRQs, FIQs, and SErrors.

Asynchronous exceptions may be *masked* by setting the A, I, and F bits in the PSTATE register to one. Additionally, when a target exception level is lower than the current exception level, the asynchronous exception is masked implicitly. The asynchronous exception will not be processed until it is unmasked by clearing the appropriate bit in the PSTATE register, and/or entering an exception level less than or equal to the exception level that the asynchronous exception is routed to. When the target exception level is higher than the current exception level, and the target exception level is EL2 or EL3, the asynchronous exception is processed, regardless of the settings in the PSTATE register. Listing 12.5 shows how to enable asynchronous exceptions by clearing the A, I, and F, bits in the PSTATE register. To set the bits, use DAIFSet instead of DAIFClr.

12.4 The boot process

In order to create a bare metal program we must understand the processor does when power is first applied or after a reset. The ARM CPU begins to execute code at a predetermined address which depends on the configuration of the CPU. In order for the system to work, the startup code must be at the correct address when the system starts up.

On the Raspberry Pi, when power is first applied, the ARM CPU is disabled and the Graphics Processing Unit (GPU) is enabled. The GPU runs a program that is stored in flash memory or ROM. That program, called the first stage boot loader, reads the second stage boot loader from a file named (bootcode.bin) on the SD card. That program enables the SDRAM, and then loads the third stage bootloader, start.elf. At this point, some basic hardware configuration is performed, and then the kernel is loaded to address 0x8000 from the kernel8.img file on the SD card. Once the kernel image file is loaded, a "b #0x8000" instruction is placed at address 0, and the ARM CPU is enabled. The ARM CPU executes the branch instruction at address 0, which immediately jumps to the kernel code at address 0x8000.

To run a bare metal program on the Raspberry Pi, it is only necessary to build an executable image and store it as kernel8.img on the SD card. Then, the boot process will load the bare metal program instead of the Linux kernel image. Care must be taken to ensure that the linker prepares the program to run at address 0x8000 and places the first executable instruction at the beginning of the image file. It is also important to make a copy of the original kernel image, so that it can be restored (using another computer). If the original kernel image is lost, then there will be no way to boot Linux until it is replaced.

12.5 Writing a bare metal program

A bare metal program should be divided into several files. Some of the code may be written in assembly, and other parts in C or some other language. The initial startup code, and the entry and exit from exception handlers, must be written in assembly. However, it may be much more productive to write the main program and the remainder of the exception handlers as C functions and have the assembly code call them.

12.5.1 Startup code

On startup or reset, the processor begins execution in EL3. The bare-metal program must include some start-up code. The startup code will:

- initialize the stack pointers for all of the modes,
- set up interrupt and exception handling,

- initialize the `.bss` section,
- configure the CPU and critical systems (optional),
- set up memory management (optional),
- set up process and/or thread management (optional),
- initialize devices (optional), and
- branch to `main()`.

The startup code requires some knowledge of the target platform, and at least part of it must be written in assembly language. The following listing shows a function named `_start` which sets up the stacks, initializes the `.bss` section, sets up the vector tables, then calls the `main` function:

Listing 12.6 AArch64 startup code.

```
1   // FILE:  start.S
2   // This is the startup code for a Raspberry Pi bare-metal
3   // program.  It assumes that only one CPU core is going
4   // to be used. The startup code for multiple CPU operation
5   // would require setting up stacks space for each CPU.
6
7   // Stack locations
8           .equ    stack_top, 0x20000000 // Raspberry Pi 2
9           .equ    EL3_stack_top, stack_top
10          .equ    EL2_stack_top, stack_top - 0x2000
11          .equ    EL1_stack_top, stack_top - 0x4000
12          .equ    EL0_stack_top, stack_top - 0x6000
13
14  // ------------------------------------------------------------
15  // The startup code should be loaded by the boot loader.
16  // The entry point is _start which performs initialization of
17  // the hardware, then calls a C function.
18          .section .text.boot      // a special section name
19          .global _start
20          .func   _start
21  _start: // On reset, we should be in EL3
22
23          // For our example, we only want to run on a single
24          // CPU core, so we cause the others to go to sleep
25          // waiting for an exception to occur. If an exception
26          // occurs, it just goes back to sleep
27          mrs     x0, mpidr_el1    // get the CPU ID
28          and     x0, x0, #3       // mask off upper bits
29          cmp     x0, #0           // if not CPU 0
30          beq     nosleep
31  sleep:  wfi                      // wait for interrupt
32          b       sleep            // go back to sleep
33
```

```
nosleep:
        // Clear the .bss segment to all zeros
        // The __bss_start__ and __bss_end__ symbols are
        // defined by the linker.
        ldr     x1,=__bss_start__ // load pointer to bss and
        ldr     x2,=__bss_end__   // to byte following bss
        mov     x3,#0             // load fill value (zero)
bssloop:cmp     x1,x2             // Start filling
        bge     bssdone
        str     x3,[x1],#8
        b       bssloop           // loop until done
bssdone:
        // Now we check the current EL, and initilaize from there down
        // to EL0
        mrs     x0, currentel  // read the current EL
        lsr     x0,x0,#2       // shift it right
        cmp     x0,#3          // are we in EL3?
        blt     el2_entry      // if not, check for EL2
        // initialize EL3 stack pointer
        ldr     x1,=EL3_stack_top
        mov     sp,x1
        // Set up vector table for EL3
        ldr     x1, =Vector_table_el1
        msr     VBAR_EL3, x1
        // Initialize SCTLR_EL2 and HCR_EL2 to safe values before
        // entering EL2.
        msr     SCTLR_EL2, XZR
        msr     HCR_EL2, XZR
        // Set the EL2 execution state.
        mrs     x0, SCR_EL3
        orr     x0, x0, #(1<<10) // RW: EL2 Execution state is AArch64
        orr     x0, x0, #(1<<0)  // NS: EL1 is non-secure
        msr     SCR_EL3, x0
        mov     x0, #0b01001     // DAIF=0000
        msr     SPSR_EL3, x0     // M[4:0]=01001 EL2h must match SCR_EL3.RW
        // Perform EL2 entry.
        adr     x0, el2_entry    // el2_entry points to the first
        msr     ELR_EL3, x0      // instruction of EL2 code.
        eret

el2_entry:
        mrs     x0, currentel  // read the current EL
        lsr     x0,x0,#2       // shift it right
        cmp     x0,#2          // are we in EL2?
        blt     el1_entry      // if not, check for EL1
        // initialize EL2 stack pointer
        ldr     x1,=EL2_stack_top
        mov     sp,x1
        // Set up vector table for EL2
```

```
83          ldr     x1, =Vector_table_el1
84          msr     VBAR_EL2, x1
85          // Initialize the SCTLR_EL1 register before entering EL1.
86          msr     SCTLR_EL1, XZR
87          // Set the EL1 Execution state.
88          mrs     x0, HCR_EL2
89          orr     x0, x0, #(1<<31) // RW: EL1 Execution state is AArch64.
90          msr     HCR_EL2, x0
91          mov     x0, #0b00101     // DAIF=0000
92          msr     SPSR_EL2, x0     // M[4:0]=00101 EL1h must match HCR_EL2.RW
93          adr     x0, el1_entry    // el1_entry points to the first
94          msr     ELR_EL2, x0      // instruction of EL1 code.
95          eret
96
97  el1_entry:
98          mrs     x0, currentel    // read the current EL
99          lsr     x0,x0,#2         // shift it right
100         cmp     x0,#1            // are we in EL1?
101         blt     el0_entry        // if not, we must be in EL0
102         // initialize EL1 stack pointer
103         ldr     x1,=EL1_stack_top
104         mov     sp,x1
105         // Set up vector table for EL1
106         ldr     x1, =Vector_table_el1
107         msr     VBAR_EL1, x1
108         // initialize EL0 stack pointer
109         ldr     x1,=EL0_stack_top
110         msr     sp_EL0,x1
111         // Initialize the SCTLR_EL0 register before entering EL1.
112         mov     x0,#(1<<16)      // Allow EL0 to use wfi instruction
113         orr     x0,x0,#(1<<18)   // Allow EL0 to use wfe instruction
114         msr     SCTLR_EL1, x0
115         // Set the EL0 Execution state.
116         mov     x0, #0b00000     // DAIF=0000 M[4:0]=00000 EL0t.
117         msr     SPSR_EL1, x0
118         adr     x0, el0_entry    // el0_entry points to the first
119         msr     ELR_EL1, x0      // instruction of EL0 code.
120         eret
121
122 el0_entry:
123         // Once we are in EL0, the only way to go back up is by causing
124         // an exception, which will invoke an exception handler.
125
126         // Call the Main function
127         bl      main
128
129         // If main ever returns, hang forever
130 hang:   b       hang     // this should never happen
131         .size   _start, . - _start
```

```
132        .endfunc
```

The first task of the startup code is to set all bytes in the .bss section to zero. Recall that the .bss section is used to hold data that is initialized to zero, but that the program file does not actually contain all of the zeros. Programs running under an operating system can rely on the C standard library to initialize the .bss section. If it is not linked to a C library, then a bare metal program must set all of the bytes in the .bss section to zero for itself.

The second task for the startup code is to initialize the vector table and stack pointer for each exception level. When an exception or interrupt occurs, the processor will automatically change into the appropriate exception level and begin executing an exception handler, using the stack pointer for that exception level. Hardware interrupts can be masked, but synchronous exceptions cannot be disabled. In order to guarantee correct operation, a stack must be set up for each exception level, and an exception handler must be provided. The exception handler does not actually have to do anything.

On the Raspberry Pi 2 and 3, memory is mapped to begin at address 0, and all models have at least 512 megabytes of memory. Therefore, it is safe to assume that the last valid memory address is 0x01FFFFFF. If each exception level is given eight kilobytes of stack space, then all of the stacks together will consume 32 kilobytes, and the initial stack addresses can be easily calculated. Since the C compiler uses a full descending stack, the initial stack pointers can be assigned addresses 0x20000000, 0x01FFFE000, 0x01FFFC000, etc.

After initializing the .bss section and setting up the stacks and vector tables, the startup code switches to EL0 and calls the main() function. The main thread will run with user privilege, and exceptions will cause the CPU to go to EL1 and begin executing the exception handler.

12.5.2 Main program

The final part of this bare metal program is the main() function. Listing 12.7 shows a very simple main program which reads from three GPIO pins which have push-buttons connected to them, and controls three other pins that have LEDs connected to them. When a button is pressed the LED associated with it is illuminated. The functions to control the GPIO device, which is introduced in Chapter 11, have been removed from the main program file. This makes the main program *portable*; it can run on any AArch4 system that has a GPIO device, with the addition of another file to implement the functions for using the GPIO device for that system.

Listing 12.7 A simple main program.

```
 1   // FILE: main.s
 2   // This program reads from three buttons connected to GPIO3-5, and
 3   // controls three leds connected to GPIO0-2. The main loop runs
 4   // continuously.
 5           .global main
 6   main:   stp     x29, x30, [sp, #-16]! // push lr and x29
 7           // Set the GPIO pins
 8           mov     x0,#22          // Pin 0
 9           bl      GPIO_dir_output // set for output
10           mov     x0,#23          // Pin 1
11           bl      GPIO_dir_output // set for output
12           mov     x0,#24          // Pin 2
13           bl      GPIO_dir_output // set for output
14
15           mov     x0,#25          // Pin 3
16           bl      GPIO_dir_input  // set for input
17           mov     x0,#26          // Pin 4
18           bl      GPIO_dir_input  // set for input
19           mov     x0,#27          // Pin 5
20           bl      GPIO_dir_input  // set for input
21   // Main loop just reads buttons and updates the LEDs.
22   loop:
23           // Read the state of the inputs and
24           // set the ouputs to the same state.
25           mov     x0,#22          // Pin 22
26           bl      GPIO_get_pin    // read it
27           mov     x1,x0           // copy pin state to x1
28           mov     x0,#25          // Pin 25
29           bl      GPIO_set_pin    // write it
30
31           mov     x0,#23          // Pin 23
32           bl      GPIO_get_pin    // read it
33           mov     x1,x0           // copy pin state to i1
34           mov     x0,#26          // Pin 26
35           bl      GPIO_set_pin    // write it
36
37           mov     x0,#24          // Pin 24
38           bl      GPIO_get_pin    // read it
39           mov     x1,x0           // copy pin state to x1
40           mov     x0,#27          // pin 27
41           bl      GPIO_set_pin    // write it
42
43           b       loop
44
45           // We should never return, but just in case...
46           ldp     x29, x30, [sp],#16 // pop lr and x29
47           ret
```

12.5.3 The linker script

When compiling the program, it is necessary to perform a few extra steps to ensure that the program is ready to be loaded and run by the boot code. The last step in compiling a program is to link all of the object files together, possibly also including some object files from system libraries. A linker script is a file that tells the linker which sections to include in the output file, as well as which order to put them in, what type of file is to be produced, and what is to be the address of the first instruction. The default linker script used by GCC creates an ELF executable file, which includes startup code from the C library and also includes information which tells the loader where the various sections reside in memory. The default linker script creates a file that can be loaded by the operating system kernel, but which cannot be executed on bare metal.

For a bare metal program, the linker must be configured to link the program so that the first instruction of the startup function is given the correct address in memory. This address depends on how the boot loader will load and execute the program. On the Raspberry Pi this address is `0x8000`. The linker will automatically adjust any other addresses as it links the code together. The most efficient way to accomplish this is by providing a custom linker script to be used instead of the default system script. Additionally, either the linker must be instructed to create a flat binary file, rather than an ELF executable file, or a separate program (`objcopy`) must be used to convert the ELF executable into a flat binary file.

Listing 12.8 is an example of a linker script that can be used to create a bare metal program. The first line is just a comment. The second line specifies the name of the function where the program begins execution. In this case, it specifies that a function named `_start` is where the program will begin execution. Next, the file specifies the sections that the output file will contain. For each output section, it lists the input sections that are to be used.

The first output section is the `.text` section, and it is composed of any sections whose names end in `.text.boot` followed by any sections whose names end in `.text`. In Listing 12.6, the `_start` function was placed in the `.text.boot` section, and it is the *only thing* in that section. Therefore the linker will put the `_start` function at the very beginning of the program. The remaining text sections will be appended, and then the remaining sections, in the order that they appear. After the sections are concatenated together, the linker will make a pass through the resulting file, correcting the addresses of branch and load instructions as necessary so that the program will execute correctly.

12.5.4 Putting it all together

Compiling a program that consists of multiple source files, a custom linker script, and special commands to create an executable image can become tedious. The make utility was cre-

Listing 12.8 A sample Gnu linker script.

```
/* FILE: bare_metal.ld - linker script for bare metal */
ENTRY(_start)

SECTIONS
{
    . = 0x8000; /* Raspbery Pi will load the image here by default */

    __text_start__ = .;
    .text :
    {
        KEEP(*(.text.boot)) /* put the start function first ! */
        *(.text)
    }
    . = ALIGN(4096); /* align to page size */
    __text_end__ = .;

    __rodata_start__ = .;
    .rodata :
    {
        *(.rodata)
    }
    . = ALIGN(4096); /* align to page size */
    __rodata_end__ = .;

    __data_start__ = .;
    .data :
    {
        *(.data)
    }
    . = ALIGN(4096); /* align to page size */
    __data_end__ = .;

    __bss_start__ = .;
    .bss :
    {
        bss = .;
        *(.bss)
    }
    . = ALIGN(4096); /* align to page size */
    __bss_end__ = .;
    _end = .;
}
```

Listing 12.9 A sample make file.

```
# source files
SOURCES_ASM := main1.s RPI_GPIO.s start1.s handlers1.s
SOURCES_C   :=

# object files
OBJS        := $(patsubst %.s,%.o,$(SOURCES_ASM))
OBJS        += $(patsubst %.c,%.o,$(SOURCES_C))

# Build flags
INCLUDES    := -I.
ASFLAGS     :=
CFLAGS      := $(INCLUDES)

# build targets
all: kernel8.img

# Build image for Raspberry Pi
kernel8.img: bare.elf
        objcopy bare.elf -O binary kernel8.img

# Build the ELF file
bare.elf: $(OBJS) bare_metal.ld
        ld $(OBJS) -Tbare_metal.ld -o $@

# Compile C to object file
%.o: %.c
        gcc $(CFLAGS) -c $< -o $@

# Compile Assembly to object file
%.o: %.s
        gcc $(ASFLAGS) -c $< -o $@

# Clean up the build directory
clean:
        rm -f $(OBJS) kernel.elf kernel8.img uImage

dist-clean: clean
        rm -f *~
```

ated specifically to help in this situation. Listing 12.9 shows a make file that can be used to combine all of the elements of the program and produce a `kernel8.img` file for the Raspberry Pi. Fig. 12.5 shows how the program can be built by typing "make" at the command line.

```
lpyeatt@Pi$ make
gcc  -c main1.s -o main1.o
gcc  -c RPI_GPIO.s -o RPI_GPIO.o
gcc  -c start1.s -o start1.o
gcc  -c handlers1.s -o handlers1.o
ld main1.o RPI_GPIO.o start1.o handlers1.o  -Tbare_metal.ld -o bare.elf
objcopy bare.elf -O binary kernel8.img
lpyeatt@Pi$ make
```

Figure 12.5: Running make to build the image.

12.6 Using an interrupt

The main program shown in Listing 12.7 is extremely wasteful because it runs the CPU in a loop, repeatedly checking the status of the GPIO pins. It uses far more CPU time (and electrical power) than is necessary. In reality, the pins are unlikely to change state very often, and it is sufficient to check them a few times per second. It only takes a few nanoseconds to check the input pins and set the output pins so the CPU only needs to be running for a few nanoseconds at a time, a few times per second.

A much more efficient implementation would set up a timer to send interrupts at a fixed frequency. Then the main loop can check the buttons, set the outputs, and put the CPU to sleep. Listing 12.10 shows the main program, modified to put the processor to sleep after each iteration of the main loop. The only difference between this main function and the one in Listing 12.7 is the addition of a wfi instruction at line 44. The new implementation will consume far less electrical power and allow the CPU to run cooler, thereby extending its life. However, some additional work must be performed in order to set up the timer and interrupt system before the main function is called.

Listing 12.10 An improved main program.

```
1  // FILE: main.s
2  // This program reads from three buttons connected to GPIO3-5, and
3  // controls three leds connected to GPIO0-2. The main loop runs
4  // continuously.
5          .global main
6  main:   stp     x29, x30, [sp, #-16]! // push lr and x29
7          // Set the GPIO pins
8          mov     x0,#22          // Pin 0
9          bl      GPIO_dir_output // set for output
10         mov     x0,#23          // Pin 1
11         bl      GPIO_dir_output // set for output
```

```
12        mov     x0,#24          // Pin 2
13        bl      GPIO_dir_output // set for output
14
15        mov     x0,#25          // Pin 3
16        bl      GPIO_dir_input  // set for input
17        mov     x0,#26          // Pin 4
18        bl      GPIO_dir_input  // set for input
19        mov     x0,#27          // Pin 5
20        bl      GPIO_dir_input  // set for input
21 // Main loop just reads buttons and updates the LEDs.
22 loop:
23        // Read the state of the inputs and
24        // set the ouputs to the same state.
25        mov     x0,#22          // Pin 22
26        bl      GPIO_get_pin    // read it
27        mov     x1,x0           // copy pin state to x1
28        mov     x0,#25          // Pin 25
29        bl      GPIO_set_pin    // write it
30
31        mov     x0,#23          // Pin 23
32        bl      GPIO_get_pin    // read it
33        mov     x1,x0           // copy pin state to i1
34        mov     x0,#26          // Pin 26
35        bl      GPIO_set_pin    // write it
36
37        mov     x0,#24          // Pin 24
38        bl      GPIO_get_pin    // read it
39        mov     x1,x0           // copy pin state to x1
40        mov     x0,#27          // pin 27
41        bl      GPIO_set_pin    // write it
42
43        // Put processor to sleep until and interrupt occurs
44        wfi
45        b       loop            // check buttons again
46
47        // We should never return, but just in case...
48        ldp     x29, x30, [sp],#16 // pop lr and x29
49        ret
```

12.6.1 Startup code with interrupt enabling

Some changes must be made to the startup code in Listing 12.6 so that after setting up the vector table, it calls a function to initialize the interrupt controller then calls another function to set up the timer. In Listing 12.11, lines 111 through 114 have been added to initialize the interrupt controller and enable the timer:

Listing 12.11 ARM startup code with timer interrupt.

```
// FILE:  start.s
// This is the startup code for a Raspberry Pi bare-metal
// program.  It assumes that only one CPU core is going
// to be used. The startup code for multiple CPU operation
// would require setting up stacks space for each CPU.

// Stack locations
        .equ    stack_top, 0x20000000 // Raspberry Pi 2
        .equ    EL3_stack_top, stack_top
        .equ    EL2_stack_top, stack_top - 0x2000
        .equ    EL1_stack_top, stack_top - 0x4000
        .equ    EL0_stack_top, stack_top - 0x6000


// ------------------------------------------------------------
// The startup code should be loaded by the boot loader.
// The entry point is _start which performs initialization of
// the hardware, then calls a C function.
        .section .text.boot    // a special section name
        .global _start
        .func   _start
_start: // On reset, we should be in EL3

        // For our example, we only want to run on a single
        // CPU core, so we cause the others to go to sleep
        // waiting for an exception to occur. If an exception
        // occurs, it just goes back to sleep
        mrs     x0, mpidr_el1   // get the CPU ID
        and     x0, x0, #3      // mask off upper bits
        cmp     x0, #0          // if not CPU 0
        beq     nosleep
sleep:  wfi                     // wait for interrupt
        b       sleep           // go back to sleep

nosleep:
        // Clear the .bss segment to all zeros
        // The __bss_start__ and __bss_end__ symbols are
        // defined by the linker.
        ldr     x1,=__bss_start__ // load pointer to bss and
        ldr     x2,=__bss_end__   // to byte following bss
        mov     x3,#0           // load fill value (zero)
bssloop:cmp     x1,x2           // Start filling
        bge     bssdone
        str     x3,[x1],#8
        b       bssloop         // loop until done
bssdone:
        // Now we check the current EL, and initialaize from there down
        // to EL0
```

```
48      mrs     x0, currentel   // read the current EL
49      lsr     x0,x0,#2        // shift it right
50      cmp     x0,#3           // are we in EL3?
51      blt     el2_entry       // if not, check for EL2
52      // initialize EL3 stack pointer
53      ldr     x1,=EL3_stack_top
54      mov     sp,x1
55      // Set up vector table for EL3
56      ldr     x1, =Vector_table_el1
57      msr     VBAR_EL3, x1
58      // Initialize SCTLR_EL2 and HCR_EL2 to safe values before
59      // entering EL2.
60      msr     SCTLR_EL2, XZR
61      msr     HCR_EL2, XZR
62      // Set the EL2 execution state.
63      mrs     x0, SCR_EL3
64      orr     x0, x0, #(1<<10) // RW: EL2 Execution state is AArch64
65      orr     x0, x0, #(1<<0)  // NS: EL1 is non-secure
66      msr     SCR_EL3, x0
67      mov     x0, #0b01001     // DAIF=0000
68      msr     SPSR_EL3, x0     // M[4:0]=01001 EL2h must match SCR_EL3.RW
69      // Perform EL2 entry.
70      adr     x0, el2_entry    // el2_entry points to the first
71      msr     ELR_EL3, x0      // instruction of EL2 code.
72      eret
73
74  el2_entry:
75      mrs     x0, currentel   // read the current EL
76      lsr     x0,x0,#2        // shift it right
77      cmp     x0,#2           // are we in EL2?
78      blt     el1_entry       // if not, check for EL1
79      // initialize EL2 stack pointer
80      ldr     x1,=EL2_stack_top
81      mov     sp,x1
82      // Set up vector table for EL2
83      ldr     x1, =Vector_table_el1
84      msr     VBAR_EL2, x1
85      // Initialize the SCTLR_EL1 register before entering EL1.
86      msr     SCTLR_EL1, XZR
87      // Set the EL1 Execution state.
88      mrs     x0, HCR_EL2
89      orr     x0, x0, #(1<<31) // RW: EL1 Execution state is AArch64.
90      msr     HCR_EL2, x0
91      mov     x0, #0b00101     // DAIF=0000
92      msr     SPSR_EL2, x0     // M[4:0]=00101 EL1h must match HCR_EL2.RW
93      adr     x0, el1_entry    // el1_entry points to the first
94      msr     ELR_EL2, x0      // instruction of EL1 code.
95      eret
96
```

```
97   el1_entry:
98           mrs     x0, currentel   // read the current EL
99           lsr     x0,x0,#2        // shift it right
100          cmp     x0,#1           // are we in EL1?
101          blt     el0_entry       // if not, we must be in EL0
102          // initialize EL1 stack pointer
103          ldr     x1,=EL1_stack_top
104          mov     sp,x1
105          // Set up vector table for EL1
106          ldr     x1, =Vector_table_el1
107          msr     VBAR_EL1, x1
108          // initialize EL0 stack pointer
109          ldr     x1,=EL0_stack_top
110          msr     sp_EL0,x1
111          // Set up the timer and enable interrupts
112          bl      enable_timer
113          bl      IC_init
114          bl      config_interrupt
115          // Initialize the SCTLR_EL0 register before entering EL0.
116          // mov   x0,#(1<<16)     // Allow EL0 to use wfi instruction
117          // orr   x0,x0,#(1<<18)  // Allow EL0 to use wfe instruction
118          // msr   SCTLR_EL1, x0
119          // Set the EL0 Execution state.
120          mov     x0, #0b00000    // DAIF=0000 M[4:0]=00000 EL0t.
121          msr     SPSR_EL1, x0
122          adr     x0, el0_entry   // el0_entry points to the first
123          msr     ELR_EL1, x0     // instruction of EL0 code.
124          eret
125
126  el0_entry:
127          // Once we are in EL0, the only way to go back up is by causing
128          // an exception, which will invoke an exception handler.
129          // Call the Main function
130          bl      main
131
132          // If main ever returns, hang forever
133  hang:   b       hang    // this should never happen
134          .size   _start, . - _start
135          .endfunc
```

12.6.2 Interrupt controllers

The Raspberry Pi has a relatively simple interrupt controller. It can enable and disable interrupt sources, and requires that the programmer read up to three registers to determine the source of an interrupt. For our purposes, we only need to manage the ARM timer interrupt. Listing 12.12 provides a few basic functions for using this device to enable the timer interrupt.

Extending these functions to provide more functionality would not be very difficult, but would take some time. It would be necessary to set up a mapping from the interrupt bits in the interrupt register controller to integer values, so that each interrupt source has a unique identifier. Then the functions could be written to use those identifiers. The result would be a software implementation that is more portable.

Listing 12.12 Functions to manage the Raspberry Pi interrupt controller.

```
// FILE: RasPiIC.s
// Functions to manage the Interrupt Controller on the
// Raspberry Pi
        // Address of Interrupt Controller
        .equ    IC,     0x3F00B000
        // Register offsets
        .equ    IRQBP,  0x200 // IRQ basic pending
        .equ    IRQP1,  0x204 // IRQ pending 1
        .equ    IRQP2,  0x208 // IRQ pending 2
        .equ    FIQC,   0x20C // FIQ control
        .equ    IRQEN1, 0x210 // IRQ enable 1
        .equ    IRQEN2, 0x214 // IRQ enable 2
        .equ    IRQBEN, 0x218 // Enable basic IRQs
        .equ    IRQDA1, 0x21C // IRQ disable 1
        .equ    IRQDA2, 0x220 // IRQ disable 2
        .equ    IRQBDA, 0x224 // Disable basic IRQs
// -----------------------------------------------------------
        .text
        .align 2
// -----------------------------------------------------------
// Initialization of the Interrupt Controller (IC)
        .global IC_init
IC_init:
        // disable all interrupts
        ldr     x0,=IC
        mov     x1,#0
        str     w1,[x0,#IRQEN1]
        str     w1,[x0,#IRQEN2]
        str     w1,[x0,#IRQBEN]
        ret
// -----------------------------------------------------------
// config_interrupt (int ID, int CPU);
// On Raspberry Pi, this just enables the timer interrupt
        .global config_interrupt
config_interrupt:
        ldr     x0,=IC
        mov     x1,#1
        str     w1,[x0,#IRQBEN]
        ret
// -----------------------------------------------------------
```

```
41    // int get_interrupt_number();
42    // Get the interrupt ID for the current interrupt.
43    // On Raspberry Pi, just read and return the pending register.
44            .global get_interrupt_number
45    get_interrupt_number: // Read the ICCIAR from the CPU Interface
46            ldr     x0,=IC
47            ldr     w0,[x0,#IRQBP]
48            ret
49    // ------------------------------------------------------------
50    // void end_of_interrupt(int ID);
51    // Notify the IC that the interrupt has been processed.
52    // On Raspberry Pi, this does nothing
53            .global end_of_interrupt
54    end_of_interrupt:
55            ret
```

12.6.3 Timers

The Raspberry Pi provides several timers that could be used, but the ARM timer is the easiest to configure. The following listing provides a few basic functions for managing this device:

Listing 12.13 Functions to manage the Raspberry Pi timer0 device.

```
1     // FILE: RasPi_timer.s
2     // The timer runs off the 250MHz APB_clock source
3             .equ    TIMER_BASE, 0x3F008400
4             .equ    LOAD,    0x00  // Load
5             .equ    VALUE,   0x04  // Value (read only)
6             .equ    CONTROL,0x08   // Control
7             .equ    IRQACK,  0x0C  // IRQ Clear/Ack (write only)
8             .equ    RAWIRQ,  0x10  // Raw IRQ (read only)
9             .equ    MSKIRQ,  0x14  // Masked IRQ (read only)
10            .equ    RELOAD,  0x18  // Reload
11            .equ    PREDIV,  0x1C  // Pre-divider
12            .equ    COUNT,   0x20  // Free-running counter
13
14            .text
15            .align 2
16    // ------------------------------------------------------------
17    // Configures and enables timer0 to generate interrupts at a
18    // fixed frequency.  Also configures the Generic Interrupt
19    // Controller (GIC) to send interrupts to CPU 0.
20            .global enable_timer
21    enable_timer:
22            ldr     x0,=TIMER_BASE
23            mov     x1,#0x7F        // divide clock to 1,953,125Hz
24            str     w1,[x0,#PREDIV]
```

```
25          ldr     x1,=954         // should give about 8Hz
26          str     w1,[x0,#LOAD]
27          ldr     w1,=0b11111000000000010101010
28          str     w1,[x0,#CONTROL]
29          ret
30  // -----------------------------------------------------------
31  // int check_timer_interrupt()
32  // Check and clear the timer 0 interrupt.  Returns 1 if the
33  // interrupt was active.  Returns 0 otherwise.
34          .global check_timer_interrupt
35  check_timer_interrupt:
36          ldr     x1,=TIMER_BASE
37          ldr     w0,[x1,#MSKIRQ]
38          ands    x0,x0,#1
39          beq     cti_ret
40          str     w0,[x1,#IRQACK]
41  cti_ret:ret
```

12.6.4 Exception handling

The final step in writing the bare metal code to operate in an interrupt-driven fashion is to modify the IRQ handlers from Listing 12.1 so that the actually does something. Listing 12.14 shows a new version of the IRQ exception handlers which check and clear the timer interrupt, then return to the location and CPU mode that were current when the interrupt occurred.

Listing 12.14 Stubs for the exception handlers.

```
1   // FILE: handlers.s
2           .text
3           .align  2
4   // Typical exception vector table code.
5   //-----------------------------------------------
6           .balign 0x800  // align to a 2 kilobyte boundary
7           .global Vector_table_el1
8   Vector_table_el1:
9   curr_el_sp0_sync:
10          // The exception handler for the synchronous
11          // exception from the current EL using SP0.
12          eret    // return from exception
13
14          .balign 0x80    // align to next 128 byte boundary
15  curr_el_sp0_irq:
16          // The exception handler for the IRQ exception
17          // from the current EL using SP0.
18          // find out which interrupt we are servicing
19          bl      get_interrupt_number // returns in r0
```

```
20          cmp     x0,#54          // is it the timer interrupt?
21          bne     i1
22          bl      check_timer_interrupt
23  i1:     bl      end_of_interrupt// tell IC we are done
24          eret    // return from exception
25
26          .balign 0x80    // align to next 128 byte boundary
27  curr_el_sp0_fiq:
28          // The exception handler for the FIQ exception
29          // from the current EL using SP0.
30          eret    // return from exception
31
32          .balign 0x80    // align to next 128 byte boundary
33  curr_el_sp0_serror:
34          // The exception handler for the system error
35          // exception from the current EL using SP0.
36          eret    // return from exception
37
38          .balign 0x80    // align to next 128 byte boundary
39  curr_el_spx_sync:
40          // The exception handler for the synchronous
41          // exception from the current EL using the
42          // current SP.
43          eret    // return from exception
44
45          .balign 0x80    // align to next 128 byte boundary
46  curr_el_spx_irq:
47          // The exception handler for IRQ exception
48          // from the current EL using the current SP.
49          bl      get_interrupt_number // returns in r0
50          cmp     x0,#54          // is it the timer interrupt?
51          bne     i2
52          bl      check_timer_interrupt
53  i2:     bl      end_of_interrupt// tell IC we are done
54          eret    // return from exception
55
56          .balign 0x80    // align to next 128 byte boundary
57  curr_el_spx_fiq:
58          // The exception handler for the FIQ exception
59          // from the current EL using the current SP.
60          eret    // return from exception
61
62          .balign 0x80    // align to next 128 byte boundary
63  curr_el_spx_serror:
64          // The exception handler for the system error
65          // exception from the current EL using the
66          // current SP.
67          eret    // return from exception
68
```

```
            .balign 0x80    // align to next 128 byte boundary
lower_el_aarch64_sync:
            // The exception handler for the synchronous
            // exception from a lower EL (AArch64).
            eret    // return from exception

            .balign 0x80    // align to next 128 byte boundary
lower_el_aarch64_irq:
            // The exception handler for the IRQ exception
            // from a lower EL (AArch64).
            bl      get_interrupt_number // returns in r0
            cmp     x0,#54          // is it the timer interrupt?
            bne     i3
            bl      check_timer_interrupt
i3:         bl      end_of_interrupt// tell IC we are done
            eret    // return from exception

            .balign 0x80    // align to next 128 byte boundary
lower_el_aarch64_fiq:
            // The exception handler for the FIQ exception
            // from a lower EL (AArch64).
            eret    // return from exception

            .balign 0x80    // align to next 128 byte boundary
lower_el_aarch64_serror:
            // The exception handler for the system error
            // exception from a lower EL(AArch64).
            eret    // return from exception

            .balign 0x80    // align to next 128 byte boundary
lower_el_aarch32_sync:
            // The exception handler for the synchronous
            // exception from a lower EL(AArch32).
            eret    // return from exception

            .balign 0x80    // align to next 128 byte boundary
lower_el_aarch32_irq:
            // The exception handler for the IRQ exception
            // from a lower EL (AArch32).
            bl      get_interrupt_number // returns in r0
            cmp     x0,#54          // is it the timer interrupt?
            bne     i4
            bl      check_timer_interrupt
i4:         bl      end_of_interrupt// tell IC we are done
            eret    // return from exception

            .balign 0x80    // align to next 128 byte boundary
lower_el_aarch32_fiq:
            // The exception handler for the FIQ exception
```

```
118        // from a lower EL (AArch32).
119        eret    // return from exception
120
121        .balign 0x80    // align to next 128 byte boundary
122  lower_el_aarch32_serror:
123        // The exception handler for the system error
124        // exception from a lower EL(AArch32).
125        eret    // return from exception
```

12.6.5 Building the interrupt-driven program

Finally, the make file must be modified to include the new source files that were added to the program. Listing 12.15 shows the modified make file. The only change is that two extra object files have been added, when make is run, those files will be compiled and linked with the program. Fig. 12.6 shows how the program can be built by typing "make" at the command line.

Listing 12.15 A sample make file.

```
1
2  # source files
3  SOURCES_ASM := main2.s RPI_GPIO.s start2.s handlers2.s RasPiIC.s RasPi_timer.s
4  SOURCES_C   :=
5
6  # object files
7  OBJS        := $(patsubst %.s,%.o,$(SOURCES_ASM))
8  OBJS        += $(patsubst %.c,%.o,$(SOURCES_C))
9
10 # Build flags
11 INCLUDES    := -I.
12 ASFLAGS     :=
13 CFLAGS      := $(INCLUDES)
14
15 # build targets
16 all: kernel8.img
17
18 # Build image for Raspberry Pi
19 kernel8.img: bare.elf
20        objcopy bare.elf -O binary kernel8.img
21
22 # Build the ELF file
23 bare.elf: $(OBJS) bare_metal.ld
24        ld $(OBJS) -Tbare_metal.ld -o $@
25
26 # Compile C to object file
27 %.o: %.c
28        gcc $(CFLAGS) -c $< -o $@
```

```
29
30   # Compile Assembly to object file
31   %.o: %.s
32          gcc $(ASFLAGS) -c $< -o $@
33
34   # Clean up the build directory
35   clean:
36          rm -f $(OBJS) kernel.elf kernel8.img uImage
37
38   dist-clean: clean
39          rm -f *~
```

```
pyeatt@pi64 $ make
gcc   -c main2.s -o main2.o
gcc   -c RPI_GPIO.s -o RPI_GPIO.o
gcc   -c start2.s -o start2.o
gcc   -c handlers2.s -o handlers2.o
gcc   -c RasPiIC.s -o RasPiIC.o
gcc   -c RasPi_timer.s -o RasPi_timer.o
ld main2.o RPI_GPIO.o start2.o handlers2.o RasPiIC.o RasPi_timer.o  ↵
    -Tbare_metal.ld -o bare.elf
objcopy bare.elf -O binary kernel8.img
pyeatt@pi64 $
```

Figure 12.6: Running make to build the image.

12.7 ARM processor profiles

Since its introduction in 1982 as the flagship processor for Acorn RISC Machine, the ARM processor has gone through many changes. Throughout the years, ARM processors have always maintained a good balance of simplicity, performance, and efficiency. Although originally intended as a desktop processor, the ARM architecture has been more successful than any other architecture for use in embedded applications. That is at least partially because of good choices made by its original designers. The architectural decisions resulted in a processor that provides relatively high computing power with a relatively small number of transistors. That results in relatively low power consumption.

Today, there are almost 20 major versions of the ARMv7 architecture, targeted for everything from smart sensors to desktops and servers, and sales of ARM based processors outnumber all other processor architectures combined. Historically, ARM has given numbers to various versions of the architecture. With the ARMv7, they introduced a simpler scheme to describe different versions of the processor. They divided their processor families into three major *pro-files*:

ARMv7-A Applications processors are capable of running a full, multiuser, virtual memory, multiprocessing operating system.

ARMv7-R: Real-time processors are for embedded systems that may need powerful processors, cache, and/or large amounts of memory.

ARMv7-M: Microcontroller processors only execute Thumb instructions and are intended for use in very small cost-sensitive embedded systems. They provide low cost, low power, and small size, and may not have hardware floating point or other high-performance features.

In 2014, ARM introduced the ARMv8 (AArch64) architecture. This is the first radical change in the ARM architecture in over 30 years. The new architecture extends the register set to thirty 64-bit general purpose registers, and has a completely new instruction set. Compatibility with ARMv7 and earlier code is supported by switching the processor into 32-bit mode, so that it executes the 32-bit ARM instruction set. This is somewhat similar to the way that the Thumb instructions are supported on 32-bit ARM cores, but the change to 32-bit code can only be made when the processor is in privileged mode, and drops back to unprivileged mode.

12.8 Chapter summary

Writing bare-metal programs can be a daunting task. However, that task can be made easier by writing and testing code under an operating system before attempting to run it bare-metal. There are some functions which cannot be tested in this way. In those cases, it is best to keep those functions as simple as possible. Once the program works on bare-metal, extra capabilities can be added.

Interrupt-driven processing is the basis for all modern operating systems. The system timer allows the O/S to take control periodically and select a different process to run on the CPU. Interrupts allow hardware devices to do their jobs independently and signal the CPU when they need service. The ability to restrict user access to devices and certain processor features provides the basis for a secure and robust system.

Exercises

12.1. What are the advantages of a CPU which supports user mode and privileged mode over a CPU which does not?

12.2. What are the *privileged modes* supported by the AArch64 architecture?

12.3. The interrupt handling mechanism is somewhat complex and requires significant programming effort to use. Why is it preferred over simply having the processor poll I/O devices?

12.4. Where does program control transfer to when a hardware interrupt occurs?

12.5. What is an `swi` instruction? What is its use in operating systems? What is the key difference between an `swi` instruction and an interrupt?

12.6. Which of the following operations should be allowed only in privileged mode? Briefly explain your decision for each one.

    a. Execute an `swi` instruction.

    b. Disable all interrupts.

    c. Read the time-of-day clock.

    d. Receive a packet of data from the network.

    e. Shutdown the computer

12.7. The programs in this chapter assumed the existence of libraries of functions for controlling the GPIO pins on the Raspberry Pi. The C prototypes for the functions are: `int GPIO_get_pin(int pin)`, `void GPIO_set_pin(int pin, int state)`, `GPIO_dir_input(int pin)`, and `GPIO_dir_output(int pin)`. Write these libraries in ARM assembly language for both platforms.

12.8. Write an interrupt-driven program to read characters from the serial port on the Raspberry Pi. The UART can be configured to send an interrupt when a character is received.

When a character is received through the UART and an interrupt occurs, the character should be echoed by transmitting it back to the sender. The character should also be stored in a buffer. If the character received is newline (\n), or if the buffer becomes full, then the contents of the buffer should be transmitted through the UART. Then, the buffer cleared and prepared to receive more characters.

Index